FRONTIERS OF CLIMATE MODELING

The physics and dynamics of the atmosphere and atmosphere–ocean interactions provide the foundation of modern climate models, upon which the chemistry and biology of ocean and land-surface processes are built. *Frontiers of Climate Modeling* captures the major developments in modeling the atmosphere witnessed during the last two decades, and their implications to our understanding of climate change, whether due to natural or anthropogenic causes. Its particular emphasis is in elucidating how greenhouse gases and aerosols are altering the radiative forcing of the climate system and the sensitivity of the system to such perturbations.

The expert team of authors addresses key aspects of the atmospheric greenhouse effect, clouds, aerosols, atmospheric radiative transfer, deep convection dynamics, largescale ocean dynamics, stratosphere–troposphere interactions, and coupled ocean–atmosphere model development. This is an important reference for those interested in understanding the forces driving the climate system and how they are modeled by climate scientists.

JEFFREY KIEHL is a senior scientist at the National Center for Atmospheric Research, where he is member of the Climate Change Research Section. For the past 25 years he has worked on a wide range of research topics in climate-change science, including the role of clouds and aerosols in the climate system, the effects of increased trace gases on climate change and the role of stratospheric ozone changes on climate. He is currently using a climate-system model to study the climates of Earth for deep geologic times. He served as the chairman of the steering committee for the national Community Climate System Model program. He is a Fellow of the American Meteorological Society and recipient of the Distinguished Achievement Award in Climate System Modeling. He has served on a number of National Academy of Sciences committees and panels studying climate issues.

V. (RAM) RAMANATHAN is an atmospheric scientist whose research focuses on global climate dynamics, the greenhouse effect, aerosols, clouds, and the Earth radiation budget. At Scripps Institution of Oceanography of the University of California, San Diego, he is a distinguished professor of atmospheric and climate sciences and the Director of the Center for Atmospheric Sciences. He is also the co-chief scientist of the Atmospheric Brown Cloud Project and the Indian Ocean Experiment, which led to the discovery of the South Asian brown haze and its radiative forcing. Dr. Ramanathan is a member of the Pontifical Academy of Sciences and the National Academy of Sciences. He is a recipient of the Buys Ballot Medal from the Royal Netherlands Academy of Sciences, the Volvo environmental prize and the Rossby Medal of the American in Meteorological Society.

FRONTIERS OF CLIMATE MODELING

Edited by

J. T. KIEHL
National Center for Atmospheric Research, Boulder, Colorado

V. RAMANATHAN
Scripps Institution of Oceanography, University of California, San Diego

CAMBRIDGE
UNIVERSITY PRESS

CAMBRIDGE UNIVERSITY PRESS
Cambridge, New York, Melbourne, Madrid, Cape Town,
Singapore, São Paulo, Delhi, Tokyo, Mexico City

Cambridge University Press
The Edinburgh Building, Cambridge CB2 8RU, UK

Published in the United States of America by Cambridge University Press, New York

www.cambridge.org
Information on this title: www.cambridge.org/9780521298681

First published 2006
First paperback edition 2011

A catalogue record for this publication is available from the British Library

ISBN 978-0-521-79132-8 Hardback
ISBN 978-0-521-29868-1 Paperback

Additional resources for this publication at www.cambridge.org/9780521298681

Contents

Preface

In a career that spans over four decades, Robert D. Cess has pioneered the study of diverse topics and disciplines. He first attacked problems dealing with conductive, convective, and radiative heat transfer in engineering systems and his contribution to these topics culminated in a classic text book on radiative transfer. The hallmark of this early work is the successful application of singular peturbation techniques to solve complex radiative heat transfer problems. His intellectual curiosity took him to the study of thermal structure of planetary atmospheres. He is one of the very select few (if not the only one) who has solved the thermal structure of almost all of the inner and outer planets of the solar system including Mercury, Mars, Earth, Venus, Jupiter, Saturn, and others, including study of the satellites. He was probably the first to obtain an analytical solution for the radiative-convective equilibrium-temperature structure of the troposphere-stratosphere of Mars and Venus.

The latter part of his career has been focused exclusively on Earth, where he has made fundamental contributions to our understanding of the physics of climate with a particular focus on processes that regulate the Earth's radiation budget and the mechanisms of cloud feedback processes. He obtained worldwide recognition for a comprehensive comparative study of the nature of water-vapor and cloud feedback processes of over 15 three-dimensional climate models, and brilliantly demonstrated that cloud feedback is the major source for the wide range in climate sensitivity of climate models. This study was pivotal in persuading funding agencies in the USA and Europe to initiate new studies in radiative processes and cloud-radiative interactions. For example, the Department of Energy's Atmospheric Radiation Program is an outgrowth of this research. He has also made insightful contributions to how black carbon and other manmade aerosols regulate the radiative heating of the earth–atmosphere system. His aerosol research played a major role in our understanding of the so-called "nuclear winter" problem, the stratospheric warming during volcanic eruptions, and the role of absorbing aerosols in the troposphere. It is fitting that his remarkable contributions to one of the most timely topics of the twentieth (and the current) century, i.e., global warming, is recognized by the publication of the book that deals with topics pioneered by Professor Robert D. Cess.

Contributors

Willam D. Collins
National Center for Atmospheric
Research
1850 Table Mesa Drive
Boulder, CO 80305
303-497-1381

Dr. Aiguo Dai
National Center for Atmospheric
Research
Climate and Global Dynamics Division
Climate Analysis Section
1850 Table Mesa Drive
Boulder, CO 80305

Dr. Vener Galin
Department of Numerical Mathematics
Russian Academy of Sciences
Leninsky Prospect, 32 A
Moscow 117334
Russia

Dr. Anand Inamdar
13517 Tiverton Road
San Diego, CA 92130-1038

James F. Kasting
Department of Geosciences
443 Deike Bldg.
Pennsylvania State University
University Park, PA 16802
814-865-3207

Yoram Kaufman
NASA Goddard Space Flight Center
Code 913
Greenbelt, MD 20771
301-614-6189

Jeffrey T. Kiehl
National Center for Atmospheric Research
1850 Table Mesa Drive
Boulder, CO 80305
303-497-1350

Dr. Gearld A. Meehl
National Center for Atmospheric Research
Climate and Global Dynamics Division
Climate Change Research Section
1850 Table Mesa Drive
Boulder, CO 80305

Dr. Valentin Meleshko
Main Geophysical Observatory
7 Karbyshev Str.
194018 St. Petersburg
Russia

Dr. Jean.-Jacques Morcette
European Centre for Medium-Range Weather
Forecasts
Shinfield Park
Reading
RG2 9AX
United Kingdom

James C. McWilliams
Department of Atmospheric Sciences
UCLA
405 Hilgard Ave.
Los Angeles, CA 90095-1565
310-206-2829

Gerald R. North
Department of Atmospheric Sciences
Room 1012a
Oceanography and Meteorology Building

Texas A&M University
3146 TAMU College Station, TX 77843
409-845-7671

Dr. S. Ramachandran
Physical Research Laboratory
Ahmedabad, India

V. Ramanathan
Center for Atmospheric Sciences
Scripps Institution of Oceanography
University of California, San Diego
9500 Gilman Drive, MC 0221
La Jolla, CA 92093-0221
858-534-8815

V. Ramaswamy
Geophysical Fluid Dynamics Laboratory
Princeton Forrestal Campus 201
Forrestal Road
Princeton, NJ 08540
609-452-6510

David A. Randall
Colorado State University
Department of Atmospheric Science
Fort Collins, CO 80523
970-491-8474

Dr. Lorraine A. Remer
NASA Goddard Space Flight Center
Code 913
Greenbelt, MD 20771

David H. Rind
NASA Goddard Institute of Space Studies
2880 Broadway
New York, NY 10025
212-678-5593

Dr. Alan Robock
Director, Center for Environmental
Prediction
Department of Environmental Sciences
Rutgers University
14 College Farm Road
New Brunswick, New Jersey 08901-8551

Dr. Michael Schlesinger
Professor of Meteorology
Department of Atmospheric Sciences
University of Illinois
S. Gregory St. Urbana, IL 61801

Brian J. Soden
Geophysical Fluid Dynamics Laboratory
Princeton Forrestal Campus 201
Forrestal Road
Princeton, NJ 08540
609-452-6575

Dr. Georgiy Stenchikov
Department of Environmental Sciences
Rutgers University
14 College Farm Road
New Brunswick, New Jersey
08901-8551

Dr. Mark J. Stevens
2600 9th St Apt 9C
Boulder, CO 80304

Dr. Didier Tanre
University of Lille 1 – Science and Technology
59655 Villeneuve D'ASCQ Cedex
France

Warren Washington
National Center for Atmospheric
Research
1850 Table Mesa Drive
Boulder, CO 80305
303-497-1321

Richard Wetherald
Geophysical Fluid Dynamics Laboratory
Princeton University Forrestal Campus
201 Forrestal Road, Princeton, NJ 08540-6649

Minghua Zhang
Institute for Terrestrial and Planetary
Atmospheres
SUNY Stony Brook
Stony Brook, NY 11794-5000
631-632-8318

Acknowledgments

Acknowledgments are due to Jay Fein, NSF, Ari Patrinos, US DOE, and to the DOE for funding the workshop on Frontiers of Climate Science, in honor of Robert D. Cess, held at Scripps Institution of Oceanography, University of California, San Diego, CA, October 19–21, 1999.

1

Overview of climate modeling

JEFFREY T. KIEHL

National Center for Atmospheric Research, Boulder, CO

1.1 Earth's climate system

The study of Earth's climate system is motivated by the desire to understand the processes that determine the state of the climate and the possible ways in which this state may have changed in the past or may change in the future. The most comprehensive tool available to reach this understanding is the global Earth-system model. Earth's climate system is composed of a number of components (e.g., atmosphere, hydrosphere, cyrosphere, and biosphere). These components are non-linear systems in themselves, with various processes, which are spatially non-local. Each component has a characteristic time scale associated with it. The entire Earth system is composed of the coupled interaction of these non-local, non-linear components. Given this level of complexity, it is no wonder that the system displays a rich spectrum of climate variability on time scales ranging from the diurnal to millions of years. Chapter 4 explores issues of climate variability in more detail. This level of complexity also implies the system is chaotic (Lorenz, 1996, Hansen *et al.*, 1997), which means the representation of the Earth system is not deterministic. However, this does not imply that the system is not predictable. If it were not predictable at some level, climate modeling would not be possible. Why is it predictable? First, the climate system is forced externally through solar radiation from the Sun. This forcing is quasi-regular on a wide range of time scales. The seasonal cycle is the largest forcing Earth experiences, and is very regular. Second, certain modes of variability, e.g., the El Nino southern oscillation (ENSO), North Atlantic oscillation, etc., are quasi-periodic unforced internal modes of variability. Because they are quasi-periodic, they are predictable to some degree of accuracy. The representation of the Earth system requires a statistical approach, rather than a deterministic one.

Frontiers of Climate Modeling, eds. J. T. Kiehl and V. Ramanathan.
Published by Cambridge University Press. © Cambridge University Press 2006.

Modeling the climate system is not concerned with predicting the exact time and location of a specific small-scale event. Rather, modeling of the climate system is concerned with understanding and predicting the statistical behavior of the system; in simplest terms, the mean and variance of the climate system. Geologic data indicate that Earth's climate has experienced a range of temperature states over its 4.6 billion-year history (Kump *et al.*, 1999). Earth's climate has experienced very warm periods such as the Cretaceous and Eocene, and very cold states such as the glacial maximum of 18 000 years ago. The geologic data also indicate that Earth's climate can fluctuate on relatively fast time scales (e.g., the Dansgaard–Oeschger events of ice rafting) (Bond *et al.*, 1995). The stability of Earth's climate is further explored in Chapter 13. Finally, the present and future state of Earth's climate must account for a new forcing factor (usually considered external) due to human industrial activity (Hansen *et al.*, 1998). The forcing arising from this activity rivals and, in the future, may exceed that due to past natural forcing. Humans have an innate interest in understanding how Nature works. Understanding Earth's climate system and being able to predict the climate poses one of the most challenging scientific questions in human history.

Climate models are based on mathematical representations, which attempt to reproduce the behavior of the Earth system (Trenberth, 1992). These representations attempt to account for the most important external and internal forcing factors. They also try to include the most important processes that create feedbacks within the system. These feedbacks act to either enhance (positive feedback) or damp (negative feedback) the forcing of the system. A detailed investigation of the various feedbacks in the climate system are discussed in Chapters 8 and 9.

The complexity of the mathematical relations and their solution requires the use of large supercomputers. The chaotic nature of the climate system implies that ensembles are required to best understand the properties of the system. This requires numerous simulations of the state of the climate. The length of the climate simulations depends on the problem of interest. The time scale of ENSO is of intraseasonal length. The time scale of global warming due to increased greenhouse gases is on the time scale of a century or more (see Chapter 2). The purpose of this chapter is to provide an overview of the modeling of the climate system, and to discuss future challenges in Earth-system modeling.

1.2 Elements of the climate system

The fundamental principal that determines the state of the climate is the amount of energy entering the system. The climate system receives shortwave radiation (wavelengths less than four microns) from the Sun. The amount of radiation is determined by the solar luminosity and the distance of Earth from the Sun. Variations in the tilt of Earth relative to the Sun affects the latitudinal distribution of shortwave energy, but not the global mean energy. The energy available to the climate system is the amount absorbed by the surface–atmosphere system. For a stable climate system, this absorbed shortwave energy must be balanced by outgoing longwave energy at the top of the atmosphere. At the surface, the net radiative energy (shortwave plus longwave) is balanced by the turbulent transfer of latent and sensible energy from and to the atmosphere. A summary of the annual global mean energy budget is given in Figure 1.1 (Kiehl and Trenberth, 1997).

For equilibrium conditions to exist, perturbations to any one of the energy fluxes must be balanced through an adjustment of the remaining terms until a new balance is established. The equilibrium response in global mean temperature, T_s, from an instantaneous forcing, ΔF, is given in Equation (1.1)

$$\Delta T_s = \lambda \Delta F \qquad (1.1)$$

where λ is the climate sensitivity factor (Dickinson, 1985). It is a measure of how the climate system responds to initial forcings. If $\lambda > 1$, then the system has a net positive feedback, which amplifies the initial forcing. If $\lambda < 1$, then the system has a net negative feedback, which decreases the magnitude of the initial forcing.

1.2.1 Physical system

The physical components of the climate system include the atmosphere, ocean, land, and cryosphere. Each of these components includes various physical processes that are in general non-local and non-linear. The circulation of the atmosphere is determined through a balance of radiative, convective, and dynamical processes (Hartmann, 1994). Spatial gradients in atmospheric heating lead to atmospheric motions that transport sensible and latent energy to balance the radiative fluxes. A simple way to consider the effects of radiation and clouds on the circulation is through the quasi-geostrophic streamfunction, which depicts the flow of airmass in terms of diabatic heating and other processes. This streamfunction is given by Equation (1.2) (Holton, 1992),

$$\overline{X} \propto -\frac{\partial \overline{Q}_{diab}}{\partial y} + \frac{\partial^2 \overline{H}}{\partial y^2} + \frac{\partial^2 \overline{M}}{\partial y \partial z} + \frac{\partial \overline{F}_D}{\partial z} \qquad (1.2)$$

where y is the latitudinal length scale; $\overline{Q}_{diab}$ is the zonal net diabatic heating (e.g., convective radiative, condensational), $\overline{M}$ is the zonal momentum flux, $\overline{H}$ is the zonal heat flux and $\overline{F}_D$ is the zonal drag force. Note it is the latitudinal *gradient* of diabatic heating that leads to atmospheric circulations.

The transport of moisture within the atmosphere with associated latent heat release through condensation leads to the formation of clouds. The presence of

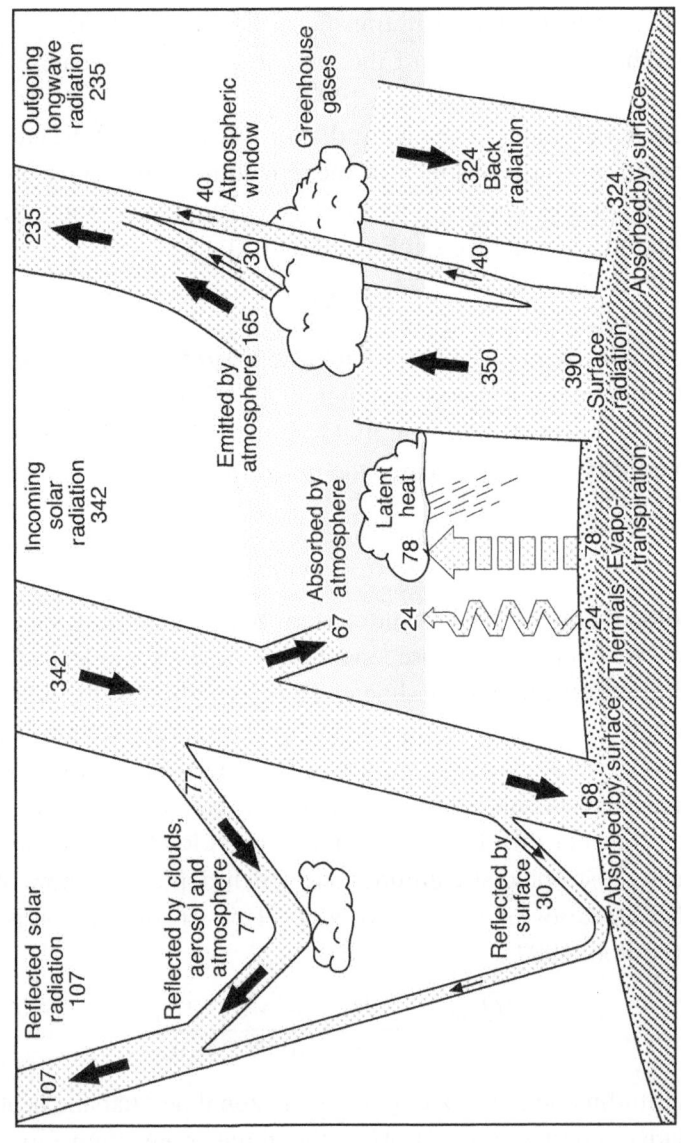

Figure 1.1. Earth's global mean annual mean energy budget as estimated by Kiehl and Trenberth (1997). Energy fluxes are in W m^{-2}.

4

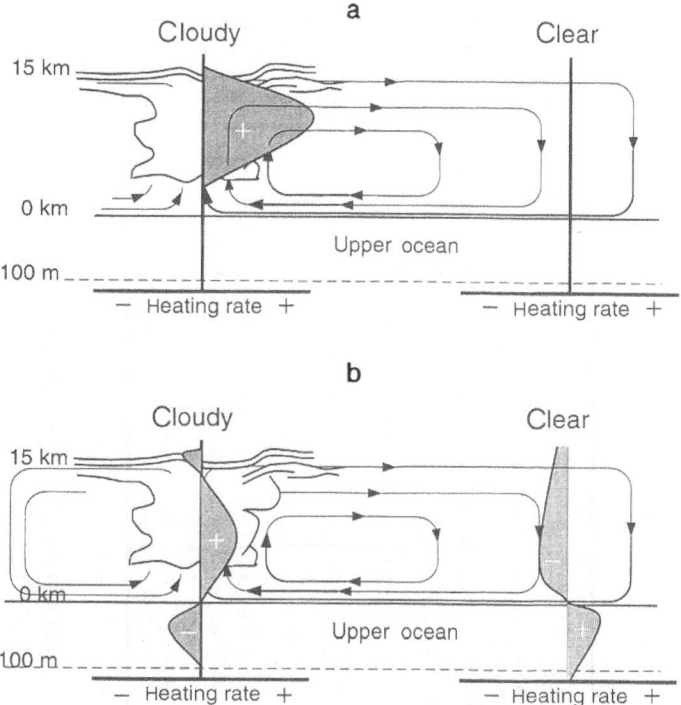

Figure 1.2. Heating profiles in convective cloud regions and clear-sky regions: (a) latent-heating gradient; (b) radiative-heating gradient. The cloud region indicates heating due to latent and radiative processes. From Webster (1994).

clouds, in turn, leads to a modulation of the gradients in energy fluxes (radiative and latent) in the atmosphere. Thus, the radiative flux gradient is supported by the gradients in latent energy release, as shown in Figure 1.2 (Webster, 1994). Both of these processes support atmospheric circulations in both the zonal direction, i.e., the Hadley circulation, and in the longitudinal direction, i.e., the Walker circulation as shown in Figure 1.3.

The ocean system is driven by surface wind stress created by atmospheric circulations. Within each ocean basin the drag of the atmospheric winds on the ocean and the influence of the Coriolis force leads to largescale gyre circulations (see Figure 1.4). These are circulations that occur in the upper ocean, and aid in the transport of energy in the poleward direction. The circulation of the deep ocean (depths greater than a few hundred meters) is driven by density gradients in the ocean. The density of seawater is determined by its temperature and salinity. Colder and more saline water is denser than warm, less saline water. In polar regions, denser water forms over less dense water, which results in a largescale overturning circulation,

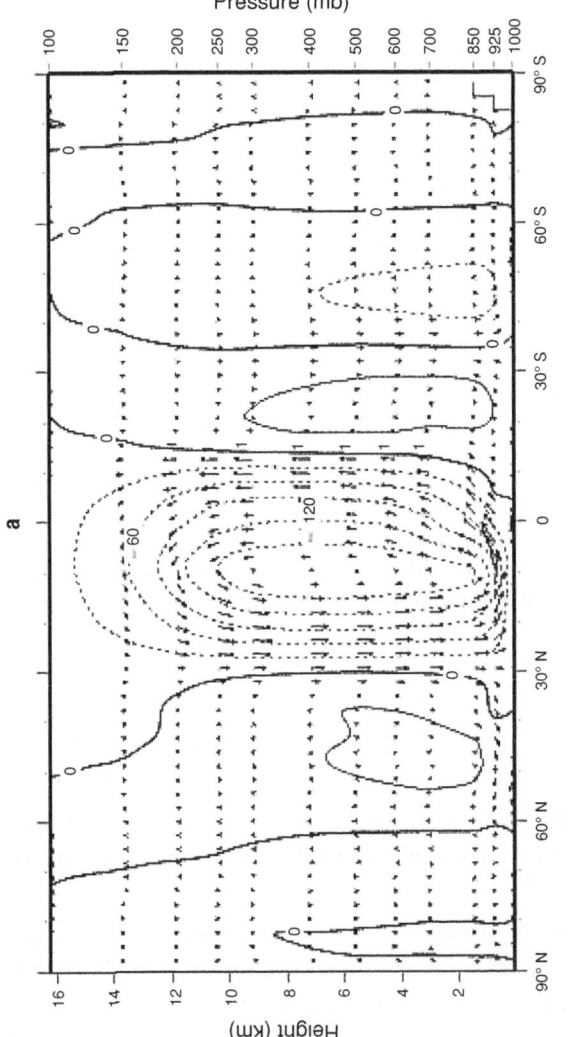

Figure 1.3. (a) Hadley circulation and (b) Walker circulation mass streamfunctions ($\times 10^9$ kg s^{-1}). From Trenberth et al. (2000).

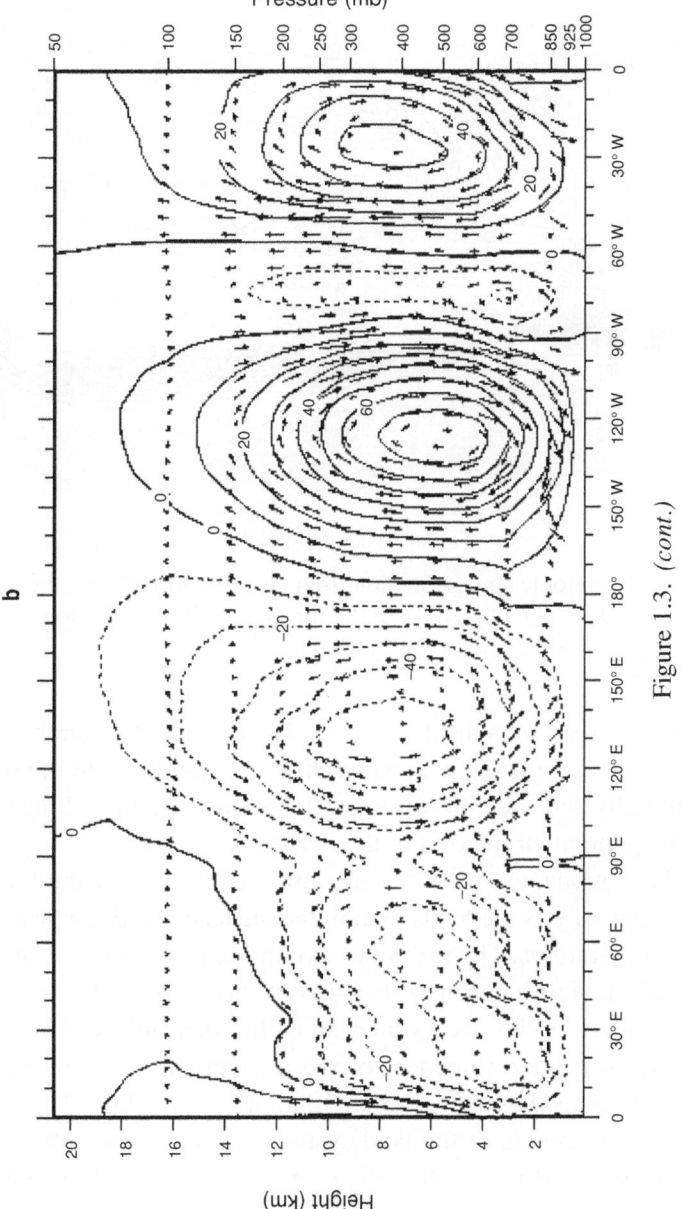

Figure 1.3. (*cont.*)

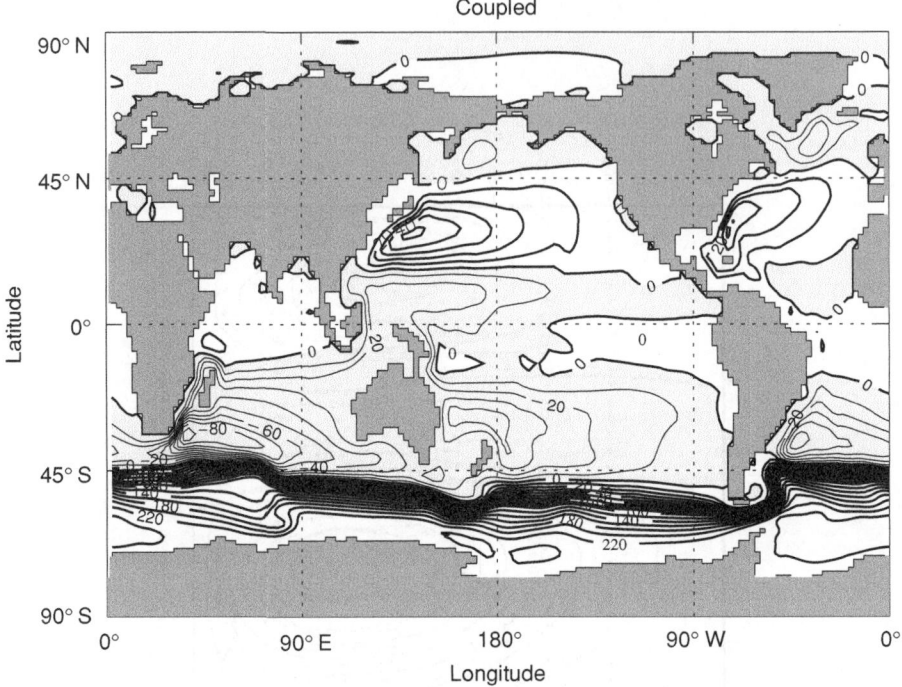

Figure 1.4. Barotropic streamfunction from an ocean model forced with observed wind stresses. From Gent *et al.* (1998). Units are Sv, Sv = sverdrup.

thermohaline circulation, in the deep ocean (Figure 1.5). This circulation transports water mass to the abyssal regions, which flows southward into the southern hemisphere. Eventually, the water moves toward the surface in upwelling regions, which completes the general circulation of the oceans.

In the polar regions, temperatures are sufficiently low for the formation of sea ice. The presence of this ice leads to a higher surface albedo, which reinforces the low surface temperatures. The ice can be dynamically advected by atmospheric and ocean circulations. Furthermore, formation of the ice leads to colder, more saline water, which supports the deep water, so-called thermohaline circulation. Thus, changes in the extent and thickness of sea ice can have a dramatic effect on not only the polar regions, but the ocean circulation as a whole.

The physical processes of the land system include the transfer of sensible energy, latent energy, and momentum between the surface and the atmosphere. The transfer of any chemical species from the surface may also be accounted for in the land system. Different vegetation results in varying efficiency of moisture exchange. The evapotranspiration process leads to important linkages between the

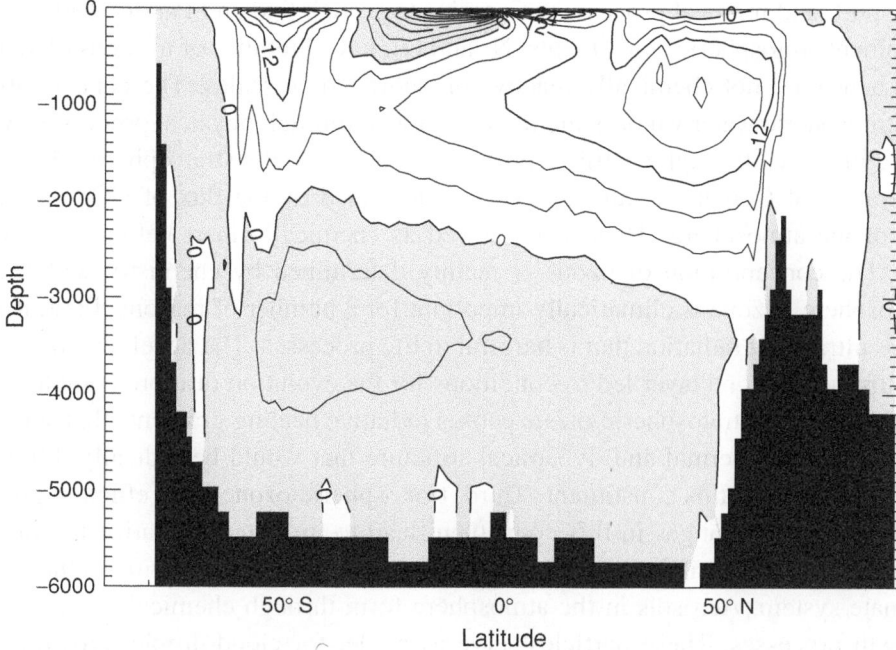

Figure 1.5. North Atlantic thermohaline circulation from an ocean model forced with observed forcing factors. From Gent *et al.* (1998). Units are Sv.

atmospheric system and the land system. Storage of water is also affected by variations in soil characteristics. Soil-moisture processes lead to long time scales in the land system (~one to two years). Water that is not stored locally runs off and eventually enters the ocean as a source of fresh water. This fresh water source alters the density of seawater, and thus affects the strength of the overturning circulation.

This brief discussion of the atmosphere, ocean, cryosphere, and land system indicates that these components are coupled through the transfer of energy, momentum, and water mass. Gradients in the flux of these quantities across boundaries lead to circulation systems within the atmosphere and ocean, where alteration in any of these fluxes leads to compensating changes to the components until a dynamical and radiative balance is restored.

1.2.2 Chemical system

The chemical composition of the atmosphere is governed by the emission of gases from Earth's surface, the chemical processes within the atmosphere, atmospheric transport, and removal processes of species (e.g., Seinfeld and Pandis, 1998). The dominant atmospheric constituents, N_2 and O_2, have lifetimes of millions of years, and hence are not chemically reactive on shorter time scales. The concentration of atmospheric water vapor is mainly determined through physical processes, with some influence through chemical processes, notably in the stratosphere. The concentration of CO_2 in the atmosphere is determined by the flux of CO_2 into and out of the atmosphere. Thus, it is viewed as chemically inert below about 100 km. The concentration of ozone is mainly determined by chemistry within the atmosphere. Ozone is climatically important for a number of reasons. First, it absorbs ultraviolet radiation that is harmful to life processes. The development of the stratospheric ozone layer led to conditions for the evolution of more complex life forms. Second, stratospheric ozone causes radiative heating gradients that support a stratospheric thermal and dynamical structure that would be radically different in the absence of this constituent. Third, tropospheric ozone is an efficient greenhouse gas, and changes in this constituent lead to important radiative forcing of the system. Chemical formation of particulates (aerosols) is also important to the climate system. Aerosols in the atmosphere form through chemical and physical growth processes. These particles serve as nuclei for cloud-droplet growth, and hence are important for cloud processes. Aerosols affect the amount of solar radiation available to the climate system by directly reflecting shortwave radiation back to space (see Chapter 11), or through modifying the reflection of clouds (the indirect effect). These particles can also act as sites for heterogeneous chemical reactions.

Ocean chemistry affects the major biogeochemical cycles of the climate system. The uptake of carbon by the ocean is a major sink that determines the amount of CO_2 left in the atmosphere. Chemical processing of carbon is regulated by biological activity in the ocean. The level of biological activity is controlled by nutrient availability, which is determined by ocean circulation. This is one reason for concern over changes in the thermohaline circulation. A slow down of this circulation means less surface water is transferred to deep water. This would lead to lower uptake of CO_2 into the ocean, which would imply a net increase in atmospheric CO_2. The ocean also plays an important role in the biogeochemical cycling of sulfur and phosphorus.

Terrestrial biogeochemical processes play a significant role in the cycling of carbon and nitrogen. The land biomass acts as a major sink for carbon, thus playing a dominant role in the determination of atmospheric CO_2 levels. Changes in the amount of biomass can have large effects on the carbon cycle. A major source of nitrous oxide, N_2O, in the atmosphere is through terrestrial nitrogen cycling. In turn, an important source of reduced nitrogen is the deposition of nitric acid, HNO_3, onto land surfaces. This process is controlled through the hydrologic cycle.

1.2.3 Ecological system

Ecological systems determine the rate at which chemical cycling occurs. Changes in terrestrial ecological systems can also affect the efficiency with which water is stored or transferred between the surface, atmosphere, and oceans (through runoff). In turn, the type, distribution, and rate of change in terrestrial and marine ecosystems depends on Earth's climate. Ecological processes function on time scales from seasonal to century in length.

1.3 Modeling the climate system

The physical components of climate models are based on the laws of conservation of energy and mass, and the second law of motion (Kiehl, 1992). Conservation of energy for the atmosphere leads to the predictive equation, Equation (1.3), for temperature, where v is the horizontal vector velocity, p is pressure, Q_{rad} is the net

$$\frac{\partial T}{\partial t} = -v \cdot \nabla T + \omega \left(\frac{kT}{p} - \frac{\partial T}{\partial p} \right) + \frac{Q_{rad}}{C_p} + \frac{Q_{con}}{C_p} + D_H \qquad (1.3)$$

radiative heating, Q_{con} is the net condensational heating, D_H is the diffusive and boundary layer heating; ω is the vertical velocity in terms of the time rate of change of pressure.

Conservation of mass implies the continuity of atmospheric circulation, Equation (1.4).

$$\frac{\partial \omega}{\partial p} = -\nabla \cdot \mathbf{v} \qquad (1.4)$$

Similar equations exist for the conservation of a constituent, such as water vapor or a chemical constituent, where the production and destruction of the substance must be included in the prediction equation for the substance. For water vapor, the predictive equation takes the form of Equation (1.5), where E represents the

$$\frac{\partial q}{\partial t} = -\mathbf{v} \cdot \nabla q - \omega \frac{\partial q}{\partial p} + E - C + D_q \qquad (1.5)$$

evaporation of liquid water to vapor, C is the condensation of vapor to liquid, and D_q is the turbulent transport of water vapor. The momentum of the atmospheric circulation is determined from Newton's second law of motion, $F = ma$, see Equation (1.6), where, $\hat{\mathbf{k}}$ is the unit vector in the vertical direction, Φ is the

$$\frac{\partial \mathbf{v}}{\partial t} = -\mathbf{v} \cdot \nabla \mathbf{v} - \omega \frac{\partial \mathbf{v}}{\partial p} + f \hat{\mathbf{k}} \times \mathbf{v} - \nabla \Phi + D_M \tag{1.6}$$

geopotential, and D_M is the turbulent transfer of momentum. The equation of state, Equation (1.7), completes the description of the atmospheric system, where R is

$$\frac{\partial \Phi}{\partial p} = -\frac{RT}{p} \tag{1.7}$$

the gas constant.

Similar equations apply to modeling the ocean, with the important difference that the ocean is an incompressible fluid. Thus the mass continuity equation for the ocean is as shown in Equation (1.8), and the ideal gas law is replaced by

$$\nabla \cdot \mathbf{v} = 0 \tag{1.8}$$

Equation (1.9),

$$\frac{\partial \Phi}{\partial z} = -\frac{\rho}{\rho_0} g \tag{1.9}$$

where ρ and ρ_0 are densities, g is the gravitational acceleration, and $\Phi = p/\rho_0$. The ocean density, as mentioned previously, is given by the temperature and salinity of the ocean, which are determined by Equations (1.10) and (1.11).

$$\frac{dT}{dt} = 0 \tag{1.10}$$

$$\frac{dS}{dt} = 0 \tag{1.11}$$

The dynamical equations for predicting sea ice include the momentum equation, an equation for thickness, the thermodynamic equation for ice temperature, and a rheology relation for the ice stress. The momentum equation includes the stress between the atmosphere and ice, and the stress between the ocean flow and ice flow, which are the dominant terms in the balance of momentum.

Equations for the land, chemical, and ecological systems are based on parametric relationships that relate key predictive quantities to largescale forcing factors (e.g., precipitation). Thus, they are not based entirely on explicit conservation laws. However, the relations for these systems are linked to fundamental understanding of physical and chemical principles, so they are not purely empirical.

1.4 Hierarchy of climate models

The science of climate modeling includes the development of models from basic principles, the comparison of climate models with observations, and the application of models to answer fundamental scientific questions. The development of models is not independent of comparison of the model simulations with observations. Typically, a process for the atmosphere, ocean, land, or cryosphere is developed by a given scientist. Development is guided by observational data, which can be from a field observation, or global satellite observations, or a combination of both. Thus, the model processes are not developed independent of physical understanding or observations. However, it is usually the case that the data or theory or understanding may be limited in some way. Thus, the modeled process is incomplete to some degree. This is why model simulations must be constantly evaluated against a diverse range of observations.

The application of models for research can occur in at least three ways. First, climate models can be used to understand the basic processes in Earth's climate system, including climate forcing, feedbacks, and response. Second, climate models can be used to test hypotheses about how the system works, e.g., to study the role of clouds in tropical intra-seasonal variability. Another example is to explore mechanisms of past climates. Third, climate models can be used to predict the state of the future climate. The use of models to study the effects of increased greenhouse gases on the climate of the next 100 years is one example. Climate models should be viewed as "numerical laboratories" where experiments can be performed to understand the Earth system better.

A hierarchy of climate models exists to look at a range of scientific questions (Harvey, 2000). These models are usually categorized in terms of the number of spatial dimensions explicitly modeled. The simplest climate models are zero-dimensional models, which represent globally annually averaged conditions. These models are based on energy balance, Equation (1.12); N_{TOA} is the net radiative flux

$$N_{TOA} = S_{abs} - F = 0 \tag{1.12}$$

at the top of the atmosphere, S_{abs} is the global mean absorbed solar radiation by the surface–atmosphere system, while F is the global mean outgoing longwave flux at the top of the atmosphere. The absorbed solar flux is formulated in terms of surface temperature to account for the effects of surface albedo on the reflection of radiation, while the outgoing longwave flux is a linear function of surface temperature. This is a steady-state model that predicts Earth's global, annual mean temperature. It was first used to investigate the stability of Earth's climate system.

Another formulation of a zero-dimensional climate model is based on the coupled atmosphere–ocean system in a linearized, time dependent form. Thus, the coupled

atmosphere, upper and deep ocean can be modeled by Equations (1.13)–(1.15)
(Dickinson, 2000); T_a, T_m, T_o represent atmospheric, mixed layer, and deep ocean

$$C_a \left(\frac{\partial T_a}{\partial t} + \frac{U_a T_a}{L_a} \right) + \lambda_{am}(T_a - T_m) + \lambda T_a = f_a \tag{1.13}$$

$$C_m \left(\frac{\partial T_m}{\partial t} + \frac{U_m T_m}{L_m} \right) + \lambda_{am}(T_m - T_a) + \lambda_{mo}(T_m - T_o) = f_m \tag{1.14}$$

$$C_o \left(\frac{\partial T_o}{\partial t} + \frac{U_o T_o}{L_o} \right) + \lambda_{mo}(T_o - T_m) = f_o \tag{1.15}$$

temperatures, respectively; U_a, U_m, U_o and L_a, L_m, L_o represent characteristic velocities and length-scales for the atmosphere, mixed layer, and deep ocean. The
second terms within brackets account for the largescale advection of sensible energy. The term λ_{am} is the atmosphere–ocean coupling coefficient, λ is a parameter
that represents the radiative damping of the atmosphere; λ_{mo} is the mixed layer–
deep ocean coupling coefficient. Stochastic forcing of the atmosphere, mixed layer,
and deep ocean are represented by f_a, f_m, and f_o, respectively. These coupled
equations can be solved to look at the transient response of the climate system to
imposed forcing. They are also useful for understanding what processes control
the overall time scale of the entire climate system, as opposed to the time scale of
individual components of the system.

One-dimensional climate models may include the vertical direction. Models that
include only the vertical balance between radiative and convective processes are
called radiative-convective models (Ramanathan and Coakley, 1978). They have
been used to study the effects of increased trace gases on climate. Energy-balance
models that neglect the vertical dimension, but include a meridional direction are
another form of one-dimensional climate models. These models have been used
to consider sea-ice feedback processes as amplifiers of climate change (Lian and
Cess, 1977).

Two-dimensional climate models include the zonally averaged momentum, thermodynamic, and continuity equations. Thus, these equations yield information on
the meridional temperature and circulation (i.e., latitude versus pressure). Contributions from the zonal circulations (eddy terms) to the zonal mean must be
included parametrically. These models are frequently used to study stratospheric
circulations and chemical interactions. Two-dimensional ocean models, based on
the zonal mean, also exist, and are used to study the sensitivity of the thermohaline circulation to external forcings. Energy-balance models have been extended to
two dimensions in latitude and longitude, where horizontal dynamics are treated
as diffusive in nature (North *et al.*, 1983) and are discussed in Chapter 3. These

models have been used to study a range of climate-change problems related to the distribution of changes to surface temperature.

Recently, a new type of model has appeared, called an intermediate climate model (e.g., Rahmstorf, 1995). These models are used to look at climate problems that include processes with long time scales, typically century to millennium in length. Intermediate climate models employ simplifying approximations for atmospheric processes (either energy balance or radiative-convective), but use more sophisticated ocean, cryosphere, and/or biosphere models. Models of this type have been applied to study the stability of the thermohaline circulation, and the evolution of the biosphere over geological time scales.

The solution of the full three-dimensional equations for momentum, energy, and mass constitute the most complete and complex form of climate models. Originally, these models were called general circulation models (GCMs), recently a more appropriate term, global climate models has been adopted. The remainder of this chapter will focus on coupled global climate models.

1.5 Practical aspects to climate modeling

Global climate models are based on the numerical and computational solution of evolution equations describing the system components and the coupling of these components (Hack, 1992). The numerical solution of the model equations is carried out on a specified spatial grid, or at a specified spatial resolution. Dynamic motions and processes with scales greater than this spatial grid-scale are *explicitly* resolved. Motions and processes smaller than this grid-scale are spatially unresolved, and must be *implicitly* included in the model. These sub-grid processes must be parametrically represented (parameterized) in terms of the largescale resolved fields, e.g., temperature, winds, etc.

Typical grid lengths for climate models are 200 to 300 kilometers. The vertical grid spacing varies from hundreds of meters to one–two kilometers. At these spatial resolutions a number of important processes are unresolved and must be parameterized. The temporal resolution, or time step, of the components is determined by the grid size, Δx, and the maximum velocity, U, in the model shown in Equation (1.16).

$$\Delta t = \frac{\Delta x}{U} \qquad (1.16)$$

Information cannot propagate any faster than U across the grid, so time steps greater than Δt lead to numerical instabilities. Various numerical techniques exist to solve the dynamic equations for the atmosphere, ocean, and sea ice. These techniques are chosen based on their accuracy and computational efficiency. All of these techniques have certain advantages and disadvantages. For example, the spectral technique, which represents the predicted fields in terms of a spherical-harmonic basis set, works well in polar regions, but introduces oscillations near steep topography (spectral-ringing effect). While finite difference techniques eliminate these oscillations, they must use artificial numerical filtering in polar regions because the grid spacing decreases, which requires a shorter time step based on Equation (1.16). Similar arguments can be made for other numerical techniques for simulating the global climate. This is one reason there is no universal approach for designing climate models. Thus, even the explicitly resolved part of global climate models has a certain level of uncertainty associated with the particular numerical technique employed.

In the atmosphere, sub-grid-scale parameterizations include cloud processes, radiation, boundary-layer effects, and turbulent mixing in the free atmosphere. Cloud processes, such as convection, vertically transport energy and moisture in the atmosphere. Convection can occur in shallow layers in the atmosphere, or as deep penetrative convection as observed in the tropical regions. This type of convection is often tied directly to the boundary-layer model, since perturbations in energy within the boundary layer can lead to a destabilization of the atmosphere. This type of convection detrains water into the upper troposphere, and dries the lower atmosphere through precipitation. These convection schemes thus produce latent heating and drying/moistening effects that are included in the thermodynamic equation and the predictive equation for water vapor, respectively. Clouds can also form through saturation of a stable layer in the atmosphere. This process usually employs a cloud microphysics scheme. The radiation model uses the predicted cloud amounts and clear-sky atmospheric profile data to calculate radiative fluxes and heating rates. The boundary-layer model represents turbulent mixing effects near the surface, and hence links the near-surface layer to the free atmosphere.

In the ocean, boundary-layer effects are also modeled with similar assumptions as in atmospheric models concerning the dependence on stability, where now salinity effects must be included. The formation of deep water through buoyancy effects is the ocean analog of atmospheric deep convection. When the near-surface water density exceeds the density of deeper water, the column is convectively adjusted to remove the instability. For sea-ice formation, the sub-grid parameterizations include the radiative effects on the ice: the thermodynamic processes that govern the thickness of ice. For the land and ecological components, all of these processes must be parameterized, since the scales are all below the resolved 200 to 300 kilometer scales.

An important aspect of Earth-system modeling is the coupling of the various processes within and across specific components. For example, within the atmosphere component, coupling between the convection model and the boundary-layer model is known to be important for simulating a realistic climate. Links between cloud processes and radiation processes have also been shown to be important for an accurate simulation of the climate system (Ramanathan *et al.*, 1983). This is not surprising given the description of how gradients in energy fluxes within the atmosphere affect the circulation. Coupling across the components is accomplished through the transfer of fluxes of momentum, energy, and mass. The effects of surface wind stress on the ocean and sea ice are accounted for by exchanging information on near-boundary winds and stability, since the drag coefficients depend on the stability of the surface layer. The flux of energy and water (liquid or solid) is needed for the land, ocean, and sea-ice models. For biogeochemical modeling, the flux of trace species across the boundary is also required. To account for all of these exchanges, many Earth-system models have developed a separate component to handle all of the flux-transfer issues (e.g., Boville and Gent, 1998).

The final and essential component to any climate model is the issue of data-input and -output. Data input usually consists of boundary conditions for the Earth system: the land configuration and bathymetry for the ocean model, properties of the soil and vegetation for the land model, constituent datasets for the atmospheric model (assuming these are not all predicted by a chemical component). Each component model requires some input information. By far the largest exchange of information occurs when data are exported (or output) from the model. Given that each component has state information to transfer, current GCMs can export giga-bytes (10^9) of data for a single simulated year. Multi-century simulations can lead to terabytes (10^{12}) of information for a single experiment. As the complexity of Earth-system models increases, so will the data volumes, placing new demands on ways of organizing, storing, and accessing these data.

1.6 Evaluation of Earth-system models

The evaluation of Earth-system models requires observational data for each of the component systems. Since models are global in extent, the best observational data should be nearly global in spatial coverage. Climate models simulate time scales that range from diurnal to millennia in length. This implies that the temporal resolution of the observational data should also be this long. Of course, extensive datasets like this do not exist. In terms of global coverage, the best available data are obtained from satellites. These data are limited to the past 20 to 25 years, and are not continuous observations. A series of different satellite instruments have flown over the past two decades, and this introduces data inter-calibration concerns. Satellite coverage (temporal and spatial) is governed by the orbit configuration, which also introduces problems when attempting to compile global data. Finally, satellites do not directly measure geophysical quantities of interest. They rely on retrieval algorithms that convert measured radiance into a geophysical quantity. These algorithms must also be evaluated to insure the "observations" are reliable. Despite these limitations, satellite data still provide an invaluable view of the Earth system, and serve as an important data source for evaluating the veracity of Earth-system models.

Other data are required, however, for a more complete analysis of model simulations. The vertical distribution of temperature, winds, or ocean currents, chemical constituents in the troposphere and ocean, require a combination of *in situ* observations and data-assimilation models. Assimilation models are similar in structure to climate models, but are strictly constrained by networks of observational data. The data from this process are composed of a hybrid of model and observations. The advantages of these data are that they are globally gridded data, which can be easily used for model evaluation. The disadvantages of these data are that they include model biases related to specific parameterizations. Secondly, they can be affected by changes in the assimilated data and methods of assimilation. These changes can affect data continuity.

Finally, actual field observations can be used for model evaluation. This is problematic since climate models are statistical by nature, not deterministic, and the spatial scales of a field program are, in general, much smaller than the scale of a GCM grid-box. Hence, data from a specific field experiment cannot be used to evaluate the ability of the model to replicate that specific event. However, the field data, if of sufficient temporal and spatial extent, can be used to determine the statistical nature of certain relevant quantities.

Comparison of models and observations is a challenging area of research. Extensive analysis and evaluation of model simulations are a valuable tool for improving models and for understanding the way the climate system works. This type of analysis goes beyond a simple comparison of two datasets, to include an in-depth diagnostic and model sensitivity approach. Model simulations are analyzed to look at the mean state of the simulated climate and the spatial and temporal variation of a specific aspect of the system.

An example of a mean state diagnostic is shown in Figure 1.6, which shows the difference between a model-simulated annual mean sea-surface temperature (SST) and that of observations. This figure indicates the spatial structure of SST biases, but does not convey any information about biases in the temporal behavior of the climate model. Figure 1.7 shows the temporal variation in the tropical SST for a specific region from a climate model and from observations. Since the climate model attempts to represent the statistical behavior of the system, the specific timing of the anomalies in warm and cold water are not relevant. What is important is the amplitude and frequency of the model variability compared to observations. This comparison also raises the issue of the chaotic nature of the climate system. This

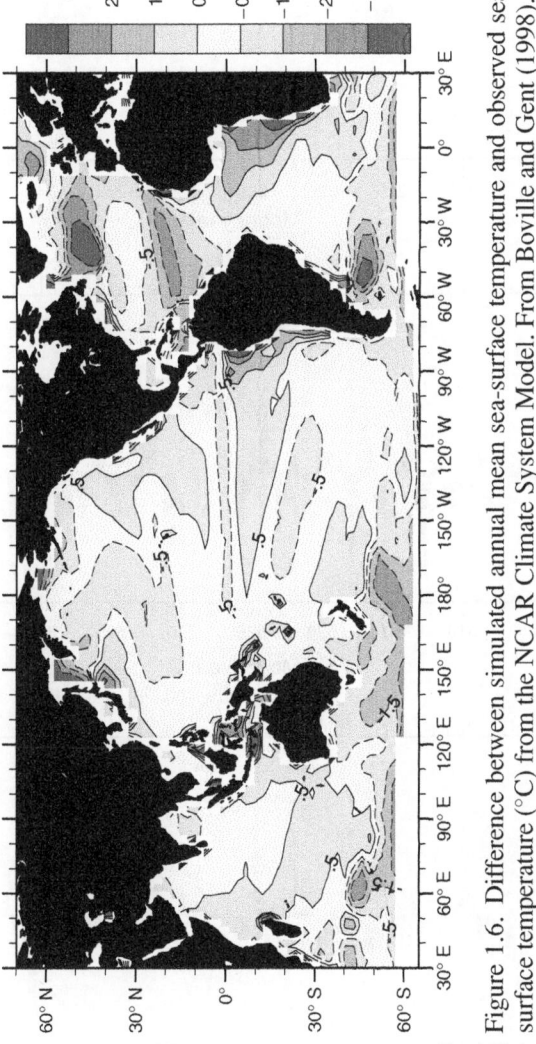

Figure 1.6. Difference between simulated annual mean sea-surface temperature and observed sea-surface temperature (°C) from the NCAR Climate System Model. From Boville and Gent (1998).

19

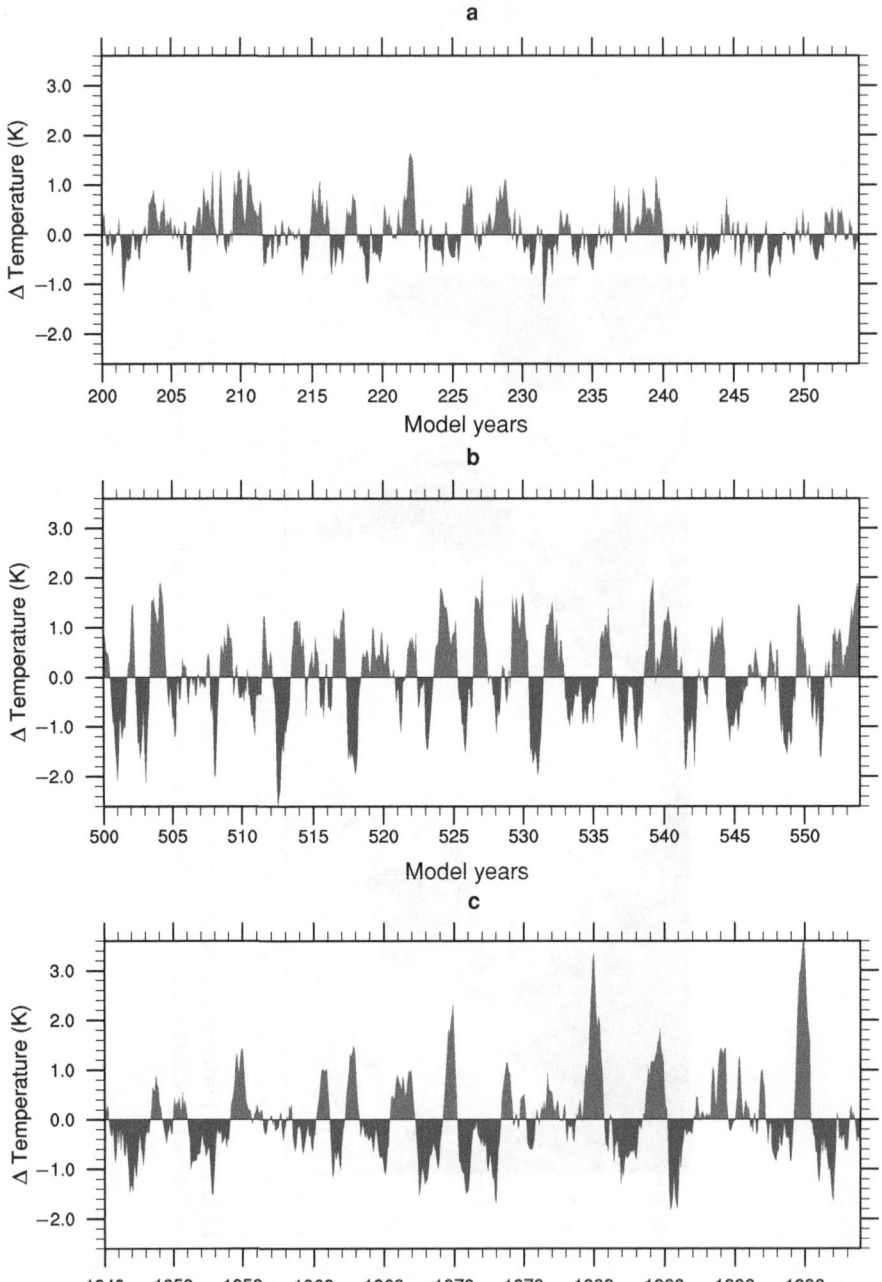

Figure 1.7. Variability in simulated tropical SSTs for the central equatorial Pacific ocean (a) CSM1.3 control; (b) CCSM2.0 control, and (c) NCEP re-analysis. From Otto-Bliesner and Brady (2001).

fact requires that an ensemble of simulations be compared to observations, since only one realization of the real world exists.

Another way of testing models is to consider climate change due to external forcing of the system. Since the response of the climate system for a given forcing is related to the sensitivity of the system, this approach is useful for evaluating climate sensitivity. Paleoclimate simulations are one example of this approach. Estimates of the forcing during the last glacial maximum (LGM) are 4 to 6 W m^{-2}. The climate is estimated to have cooled by two to three degrees during the LGM. Models with appropriate boundary conditions for the LGM can simulate this change in climate. This forced response can then be compared with the observed response. On a much shorter time scale, the response of the climate system to volcanic eruptions can also be used to evaluate climate models and is discussed in Chapter 6.

There is some degree of skepticism concerning the predictive capabilities of climate models. These concerns center on the ability to represent all of the diverse processes of nature realistically. Since many of these processes (e.g., clouds, sea ice, water vapor) strongly affect the sensitivity of climate models, there is concern that model response to increased greenhouse-gas concentrations may be in error. For this reason alone, it is imperative that climate models be compared to a diverse set of observations in terms of the time mean, the spatio-temporal variability and the response to external forcing. To the extent that models can reproduce observed features for *all* of these features, belief in the model's ability to predict future climate change is better justified.

1.7 Future challenges for Earth-system modeling

The future of climate-system modeling ultimately rests on continued scientific development and understanding. The future of this field also depends on technological advances in computation and mass data storage. Finally, the increasing complexity of this field introduces social issues.

1.7.1 Science issues

The level of complexity of an Earth-system model is determined by the scientific questions to be addressed. Over the past 30 years, emphasis has focused on the physical components of these models. Continued research and development is required to improve these basic components of Earth-system models.

In atmospheric models, improvement in the treatment of cloud processes is perhaps the most important issue. Cloud parameterizations have increased in complexity. Many models include more detailed microphysical treatments to model the formation and dissipation of stable clouds. Physical links between aerosols and clouds are required in these models, which leads to coupling between chemistry and climate processes. Unfortunately, there are too few observational data to evaluate if these more complex parameterizations are more accurate. Clouds play a fundamental role in determining the sensitivity of climate models as shown in Chapter 8. It is imperative that improvements occur in this area of model development.

In ocean models, improvements in the treatment of flow along the bottom topography are needed. The treatment of sub-grid mixing processes also requires improvement. Certain mixing processes that are known to occur in the ocean are currently not accounted for in models. Deep water formation is parameterized through a simple convective adjustment mechanism. Perhaps more sophisticated methods are required to represent more accurately this ocean process. Improved treatment of coastal processes is also needed. In sea-ice models, ice thermodynamic treatment is an area that will require more emphasis. It is very simplistic to assume sea-ice properties are homogeneous. Models of glacial growth are required to study longer time-scale problems, such as glacial to inter-glacial transitions. Modeling these effects requires the addition of land-ice models within climate-system models.

To address issues on biogeochemical cycles, a more complex treatment of biological processes is required to determine the rates of cycling. This in turn requires ecological models that dynamically determine the population distribution and type of organisms that are important to the biology. Each of these components has associated with them a particular time scale; multiple time scales lead to increasing demands on computational resources to carry out meaningful simulations of the climate system.

Added complexity to climate models implies the prediction of new climate variables. Evaluation of these new fields and continued evaluation of current model output requires extended and improved observational data. New satellite data on clouds and aerosols are required to evaluate these climatically important processes. Improved analysis products will continue to help evaluate the dynamic and thermodynamic simulations of the models. Increased information is needed on the state of the ocean, both circulation and thermal properties. Additional observations on quantities related to the chemical and biogeochemical states are also needed for model evaluation.

Most important to the future of climate-system model research is the education and training of the next generation of Earth-system modelers. Continued improvement in modeling the Earth system requires new and improved approaches to modeling components of the system, and the addition of whole new components. Education of this new generation rests on the dedication of both university faculty and scientists at national laboratories. Indeed, a partnership between these two sources of mentors is perhaps the best way to accomplish the education and training of a new generation of climate scientists.

1.7.2 Technology issues

Increased complexity of climate models creates new technological demands for the field of climate research. As computer codes grow in complexity and as more development occurs over greater distances, management of Earth-system-model codes will require a new infrastructure. This infrastructure will need to deal with issues related to documentation of the model, coding standards, and enhanced communication among the scientists and engineers working on the Earth-system model. These are issues that modeling projects have rarely had to face. Yet, they are very real challenges that must be met with creative management solutions.

The development and implementation of future climate models will require more flexible and efficient coding structures to meet the ever-changing architectures of computer hardware. Significant changes in the paradigm of supercomputing have occurred over recent years. Changes in this paradigm may well continue into the future. Yet, climate research requires some level of stability to carry out long-term, reproducible investigations. Matching this goal of the climate-scientists with the reality of changing computer hardware will be a major challenge in the field of climate-system modeling.

As stated previously, the growing complexity of climate models implies increasing volumes of simulated data. The management and access of these large datasets will also be a challenge.

1.7.3 Social issues

There are three social issues that are relevant to the future of climate modeling: social interaction within a diverse scientific community, interaction between the climate-modeling community and the social-science community, and interactions between the science community and society at large.

As Earth-system models grow in complexity, the diversity of science disciplines required to create the models will also increase. Each science discipline has evolved in its own fashion. Scientists in these disciplines have their own terminology and methods of carrying out research. Yet, they are all working on building a single scientific research tool. Overcoming the "Tower of Babel" communication problems in building an Earth-system model requires a new level of social interaction. Although "interdisciplinary research" is a well-worn phrase, an activity like Earth-system modeling can only succeed if effort is made to work in a strongly collaborative manner.

As the predictive capability of climate models is exercised, interest from the social, policy, and impacts community will continue to grow. Thus, mechanisms will be required to enhance communication between climate scientists and social scientists. It is important that this mechanism exists, since both sides need assurance that the most accurate and valuable information is exchanged between these groups.

Finally, it is incumbent upon climate scientists to convey their science to the public. The relevance of climate science to society is immense. Throughout history, civilizations have dealt with changes in climate. Future climate change may have significant effects on society. If for no other reason than this, the science community should strive to convey their discoveries to society.

1.8 Acknowledgments

Much of what I have learned about climate modeling came from reading the papers of Professor R.D. Cess, and my collaborations with Bob over the past 20 years. His unrelenting care and integrity in carrying out research have been a model for me. I thank him for all he has given me.

References

Bond, G. C. and R. Lotti (1995). Iceberg discharges into the North Atlantic on millennial time scales during the Last Glaciation. *Science* **267**, 1005–10.

Boville, B. A. and P. R. Gent (1998). The NCAR Climate System Model, version one. *J. Climate* **11**, 1115–30.

Dickinson, R. E. (1985). Climate sensitivity. *Adv. Geophys.* **28**, 99–130.

(2000). How coupling of the atmosphere to ocean and land helps determine the time scales of interannual variability of climate. *J. Geophys. Res.* **105**, 20115–9.

Gent, P. R., F. O. Bryan, G. Danabasoglu *et al.* (1998). The NCAR Climate System global ocean component. *J. Climate* **11**, 1287–1306.

Hack, J. J. (1992). Climate system simulation: basic numerical & computational concepts. In *Climate System Modeling,* ed. K.E. Trenberth, Cambridge, Cambridge University Press.

Hansen, J., M. Sato, R. Ruedy *et al.* (1997). Forcings and chaos in interannual to decadal climate change. *J. Geophys. Res.* **102**, 25679–720.

Hansen, J., M. Sato, A. Lacis, R. Ruedy, I. Tegen, and E. Matthews (1998). Perspective: Climate forcings in the industrial era. *Proc. Natl. Acad. Sci.* **95**, 12753–8.

Hartmann, D. L. (1994). *Global Physical Climatology*. New York, Academic Press.

Harvey, D. D. L. (2000). *Global Warming: the Hard Science*, New York, Prentice Hall.

Holton, J. R. (1992). *An Introduction to Dynamic Meteorology*, 3rd edn. San Diego, Academic Press.

Kiehl, J. T. (1992). Atmospheric general circulation modeling. In *Climate System Modeling*, ed. K. E. Trenberth, Cambridge, Cambridge University Press.

Kiehl, J. T. and K. E. Trenberth (1997). Earth's annual global mean energy budget. *Bull. Amer. Meteor. Soc.* **78**, 197–208.

Kump, L. R., J. F. Kasting, and R. G. Crane (1999). *The Earth System*. Prentice Hall, New York.

Lian, M. S. and R. D. Cess (1977). Energy balance climate models: A reappraisal of ice-albedo feedback. *J. Atmos. Sci.* **34**, 1058–1062.

Lorenz, E. N. (1996). *The Essence of Chaos*. Seattle, University of Washington Press.

North, G. R., J. G. Mengel, and D. A. Short (1983). A simple energy balance model resolving the seasons and the continents: Application to the astronomical theory of the Ice Ages. *J. Geophys. Res.* **88**, 6576–86.

Otto-Bliesner, B. L. and E. C. Brady (2001). Tropical Pacific variability in the NCAR Climate System Model. *J. Climate* **14**, 3587–607.

Ramanathan, V. and J. A. Coakley, Jr. (1978). Climate modeling through radiative-convective models. *Rev. Geophys. Space Phys.* **16**, 465–89.

Ramanathan, V., E. J. Pitcher, R. C. Malone, and M. L. Blackmon (1983). The response of a spectral general circulation model to refinements in radiative processes. *J. Atmos. Sci.* **40**, 605–30.

Ramhstorf, S. (1995). Bifurcations of the Atlantic thermohaline circulation in response to changes in the hydrological cycle. *Nature* **378**, 145–9.

Seinfeld, J. H. and S. N. Pandis (1998). *Atmospheric Chemistry and Physics: from Air Pollution to Climate Change.* New York, John Wiley & Sons.

Trenberth, K. E., ed. (1992). *Climate System Modeling.* Cambridge, Cambridge University Press.

Trenberth, K. E., D. P. Stepaniak, and J. M. Caron (2000). The global monsoon as seen through the divergent atmospheric circulation. *J. Climate* **13**, 3969–93.

Webster, P. J. (1994). The role of hydrological processes in ocean–atmosphere interactions. *Rev. Geophys.* **32**, 427–76.

2

Climate-change modeling: a brief history of the theory and recent twenty-first-century ensemble simulations

WARREN M. WASHINGTON, AIGUO DAI, AND GERALD A. MEEHL

National Center for Atmospheric Research, Boulder, CO

2.1 Introduction

There is a long scientific history about the effects of increasing CO_2 on the climate system. The first scientist to provide a scientific basis and concern was the Swedish chemist Svante Arrhenius (Figure 2.1), who published a paper in 1896. He was the first researcher to explain in a qualitative way the possible climatic effects of increased concentrations of greenhouse gases. The Industrial Revolution had just started in the 1890s and in the newly developed countries of Europe and the United States there was an increase in the use of fossil fuels, at that time mostly coal. As we now know this was the start in the wide-spread use of fossil fuels as an energy source that is a major factor in the observed increases of CO_2 concentrations in the atmosphere. Arrhenius concluded that atmospheric CO_2 was important to Earth's heat balance and that increases in this gas would lead to increased atmospheric temperature. Interestingly, Arrhenius estimated that an increase of 2.5 to 3 times the CO_2 concentration would result in a globally averaged temperature increase of 8–9 °C, a climatic global warming effect not too different from that estimated by the present generation of complex and comprehensive computer climate models. In Arrhenius's simple scientific argument only some elements of the complex interactions of the climate system were known and included in his estimate. The latter does not give a detailed analysis of the changes in the climate system. Such an analysis can only be derived from advanced computer climate models.

Somewhat earlier, the Irish scientist John Tyndall (1861) (Figure 2.2), in his classic paper, "*On the absorption and radiation of heat by gases and vapors, and on the physical connection of radiation, absorption and conduction,*" described the effect of increasing CO_2 and other gases, including water vapor, on Earth's radiative

Frontiers of Climate Modeling, eds. J. T. Kiehl and V. Ramanathan.
Published by Cambridge University Press. © Cambridge University Press 2006.

Figure 2.1. Svante August Arrhenius predicted in 1896 that burning fossil fuels like coal would increase atmospheric carbon dioxide and in turn warm the climate system.

Figure 2.2. John Tyndall studied the radiative effects of increased greenhouse gases in 1861, and inferred a warmer climate.

balance. In a greenhouse analogy, solar radiation would be transmitted largely through greenhouse glass while significant terrestrial or infrared radiation would be trapped by the glass. This is an overly simple description of the fundamental essence of the greenhouse warming phenomena. In several ways the greenhouse analogy is not correct and it is somewhat of a misnomer; however, the name is widely used.

With simple idealized models, Callendar (1938), Plass (1961), Mitchell (1961), and others tried to link CO_2 increases to temperature changes. From these simple beginnings the use of more and more advanced models has led to a more holistic understanding of the complex interactions involved in global warming. This chapter will review some of the history of climate modeling and discuss in particular the use of a National Center for Atmospheric Research (NCAR) climate model for investigating the effect of increasing greenhouse gases and sulfate aerosols on the climate system. These models are sometimes referred to as general circulation models (GCMs) in that they refer to model components that compute the circulation characteristics, such as for atmospheric temperature, water vapor, wind, and pressure. In a similar manner the ocean circulation system includes the currents, temperature, and salinity distributions. As discussed in Chapter 1, the interactions between atmosphere, land, ocean, and sea ice are complex and involve the transfers of momentum, mass (e.g., water), and energy. For example, the atmosphere exchanges momentum, heat, and water with the ocean and sea ice mostly due to wind stress, heat fluxes, and water fluxes.

The first coupled model simulations of climate change were equilibrium experiments in which the amount of CO_2 in the model atmosphere was simply doubled and the differences in resulting climate were examined. Another type of simulation is the so-called transient experiments in which the amount of CO_2 and other greenhouse gases are increased gradually in a realistic fashion to examine how the climate adjusts to the slowly increasing greenhouse effect. In these early simulations, it is not just the direct changes in greenhouse radiative heating that are important, but also the myriad of complex feedback effects that must be dealt with and that make this a difficult problem even for present-day simulations of climate change. For a detailed review of the early and present climate modeling studies of the greenhouse effect, we refer the reader to the most recent report of the Intergovernmental Panel on Climate Change (IPCC, 2001).

2.2 Fundamental radiative effects of increased greenhouse gases

Each gas in the atmosphere has unique radiative physical properties. The greenhouse gases are strong absorbers and emitters of thermal infrared radiation in the 5- to 100-μm range. One curious finding is that oxygen and nitrogen – the main constituent gases in the Earth's atmosphere – are not radiatively active in the infrared wavelength range. The structure of these particular diatomic molecules does not allow them to absorb or emit infrared radiation. Here we point out that triatomic molecules such as CO_2, water vapor (H_2O), nitrous oxide (N_2O), and ozone (O_3) are all strong absorbers and emitters in the infrared wavelength range. These effects are even more strongly manifested in molecules of more than three atoms, such as methane (CH_4) and chlorofluorocarbons (CFCs). For polyatomic molecules, the atoms can (1) vibrate toward and away from each other so as to absorb and emit radiational energy in specific wavelength ranges (vibrational energy), and (2) rotate around each other in a complex manner (rotational energy). These infrared absorption and emission properties can be calculated for simple molecules; for complex molecules, however, the absorption spectra are measured experimentally.

Absorption spectra in the 8- to 12-μm range are especially sensitive to greenhouse-gas concentrations, and climate models used in climate-change scenario experiments must provide a reasonably accurate treatment of these gases in this wavelength region. Present day climate models use very detailed and accurate methods for computing water vapor, CO_2, and other greenhouse-gas absorption properties. The computational method for estimating the radiative effects makes use of grid-scale parameters such as temperature and water vapor to represent radiative effects taking place at scales much smaller than the grid-scales. This process is often called parameterization in the literature. An increase of CO_2 alone causes an increase in radiative heating which is not especially large, thus recent increases of CO_2 do not yield a sizable climate response. It is the water-vapor-increase feedback that amplifies the direct radiative effect of CO_2 or other greenhouse trace gases. Without increased evaporation and water-vapor feedback (see Chapter 9), the direct greenhouse warming response from increases in trace gases would be small.

While most climate-sensitivity experiments have involved models with only increased CO_2, it is true that other radiatively active gases are also increasing. To first order, the following percentages of yearly increases in the atmosphere are: $CO_2 \approx 0.5\%$, $O_3 \approx 1\%$ in the troposphere, $CH_4 \approx 0.9\%$, $N_2O \approx 0.2\%$, and CFCs $\approx 4\%$. The CFCs will be decreasing relatively soon because of the Montreal Protocol, which will limit the production of this greenhouse gas. It is now known with certainty that water vapor, the principal greenhouse gas, is increasing in concentration (IPCC, 2001) in a manner that is consistent with the increase of temperatures. Water-vapor observations are discussed in Chapter 10. In fact, to a first-order approximation, as the atmosphere warms it is capable of holding more water vapor and the relative humidity is roughly constant.

2.3 Water-vapor and cloud feedback mechanisms

Some aspects of water-vapor and cloud feedback to the climate system are covered in Chapters 8, 9, and 10. The major emphasis here is to review several feedback-sensitivity mechanisms in relation to increased greenhouse gases.

Möller (1963) demonstrated the critical role of water-vapor feedback in the climate system and was the first to point out the *large* effect of water-vapor feedback on surface temperature. He assumed that the climatological distribution of relative humidity remains essentially the same, although the surface temperature is allowed to change. In effect, when the surface temperature increases, absolute humidity increases since relative humidity is approximately the same. Downward infrared radiation also increases because of the increased water vapor, especially in the lower atmosphere. Also, because of increased water vapor, the solar-radiation absorption increases in the atmosphere and solar absorption decreases at Earth's surface. In a simple model with many assumptions (some questionable), Möller found that, for a surface temperature near 15 °C, a doubling of CO_2 would result in a 10 °C temperature increase. The main point of his research was to show that water vapor, the most important greenhouse gas in the atmosphere, amplifies the effect of other increasing greenhouse gases.

Cloud feedback remains a difficult aspect of greenhouse-gas climate modeling and it is an area of intense research because it can act to enhance or diminish the global warming effect. Also, it is one of the most poorly understand aspects of climate change. In very simple terms an increase in cloud cover over an area can cause more reflection of solar radiation and cooling of the climate system, but on the other hand it can lead to more "trapping" of infrared radiation which under certain circumstances can cause the climate system to warm. Over the recent years, through *in situ* field programs and satellite-based studies, the various feedback mechanisms are becoming better understood. From that understanding, they are being better modeled in climate models. The most common schemes in the atmospheric components of climate models are to tie cloudiness to a dependence upon relative humidity and the convective stability of the atmosphere. In some models the complex cloud physical processes involving cloud liquid-water and ice-water content, and cloud and ice-crystal size, are being included albeit with many simplifications and assumptions.

2.4 Snow, sea, and land-ice albedo feedbacks

Most modeling studies have shown a large positive feedback of the climate system due to albedo changes of snow and sea ice. Though the basic mechanism is thought to be simple, it has many complex aspects. Surfaces covered by snow, sea ice or permanent land ice (e.g., Greenland) all have a high albedo. Any warming resulting in melting of the snow or sea ice exposes bare ground or open seawater, which will drastically reduce the albedo of that region. This will increase the absorption of solar radiation at the surface and thus enhance the surface warming, which in turn melts more snow and ice; soon there is a strong feedback that results in enhanced warming at high latitudes. Greenhouse-warming experiments that began with small quantities of snow and ice in the control experiment or unperturbed state will have less warming with increased greenhouse gases than model experiments with extensive snow and ice (e.g., Meehl and Washington, 1990). The dependence of the amount of greenhouse warming on the sea-ice amount in the control experiment is another reason for the differences in climate simulations.

Cess *et al.* (1989) have shown that the view of the role of snow and sea-ice albedo is too simple. In some models, although snow and sea ice decrease, clouds may increase so as to increase the planetary albedo. This is plausible because increased open water at the surface, increased evaporation, and increased atmospheric water vapor can result in increases in low-level stratus clouds. Models, however, do not handle these interactions well. Sea ice has additional important feedbacks because of seasonal heat storage. For example, the high albedo of sea ice will lead to less ocean heating in that region. If sea ice disappears, then heat can be stored in the upper ocean during the summer months and retard freezing in the subsequent fall and winter months. This feedback mechanism will likely further retard sea-ice formation in subsequent seasons, and cause a stronger positive feedback. Even with our present knowledge of the key mechanisms involved in the feedbacks, the treatment of snow and sea-ice processes in climate models is undergoing continued improvement in concert with field experiments that are providing better observed data and insights into physical processes.

2.5 Energy-balance climate-model estimates

Energy-balance models (EBMs) were the first simple climate models to incorporate details of the important feedback processes involved in the greenhouse effect. These models were used in the 1960s and 1970s by Budyko (1974) and Sellers (1974) and are discussed in Chapter 3. Energy-balance models compute a surface temperature that balances energy fluxes. Depending on the model's details, terms are added to account for meridional (i.e., north–south) fluxes of energy by the atmosphere and ocean system.

The effect of increased greenhouse gases with such models is investigated by changing the infrared-radiation aspects of the model. More infrared flux toward the surface caused by increased greenhouse gases will result in a warmer surface. Because the surface and lower atmosphere in EBMs are closely coupled, the system responds by increasing surface temperature. It is also possible to include the change of the reflectance or albedo of snow and ice surfaces in the zonally averaged EBMs as a result of warmer surface temperatures. This positive feedback can amplify the warming, especially in the higher latitudes.

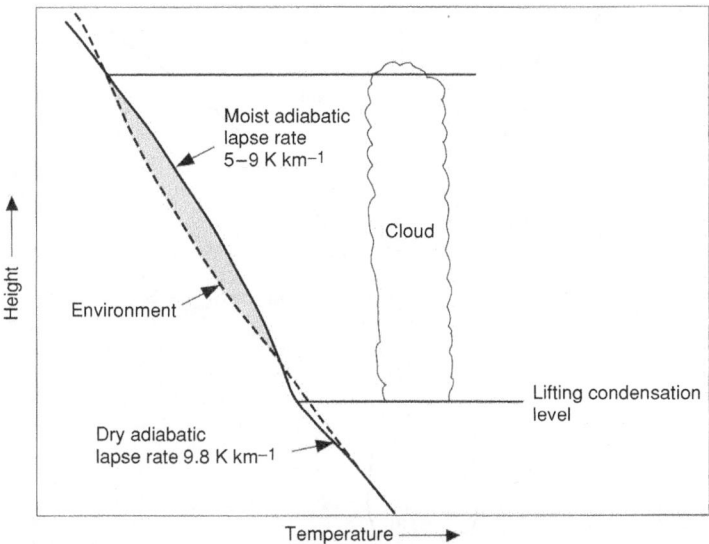

Figure 2.3. A schematic of cumulus-cloud formation showing the lifting conden-sation level (LCL), dry and moist adiabatic lapse rates, and lapse rate for air outside cloud which is labeled as environment. Note the adiabatic lapse rate is less than the environmental lapse rate indicating that condensation of water vapor into liquid water has taken place, which locally warms the atmosphere.

Reasonable assumptions in the surface physical processes of an EBM yield a global warming of about 3 °C for doubled CO_2 concentrations. In this type of model, ad hoc assumptions must be made about the feedback mechanisms and the role of dynamics in the redistribution of heat. Such models have proven useful in understanding the fundamental physics of climate change.

2.6 Radiative-convective model estimates

The next level of model complexity in studies of the greenhouse effect are radiative-convective models (RCMs), first pioneered by Manabe and collaborators at the Geophysical Fluid Dynamics Laboratory (GFDL) in the early 1960s. These mod-els are one-dimensional in the vertical for which column radiative heating and cooling rates can be computed. The convective aspect of the model is shown in Figure 2.3, where heating by convection causes the atmosphere to warm and adjust to a moist adiabatic lapse rate. In adjusting to a new stable lapse rate, convection by moist processes warms the upper troposphere and cools the lower troposphere. The relative humidity is specified as well as the levels and radiative properties of the clouds.

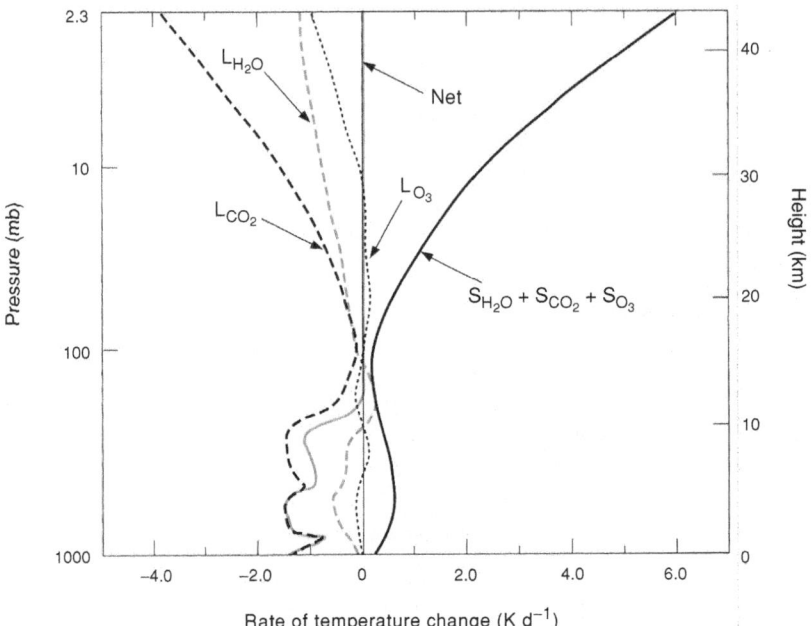

Figure 2.4. Heating rates in the atmosphere due to absorption of solar radiation (S) by atmospheric H_2O, CO_2, and O_3, and heating and cooling rates due to absorption and emission of longwave (L) or infrared radiation by H_2O, CO_2, and O_3. From Manabe and Strickler (1964).

Figure 2.4 from Manabe and Strickler (1964) shows the vertical distribution of the heating and cooling rates of various atmospheric gases, including water vapor, CO_2, and ozone. It is evident from Figure 2.4 that these gases are the largest factors in radiatively heating and cooling the atmosphere. In the figure, L denotes longwave or infrared and S denotes solar radiation. The net radiative heating and cooling are near zero in the stratosphere, but in the troposphere the net radiative cooling is compensated by vertical transfers of heat from the surface by moist and dry convection.

Solar and infrared heating and cooling rates are strongly affected by CO_2 concentrations, especially in the stratosphere. Manabe and Wetherald (1967) were the first to use a RCM for CO_2 climate-change experiments. They set solar forcing to a global annual mean, assumed that the convective adjustment of temperature has a fixed critical lapse rate that cannot be exceeded, and specified relative humidity and cloudiness. The temperature structure in the stratosphere is based on radiative equilibrium between solar and infrared fluxes, whereas in the troposphere it is based on a balance of radiative fluxes and convective processes. Manabe and Wetherald show the distribution of temperature for three different concentrations of CO_2 (Figure 2.5). Relative to the 300 ppm case at 150 ppm, the stratosphere

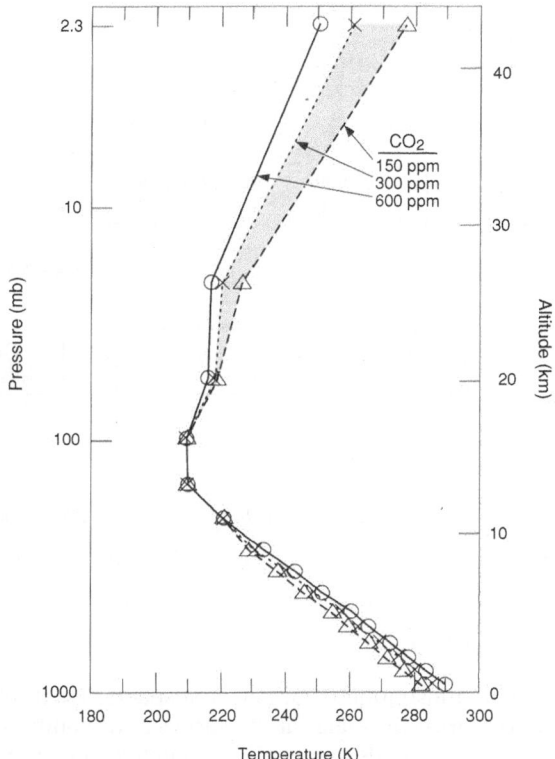

Figure 2.5. Vertical distributions of temperature in a radiative-convective model for fixed relative humidity and fixed cloud cover. Surface temperature changes are 2.88 °C for a CO_2 doubling from 150 to 300 ppm and 2.36 °C for a change from 300 to 600 ppm. From Manabe and Wetherald (1967).

warms and the troposphere cools, and at 600 ppm, the stratosphere cools and the troposphere warms. Doubling the CO_2 from 300 to 600 ppm results in a surface temperature increase of 2.4 °C, and halving the concentration results in a surface temperature decrease of 2.9 °C. The simple physics shows that increasing greenhouse gases makes it difficult for the bottom of the atmosphere to radiate energy to space so that in order for the system to reach a new equilibrium, the bottom of the atmosphere warms and the top of the atmosphere cools such that a new equilibrium is established.

Because the RCMs are so highly parameterized, they should not be expected to give reliable quantitative estimates of greenhouse-gas-caused climate change, but they are instructional for understanding the role of convective and radiative processes.

In study of radiative and water-vapor feedback, Ramanathan (1981) used an RCM to aid our understanding of the basic mechanisms. His simple one-dimensional

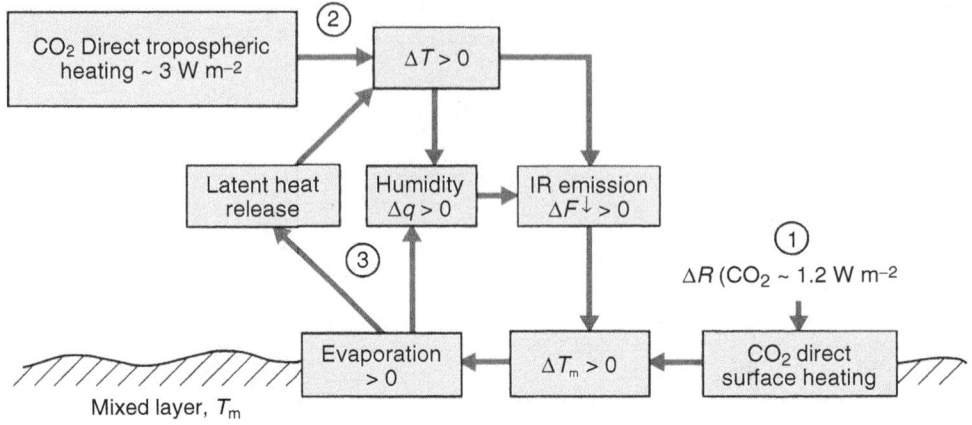

Figure 2.6. Schematic illustration of ocean–atmospheric feedback processes by which CO_2 increase warms the water at the surface. All numbers correspond to hemispheric averaged conditions and apply to doubled CO_2. From Ramanathan (1981).

atmospheric RCM was coupled to a simple, uniformly mixed, 50-m layer ocean. This type of model does not have all the feedbacks of the real climate system, but it does include many fundamental first-order processes, especially those dealing with water-vapor feedback with changes in ocean temperature. Figure 2.6 shows changes in surface temperature and fluxes of various quantities involved in the RCMs. Processes (1) and (2) are direct infrared radiative heating of the ground and the troposphere by a doubling of the atmospheric CO_2 concentrations. The total additional heating of the troposphere is about 4 W m^{-2}, of which 3 W m^{-2} is involved in process (2) that heats the troposphere, and 1 W m^{-2} in process (1) is direct radiative heating (ΔR) of Earth's surface. Process (3) enhances the direct heating effect by a series of feedbacks. The ocean surface is warmed by direct heating which causes more evaporation and a larger absolute humidity (Δq) above the surface. This, in turn, increases latent-heat release and infrared flux back toward the ocean surface ($\Delta F\downarrow$), causing increased surface temperature ($\Delta T_m > 0$). In a cyclic manner, temperature increases cause evaporation increases, etc. The process does not cause a "runaway" greenhouse effect (see also Chapter 13). The system settles down to a new equilibrium. These simple-model examples demonstrate the basic, qualitative aspects of the connection between increasing greenhouse gases and global warming.

2.7 Simulations with comprehensive three-dimensional climate models

The IPPC (2001) gives an extensive discussion of the state-of-the-art climate modeling worldwide. The discussion presented here will demonstrate the capability of NCAR-developed climate models in simulating climate change. The NCAR has had a long history of development of climate models, starting in the early 1960s and continuing to the present with comprehensive climate models. The most recent developments are due to collaborations among NCAR, universities, Department of Energy (DOE), National Aeronautics and Space Administration (NASA), and National Oceanic and Atmospheric Administration (NOAA) scientists. In the early 1990s the NCAR Climate System Model (CSM) was developed which had components contributed to by many researchers. This model is designed to be used by a large number of scientists as a tool for climate research as well as an organizing project for continued improvements in model components. The model was first written for vector supercomputer systems and later a DOE-supported version was developed, called the Parallel Climate Model (PCM) that was specifically designed for parallel computer systems. In 2000, it was decided to merge the two model approaches into a new model called the Community Climate-System Model (CCSM). This new merged model will execute on many different computer systems and it will have improved components of the climate system.

It is well known that the world's population will grow in the twenty-first century hand-in-hand with the world's economy. This will result in increasing greenhouse gases such as CO_2 in large part due to the corresponding increasing pace of fossil-fuel burning as an energy source. There are many projections of possible future climates from various modeling groups, many of which are featured in the IPPC (2001) report. These previous simulations of such changes using coupled climate models include: Manabe *et al.* (1991); Cubasch *et al.* (1992); Mitchell *et al.* (1995); Roeckner *et al.* (1999); Russell and Rind (1999); Boer *et al.* (2000); Meehl *et al.* (2000a;b); and Dai *et al.* (2001b). These simulations have been carried out using only a few realizations for each emission scenario. Individual realizations show considerable differences from one simulation to another (Delworth and Knutson, 2000), because each simulation has its own realization of the model's internal variability. However, it should be kept in mind that this is *not chaos* because of the constraints on the system from external forcing such as the solar radiation forcing and dissipative mechanisms. The differences between individual realizations cause uncertainties in model-simulated climate changes that are in addition to uncertainties associated with a given model's climate sensitivity. The other sources of uncertainties are the inter-model differences (Allen *et al.*, 2000) caused by differing methods of modeling and those that involve different future emissions scenarios (Kattenberg *et al.*, 1996). Areal and temporal averaging can reduce uncertainties in a single realization.

The preferred solution to this problem is to carry out multiple realizations (an ensemble) for each emission case (starting from different initial conditions), and to average these. Ensembles are being used increasingly in climate studies with climate model simulations (e.g., Cubasch *et al.*, 1994; Zwiers, 1996; Hansen *et al.*, 1997; Rowell, 1998; Delworth and Knutson, 2000; Wehner, 2000). Most of the previous research has employed only relatively short ensemble simulations and has focused on ensemble variability (relative to climate-change signal) rather than on the prediction of future climate changes under possible future emissions scenarios. An exception to this method is the recent study by Mitchell *et al.* (2000) who used the Hadley Centre climate model to examine the effect of stabilizing atmospheric CO_2 on global and regional climate changes. These authors used, as an idealized baseline, an ensemble of four integrations under a 1% per year CO_2 increase (with no sulfate-aerosol or other greenhouse-gas forcing) and compared the simulation with single integrations under two different CO_2 stabilization scenarios.

From Dai *et al.* (2001a) we show some results from two ensembles of five integrations of a coupled ocean–atmosphere GCM forced with projected concentrations of greenhouse gases and sulfate aerosols for the twenty-first century. A business-as-usual (BAU) scenario is used as the baseline and is compared with a CO_2 stabilization (STA550) scenario (the scenario details are given in Dai *et al.*, 2001b). This study is different from previous studies in the following ways: ensemble simulations for both the BAU and STA550 cases are used and a full range of greenhouse-gas changes as well as the sulfate-aerosol effects are included. Furthermore, these are plausible assessments of the effect of CO_2 stabilization on future climate in terms of comparing what climate changes are different in a stabilization scenario compared to that where there is no future stabilization of CO_2.

The coupled ocean–atmosphere model used here is the PCM (Washington *et al.*, 2000). The PCM is global in domain and consists of an atmospheric GCM (T42 truncation, $\sim 2.8°$ latitude/longitude resolution, with 18 vertical layers), an ocean GCM ($\sim 2/3°$ average resolution with 32 vertical layers), a land-surface model, and a sea-ice model. The PCM does not use flux adjustments. It produces a stable atmospheric and upper-ocean climate, though there is a small cooling trend of the deep ocean. The PCM is capable of simulating many of the regional climate features such as the observed El Niño amplitude (Meehl *et al.*, 2001a), the North Atlantic oscillation, and the India–Asia monsoon systems.

As noted above the development of the two future scenarios (BAU and STA550) of greenhouse-gas (CO_2, CH_4, N_2O, O_3, and CFCs) concentrations and sulfate-aerosol distributions is described in Dai *et al.* (2001b). The two scenarios were designed to be a "matched" pair. The only difference between the two scenarios is in the CO_2 concentration under the STA550 scenario: the CO_2 increase rate is substantially lower and CO_2 concentrations are projected to stabilize at 550 ppm by 2150 following Wigley *et al.* (1996). Internally-consistent emissions for CO_2 and SO_2 were generated (under the BAU scenario) using an energy-economics model (Edmonds *et al.*, 1997) driven by regionally specific assumptions with regard to population growth, economic growth, energy use per capita, technological development, etc.

The CO_2 level in 2100 is approximately 710 ppm in the BAU scenario (which is similar to the average of all concentration projections under the IPCC Special Report on Emission Scenarios (SRES) emissions scenarios; see Nakicenovic and Swart, 2000) and is approximately 540 ppm under the STA550 scenario. Global SO_2 emissions peak near 2005 (at 81 Tg S yr^{-1}, 1 Tg = 10^{12} g) and then decline steadily until 2080 when they stabilize at approximately 30 Tg S yr^{-1}. These SO_2 emissions are within the range of SO_2 emissions in the SRES scenarios (Nakicenovic and Swart, 2000). Atmospheric-sulfate loadings under the SO_2 emissions were taken from earlier simulations using the NCAR CSM (Dai *et al.*, 2001b).

The ensemble simulations started from an 1870 control simulation that was run for 230 years and used constant greenhouse gases and sulfate-aerosol concentrations appropriate for that year. Each member of the ensemble started from a different point (separated by ten or more years) in the control simulation. The simulations continued through 1999 using greenhouse-gas concentrations and sulfate loadings based on observations (Dai *et al.*, 2001a). The climate conditions at the end of the historical simulations are comparable to recent observations. The simulations into the future were extended to the end of the twenty-first century using the projected greenhouse-gas concentrations. Two five-member ensembles of runs from 1870 to 2099 were used to derive the ensemble mean and ensemble range for the historical, BAU, and STA simulations. The ensemble range was defined as the largest value minus the smallest value in the ensemble at each year. For a Gaussian process, this minimum-to-maximum range is 2.475 times the standard deviation for five-member ensembles (Dai *et al.*, 2001a).

Figure 2.7 shows the ensemble mean and range for globally averaged, annual-mean surface air temperature. For the twentieth century, the rapid warming since the late 1970s was well simulated, however, the warming around 1940 is not captured by any of the model ensemble members with the greenhouse-gas, sulfate-aerosol, and ozone-forcing discussed above. Increasing the number of the ensemble simulations to ten for the twentieth century also failed to capture the peak warming near 1940. This is in contrast to an earlier study (Delworth and Knutson, 2000), which found that one of five ensemble runs using the GFDL model forced with

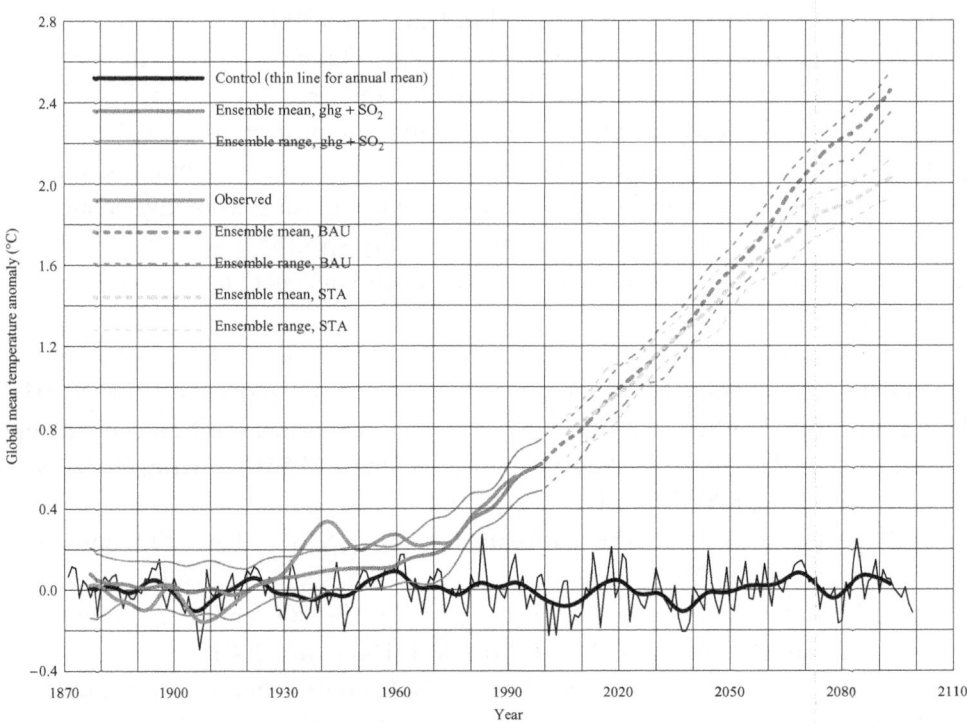

Figure 2.7. Globally averaged annual-mean surface air temperature change from 1870 to 2099 simulated by the PCM under the historical greenhouse-gas (ghg) and sulfur dioxide emissions, which in turn become sulfate aerosols forcing (red solid lines) and the BAU (red dashed lines) and STA550 (green dashed lines) future scenarios. The smoothed thin lines are ensemble ranges whereas the thick lines are ensemble averages. The thick purple line is observed surface temperature from Nicholls *et al.* (1996 and updates). The black lines are from the control simulation. From Dai *et al.* (2001a).

This figure is available for download in colour from
www.cambridge.org/9780521791328

greenhouse gases and sulfate aerosols reproduced both the peak warming around 1940 and the warming after the late 1970s. Figure 2.8 shows the time evolution of globally averaged surface air temperature from multiple ensemble simulations of twentieth century climate from the PCM compared to observations. The simulations start in the late nineteenth century, and continue to the year 2000. Temperature anomalies are calculated relative to a reference period averaged from 1890 to 1919. The thick dashed line near the bottom of the figure shows the observed data, or the actual, recorded globally averaged surface air temperatures from the past century. The blue and red lines are the average of four simulations each from the computer model. The pink and light blue shaded areas depict the range of the four simulations for each experiment, giving an idea of the uncertainty of a given realization of twentieth-century climate from the climate model. The blue line shows the average

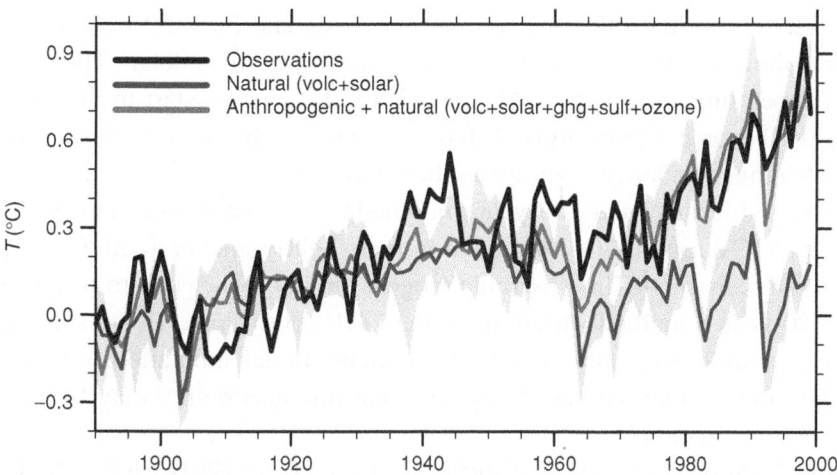

Figure 2.8. Four-member ensemble mean (red line) and ensemble member range (pink shading) for globally averaged surface air temperature anomalies (°C, anomalies are formed by subtracting the 1890–1991 mean for each simulation from its time series of annual values) for "natural" forcings, which are solar plus volcanic, plus the "anthropogenic," which has the observed increased greenhouse-gas and sulfate-aerosol effect. Also plotted are the "natural" forcings alone (dark blue line) and the ensemble member range (light blue shading). The observed is also shown (heavy black line). Note the "natural" forcing simulations do not show the late twentieth-century warming. See Meehl *et al.*, 2004 and Ammann *et al.*, 2003.
This figure is available for download in colour from
www.cambridge.org/9780521791328

from the four-member ensemble of the simulated time evolution of globally average surface air temperature when only "natural" influences (solar variability and volcanic eruptions) are included in the model. Therefore, the blue line represents what the model says global average temperatures would have been if there had been no human influences. The red line shows the average of the four-member ensemble experiment when natural forcings *and* anthropogenic influences (greenhouse gases including carbon dioxide, sulfate aerosols from air pollution, and ozone changes) are included in the model. Note that this model can reproduce the actual, observed data very well only if the combined effects of natural and anthropogenic factors are included. The conclusion that can be drawn is that naturally occurring influences on climate contributed to most of the warming that occurred before World War II, but that the large observed temperature increases since the 1970s can only be simulated in the model if anthropogenic factors are included (see Meehl *et al.*, 2004 and Ammann *et al.*, 2003). This confirms the conclusion of the IPCC Third Assessment Report that most of the warming we have observed in the latter part of the twentieth century has been due to human influences.

The ensemble-averaged, global mean surface warming from 1990–1999 to 2090–2099 is 1.9 °C under the BAU scenario and 1.5 °C under the STA550 scenario

(Figure 2.7 shows the full time series). The BAU and STA550 warming is very similar to those simulated by the CSM version1.3 (Dai *et al.*, 2000a). The ensemble-mean temperatures under the BAU and STA550 scenarios start to diverge in the 2040s, but become significantly different only after the middle 2060s when the ensemble ranges no longer overlap as shown in Figure 2.7.

The ensemble range of the 20-year smoothed global mean temperature is approximately 0.25 °C during 1870–1980 and becomes slightly smaller (0.20 °C) after 2050 (under both scenarios). This ensemble range is similar to the peak-to-trough amplitude of 20-year smoothed variations in the 1870 control run. This result suggests that the ensemble uncertainty of coupled climate-model simulations arises largely from the model's internal variability, and that this uncertainty may be estimated using control-run data.

Figure 2.9 shows the pattern of surface warming from 1961–1990 to 2070–2099 under the BAU scenario for the northern hemisphere winter, northern hemisphere summer, and the annual mean. The warming ranges between 1 and 2 °C over the oceans and is above 2 °C over many land areas, especially in northern high latitudes during winter where the warming is above 5 °C. Thirty-year averaging used here is to reduce noise in the spatial patterns. The ensemble-averaged spatial patterns of seasonal- and annual-mean surface warming are very similar under the BAU and STA550 scenarios with spatial correlation coefficients around 0.990. The spatial patterns are generally similar to those from other transient experiments using coupled climate models forced by CO_2 or CO_2 plus aerosol changes (Kattenberg *et al.*, 1996; Mitchell *et al.*, 1998; Roeckner *et al.*, 1999; Boer *et al.*, 2000; Dai *et al.*, 2001b). The PCM produces a moderate surface cooling (1–2 °C, mostly during winter) from 1961–1990 to 2070–2099 over the central mid-latitude North Atlantic Ocean. This cooling is larger at the ocean surface than in the air and extends over 1 km depth into the ocean. The cooling is caused primarily by changes in ocean currents.

The globally averaged precipitation rate generally follows global mean temperature on decadal to longer time scales (Figure 2.10). From 1990–1999 to 2090–2099 under the BAU scenario, global mean precipitation increases from approximately 3.094 mm d^{-1} to approximately 3.195 mm d^{-1}, or by 3.3%. Under the STA550 scenario, the global mean precipitation rises to only 3.168 mm d^{-1}, or by 2.4%. The global mean precipitation to temperature-change ratio (P/T) is approximately 0.050 mm d^{-1} K^{-1}, or 1–7% K^{-1}, for the 1990–2099 period under both scenarios. This may be compared with the average of 11 other models listed by Wigley, 1999 (Table 2), which is 2.2% K^{-1} (range 1.1–3.0% K^{-1}).

The ensemble range of the 20-year smoothed global mean precipitation rate is approximately 0.020 mm d^{-1} (0.7%) through the entire integration period for both scenarios (Figure 2.10). This range is comparable to (although slightly larger than) the amplitude of the 20-year smoothed variations in the control run. Again this

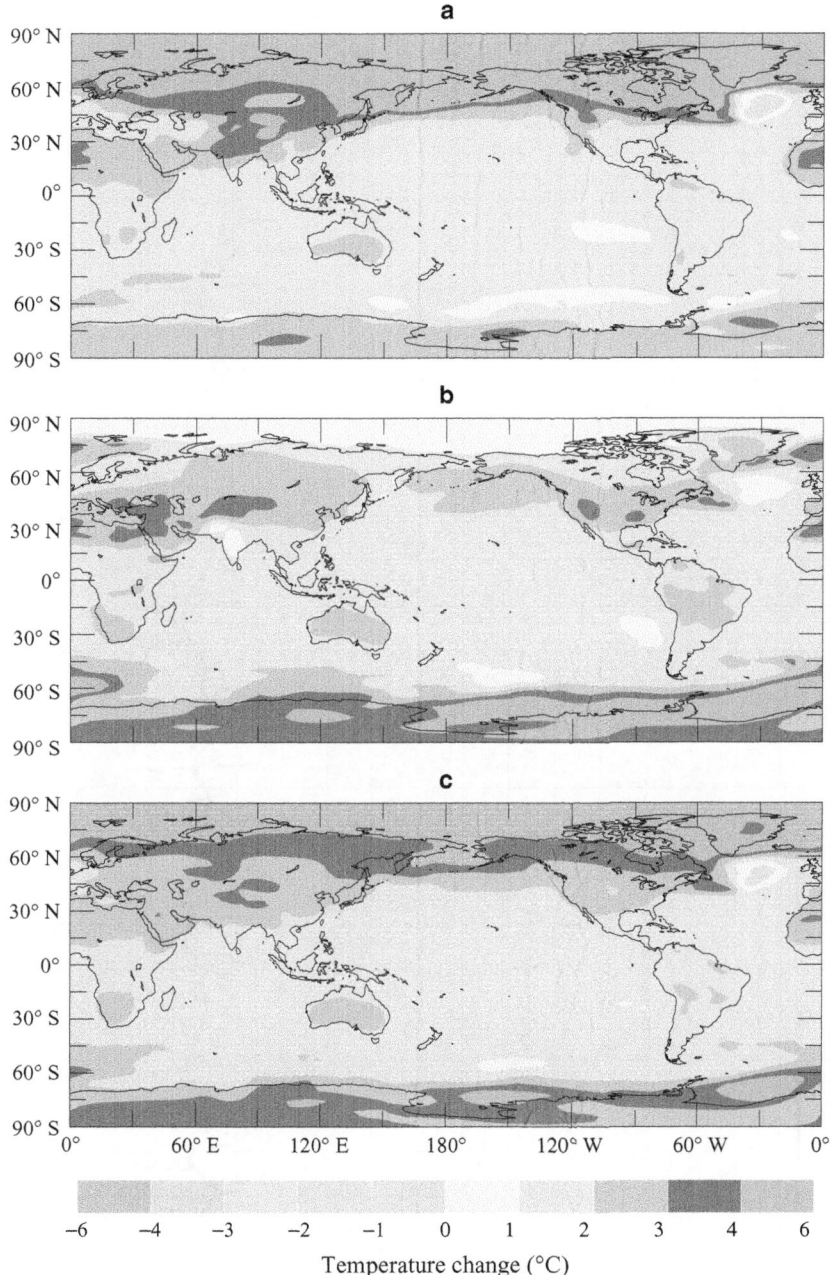

Figure 2.9. Ensemble-averaged surface air temperature changes (°C) from 1961–90 to 2070–99 under the BAU scenario for (a) December, January, February, (b) June, July, August, and (c) annual mean. Almost all the changes are statistically significant at the 5% level. From Dai *et al.* (2001a).

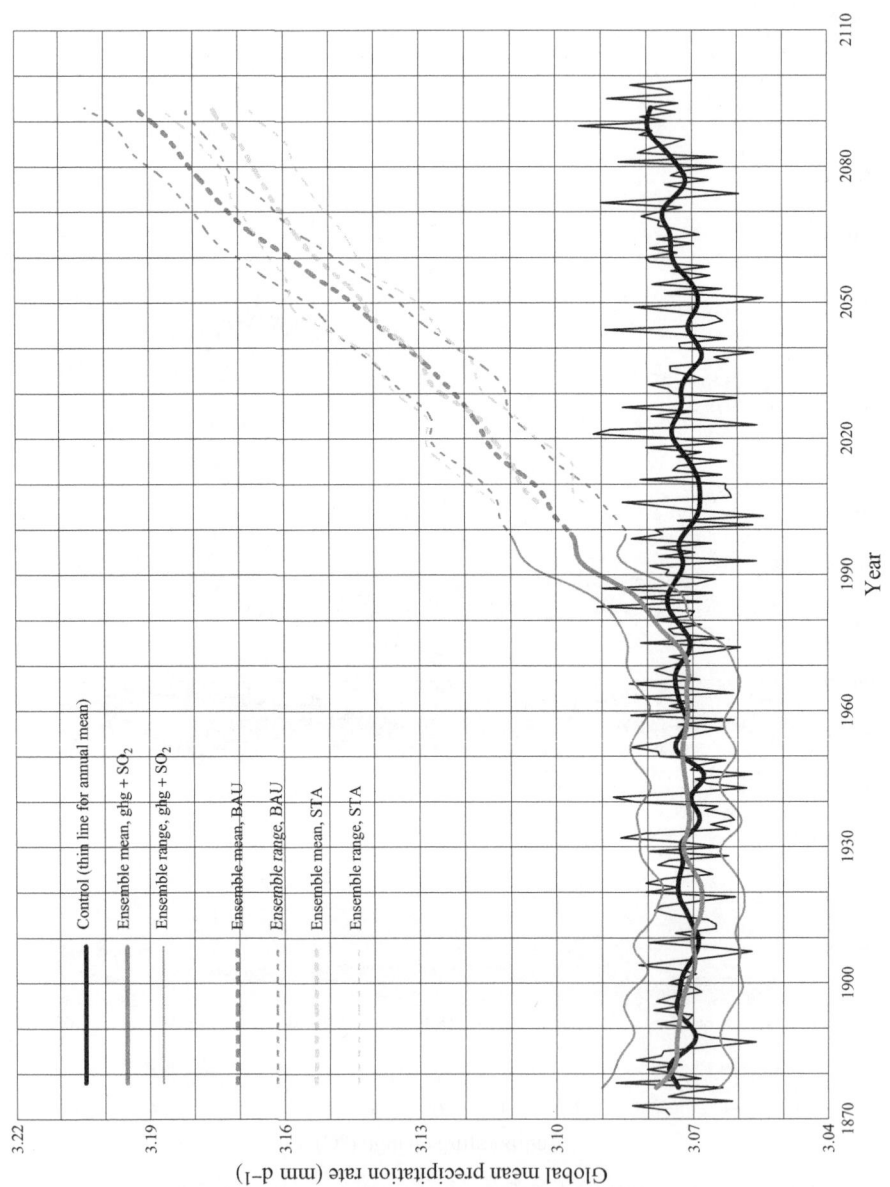

Figure 2.10. Same as Figure 2.7 except for precipitation (with no observations).

This figure is available for download in colour from
www.cambridge.org/9780521791328

44

result suggests that intra-ensemble uncertainties may be approximated by internal variations in control runs. At the end of the runs, the ensemble ranges overlap only slightly, which suggests that the difference in the global mean precipitation signals under the BAU and STA550 scenarios does not become statistically significant until late in the twenty-first century.

Regional precipitation changes can differ substantially among the ensemble runs, especially in the tropics and subtropics and for seasonal precipitation. The regional seasonal patterns of ensemble-averaged precipitation changes (Figure 2.11) are very similar between the BAU and STA550 scenarios (with slightly smaller magnitudes for the STA550 simulation). The pattern correlation coefficients between the BAU and STA550 geographical distributions in Figure 2.11 are 0.908 for December, January, February (DJF) and 0.943 for June, July, August (JJA). Our results show that ensemble-averaged simulations can provide reproducible estimates of regional precipitation changes and that lowering the rate of future CO_2 increases would not alter regional precipitation-change patterns significantly.

In general, the ensemble-averaged precipitation in the BAU case shows a 20–40% increase from 1961–1990 to 2070–2099 at high latitudes during winter (consistent with other models; see Kattenberg *et al.*, 1996) and a 10–30% decrease over the subtropical dry areas (around 30° latitude). Other notable regional changes (Figure 2.11) include a 20–40% reduction in JJA precipitation in the western United States, a 20–40% decrease in both DJF and JJA precipitation over the central mid-latitude North Atlantic, and a 20–40% increase in JJA precipitation over eastern Australia and over a region including western India, Saudi Arabia and Egypt. Similar, but slightly smaller changes occur in the STA ensemble mean. The intra-ensemble variability at the grid-box level can, however, be large even for annual mean values of temperature and precipitation. For example, annual temperature changes can differ by 2–3 °C over the sea-ice areas at northern high latitudes among ensemble members (5 °C for DJF temperature changes).

Soil-moisture changes in the BAU ensemble mean (from 1961–1990 to 2070–2099; not shown) are generally small (±3%) over most areas, with a 2–10% decrease over Europe and western Asia, western Canada and Alaska, a 5–10% increase over eastern China during JJA, a 2–5% increase in northern mid- and high-latitudes, and a 2–10% decrease over the southern United States and southeastern Australia during DJF. Changes in the STA550 simulation are similar to these over many areas.

Total cloud cover changes are within ±5% (of sky cover) over most regions in both the BAU and STA550 ensemble means, with a 2–6% decrease over northern mid- and high-latitudes during DJF, and a 2–6% increase over northern high latitudes, eastern Asia, and the tropical Indian Ocean during JJA. Mid-level clouds decrease while high-level clouds increase, as in previous climate-model simulations (Kattenberg *et al.*, 1996).

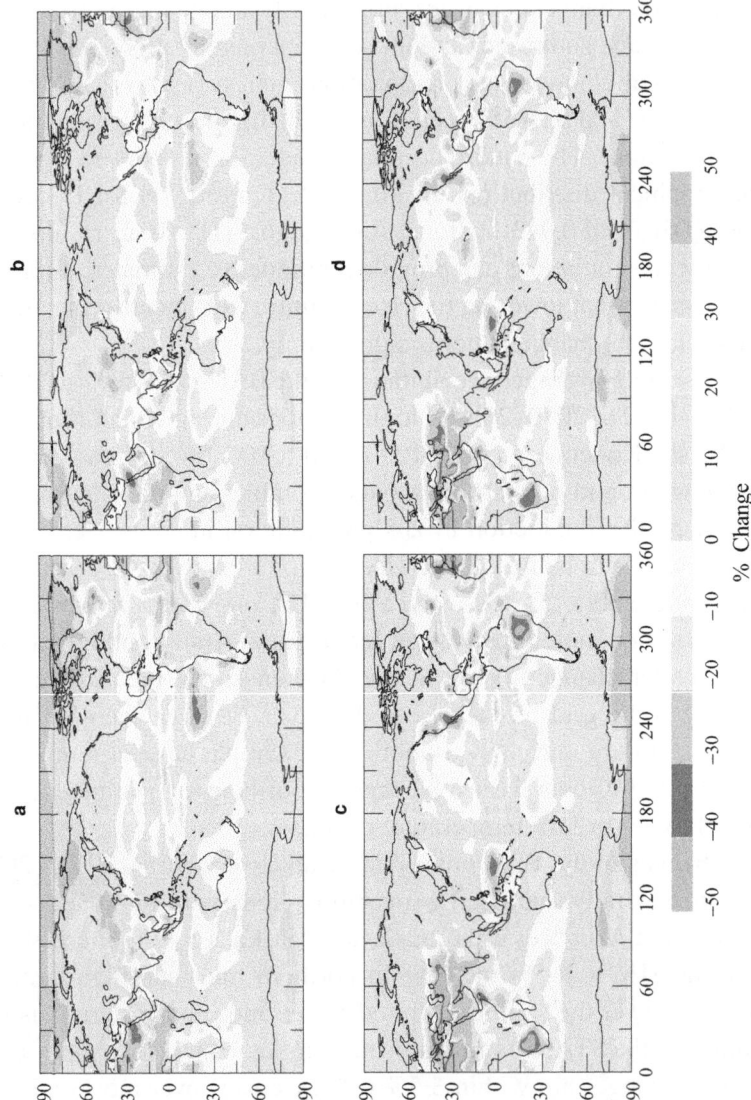

Figure 2.11. Ensemble-averaged precipitation changes (%) from 1961–90 to 2070–99 under the BAU (a and c) and STA550 (b and d) scenarios for December, January, February and June, July, August, respectively. Change with ± 10% are not significant statistically.

This figure is available for download in colour from
www.cambridge.org/9780521791328

46

2.8 Summary and discussion

These NCAR results show that ensemble averaging reduces the noise level in model-simulated climate changes, especially on regional and smaller scales. However, the results of this study still contain uncertainties resulting from model deficiencies in simulating the climate response to any given forcing and from uncertainties in future emissions. The PCM has a climate sensitivity of approximately 2.1 °C global mean warming for a doubling of atmospheric CO_2. This sensitivity, which is in the lower half of the 1.5–4.5 °C range of climate models used in the IPCC (2001), largely determines the magnitude of the simulated global mean warming by the end of twenty-first century. It should be noted that the simulated global warming over 1870–2099 is only $\sim$70% of the total warming that will eventually occur due to the forcing over this period (Dai *et al.*, 2001a).

If the climate changes obtained in these simulations are realistic, the PCM results show that the global warming trend since the late 1970s (see, e.g., Wigley, 2000) is likely to continue through much of the twenty-first century. Efforts to stabilize atmospheric CO_2 concentration at 550 ppm will only slow down the warming moderately, and the slow down will only become apparent in the middle to late twenty-first century. Compared with the BAU case, the surface warming under the STA550 scenario is reduced by a few tenths of a degree over the oceans and tropical land areas, and by 0.5–1.0 °C over mid- and high-latitude land areas.

There are no significant changes in spatial and seasonal patterns of change for global precipitation under the BAU scenario, which is likely to increase by about 3% during the next 100 years, with the largest (percentage) increases over northern mid and high latitudes during winter. Soil-moisture changes by the end of the twenty-first century are small over most areas and stabilization will reduce these changes by only a small amount.

In summary, we show in this chapter the development of simple concepts involved in the greenhouse effect and related global warming. These simple concepts have led the way towards a more comprehensive knowledge of how the climate system works and how to model it. Policymakers around the world are now becoming increasingly concerned about what the scientific community is saying about climate change. The most recent IPCC report (2001) has made an even more convincing case of observable climate change and the gap between the observed trends and the model-simulated trends is closing. This suggests that climate models are capable of making useful projections of future climate change. Some of the model simulations used in the IPCC report were the models developed at NCAR with National Science Foundation (NSF) and US DOE support and the computer time was provided by those two government agencies. These state-of-the-art modeling efforts have more become part of the CCSM. It is expected that this will result in

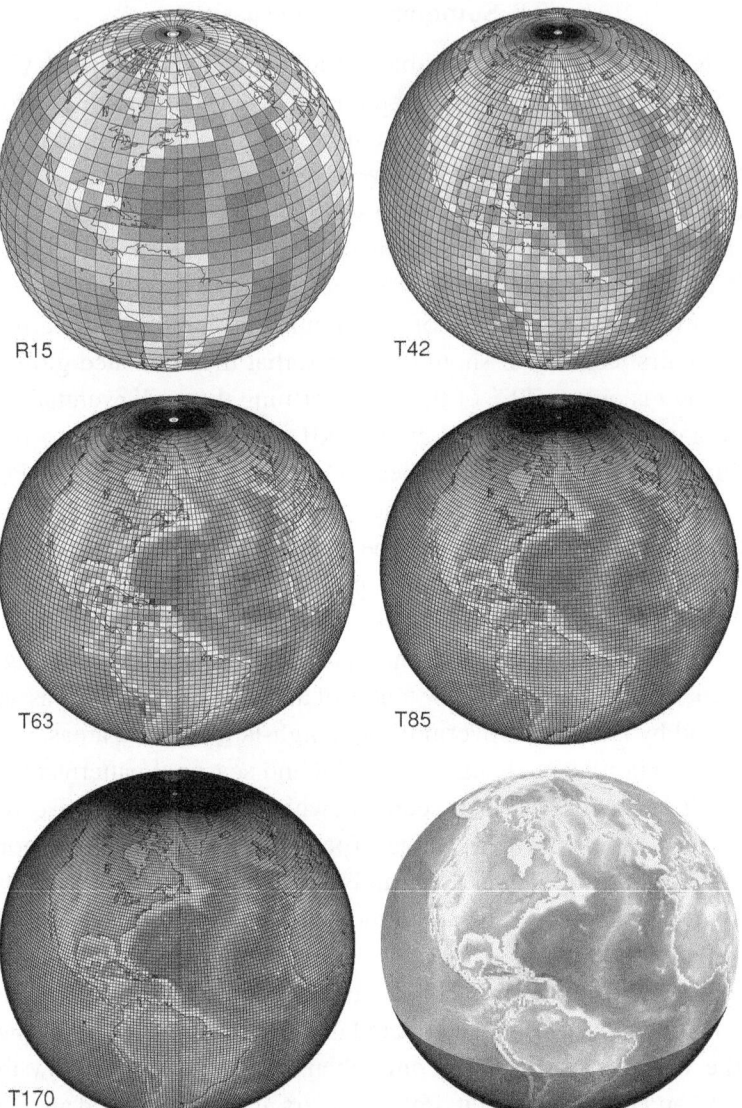

Figure 2.12. R stands for rhomboidal truncation and T stands for triangular truncation (see Washington and Parkinson, 2005 for definitions of resolution); T85 is
a horizontal resolution of about 150 km. The other resolutions can be scaled to
85. Past, present, and future climate model resolutions. Note T42 is about 2.8°
latitude–longitude resolution, which is about 300 km. With increasing resolution
the mountain, coastal, and ocean-bottom features become closer to the observed.

even more reliable models for prediction of future climates. One of the dreams of the climate-modeling community is that we will have models that have sufficient resolution in the atmosphere, oceans, and sea ice. Figure 2.12 shows the horizontal resolution of past, present, and future climate models. In the 1980s, the models had the spectral resolution of R15 and in the 1990s the resolution was T42. We will require resolutions up to T170, which is the lower left figure. To obtain such resolution will require computers more than 100 times faster than we presently have. Hopefully, in the not too distant future improvements in modeling capability and resolution will allow for improved climate prediction. Given the nature of climate change and the fact that some of the greenhouse gases such as CO_2 have very long life times (approximately 100 years), policy makers will have to make decisions with the help of imperfect but improved climate-model projections far before all of the scientific questions have been completely settled. If they delay in making these decisions, planet Earth will have carried out the experiment and the changes will not be reversible.

Acknowledgments

J. W. Weatherly, T. W. Bettge, W. G. Strand, V. B. Wayland, and A. P. Craig contributed to the PCM simulations, and J. T. Kiehl and B. A. Boville provided sulfate-loading data. We also thank K. E. Trenberth, B. D. Santer, and K. E. Taylor for their constructive comments. The US DDE and US NSF supported the PCM simulations. The input concentration scenarios were developed under the ACA-CIA (A Consortium for the Application of Climate Impact Assessments) program, funded by the Electrical Power Research Institute, the Central Research Institute of Electric Power Industry, KEMA, and NCAR. The NSF sponsors NCAR. The Tyndall picture was obtained from the BBC Huton Picture Library and the Arrhenius picture was obtained from the Chemical Heritage Foundation.

References

Allen, M. R., P. A. Stott, J. F. B. Mitchell, R. Schnur, and T. L. Delworth (2000). Quantifying the uncertainty in forecasts of anthropogenic climate change. *Nature* **407**, 617–20.

Ammann, C. M., G. A. Meehl, W. M. Washington, and C. Zender (2003). A monthly and latitudinally varying volcanic forcing dataset in simulations of 20th century climate. *Geophys. Res. Lett.* **30**, doi:309:10.1029/2003GL016875RR.

Boer, G. J., G. Flato, and D. Ramsden (2000). A transient climate change simulation with greenhouse gas and aerosol forcing: projected climate for the 21st century. *Clim. Dyn.* **16**, 427–50.

Budyko, M. I. (1974). *Climate and Life*. International Geophysical Series 18, New York, Academic Press.

Callendar, G. S. (1938). The artificial production of carbon dioxide and its influence on temperatures. *Quart. J. Roy. Meteor. Soc.* **64**, 223–37.

Cess, R. D., G. L. Potter, J. P. Blanchet *et al.* (1989). Interpretation of cloud–climate feedback as produced by 14 atmospheric general circulation models. *Science* **245**, 513–16.

Cubasch, U., K. Hasselmann, H. Hock *et al.* (1992). Time-dependent greenhouse warming computations with a coupled ocean-atmosphere model. *Clim. Dyn.* **8**, 55–69.

Cubasch, U., B. D. Santer, A. Hellbach *et al.* (1994). Monte Carlo climate change forecasts with a global coupled ocean–atmosphere model. *Clim. Dyn.* **10**, 1–19.

Dai, A., G. A. Meehl, W. M. Washington, T. M. L. Wigley, J. M. Arblaster (2001a). Ensemble simulation of twenty-first century climate changes: business-as-usual versus CO_2 stabilization. *Bull. Amer. Meteor. Soc.* **82**, 2377–88.

Dai, A., T. M. L. Wigley, B. A. Boville, J. T. Kiehl, and L. E. Buja (2001b). Climates of the 20th and 21st centuries simulated by the NCAR Climate System Model. *J. Clim.* **14**, 485–519.

Delworth, T. L. and T. R. Knutson (2000). Simulation of early 20th century global warming. *Science* **287**, 2246–50.

Edmonds, J., M. Wise, H. Pitcher, R. Richels, T. M. L. Wigley, and C. MacCracken (1997). An integrated assessment of climate change and the accelerated introduction of advanced energy technologies: an application of MiniCAM 1.0. *Mitig. Adapt. Strat. Global Change* **1**, 311–39.

Hansen, J., M. Sato, R. Ruedy, *et al.* (1997). Forcings and chaos in interannual to decadal climate change. *J. Geophys. Res.* **102**, 25679–720.

Houghton, J. T., Y. Ding, D. J. Griggs, *et al.*, eds. (2001) *Climate Change 2001: the Scientific Basis.* Cambridge, Cambridge University Press.

IPCC (2001). *Climate Change 2001: the Scientific Basis. Contributions of Working Group I to the Third Assessment Report*, eds. J. T. Houghton, Y. Ding, D. J. Griggs, M. Noguer, P. J. van der Linden, and D. Xiaosu, Cambridge, Cambridge University Press.

Kattenberg, A., F. Giorgi, H. Grassl *et al.* (1996). Climate models – projections of future climate. In *Climate Change 1995: the IPCC Second Assessment.* eds. J.T. Houghton *et al.*, Cambridge, Cambridge University Press, pp. 285–358.

Manabe, S. and R. F. Strickler (1964). On the thermal equilibrium of the atmosphere with a convective adjustment. *J. Atmos. Sci.* **24**, 241–59.

Manabe, S., R. J. Stouffer, M. D. Spelman, and K. Bryan (1991). Transient response of a coupled ocean–atmosphere model to gradual changes of atmospheric CO_2. Part I: annual mean response. *J. Clim.* **4**, 785–818.

Meehl, G. A. and W. M. Washington (1990). CO_2 climate sensitivity and snow–sea-ice albedo parameterization in an atmospheric GCM coupled to a mixed-layer ocean model. *Climatic Changes* **16**, 283–306.

Meehl, G. A., G. J. Boer, C. Covey, M. Latif, and R. J. Stouffer (2000a). The Coupled Model Intercomparison Project (CMIP). *Bull. Amer. Meteor. Soc.* **81**, 313–18.

Meehl, G. A., W. D. Collins, B. Boville *et al.* (2000b). Response of the NCAR Climate System Model to increased CO_2 and the role of physical processes. *J. Clim.* **13**, 1879–98.

Meehl, G. A., P. R. Gent, J. M. Arblaster *et al.* (2001a). Factors that affect amplitude of El Niño in global coupled climate models. *Clim. Dyn.* **17**, 515–526.

Meehl, G. A., W. M. Washington, C. M. Ammann *et al.* (2005). Combinations of natural and anthropogenic forcings in 20th century climate. *J. Clim.*, in press.

Mitchell, J. F. B., T. C. Johns, J. M. Gregory, and S. F. B. Tett (1995). Climate response to increasing levels of greenhouse gases and sulfate aerosols. *Nature* **376**, 501–4.

Mitchell, J. F. B., T. C. Johns, W. J. Ingram, and J. A. Lowe (2000). The effect of establishing atmospheric carbon dioxide concentration on global and regional climate change. *Geophys. Res. Lett.* **27**, 2977–80.

Mitchell, J. M., Jr. (1961). Recent secular changes of global temperature. *Annals New York Acad. Sci.* **95**, 235–50.

Möller, F. (1963). On the influence of changes in the CO_2 concentration in air on the radiation balance of the Earth's surface and on the climate. *J. Geophys. Res.* **68**, 3877–86.

Nakicenovic, N. and R. Swart, eds. (2000). *IPCC Special Report on Emission Scenarios*. Cambridge, Cambridge University Press.

Nicholls, N., G. V. Gruza, J. Jouzel, *et al.* (1996). Observed climate variability and change. In *Climate Change 1995: The IPCC Second Assessment*, eds. J.T. Houghton, Y. Ding, D. J. Griggs, *et al.*, Cambridge, Cambridge University Press, pp.133–92.

Plass, G. N. (1961). Comments on the influence of carbon dioxide variations on the atmospheric heat balance by L. D. Kaplan. *Tellus* **13**, 296–300.

Ramanathan, V. (1981). The role of ocean–atmosphere interactions in the CO_2 climate problem. *J. Atmos. Sci.* **38**, 918–30.

Roeckner, E., L. Bengtsson, J. Feichter, J. Lelieveld, and H. Rodhe (1999). Transient climate change simulations with a coupled atmosphere-ocean GCM including the tropospheric sulfur cycle. *J. Clim.* **12**, 3004–32.

Rowell, D. P. (1998). Assessing potential seasonal predictability with an ensemble of multidecadal GCM simulations. *J. Clim.* **11**, 109–20.

Russell, G. L. and D. Rind (1999). Response to CO_2 transient increase in the GISS coupled model: regional coolings in a warming climate. *J. Clim.* **12**, 531–9.

Seller, W. D. (1974). A reassessment of the effect of CO_2 variations on a simple global climate model. *J. Appl. Meteorol.* **13**, 831–3.

Tyndall, J. (1861). On the absorption and radiation of heat by gases and vapours, and on the physical connexion of radiation, absorption and conduction. *Philosophical Magazine and J. Science, Series 4*, **22**(146), 169–94 and **22**(147), 273–85.

Washington, W. M. and C. Parkinson (2005) *An Introduction to Three-Dimensional Climate Modeling*. Sausalito, CA, University Science Books.

Washington, W. M., J. W. Weatherly, G. A. Meehl *et al.* (2000). Parallel Climate Model (PCM) control and transient simulations. *Clim. Dyn.* **16**, 755–74.

Wehner, M. F. (2000). A method to aid in the determination of the sampling size of AGCM ensemble simulations. *Clim. Dyn.* **16**, 321–31.

Wigley, T. M. L. (1999). *The Science of Climate Change: Global and US Perspectives*, Arlington, VA, Pew Center on Global Climate Change.

(2000). Volcanoes and record-breaking temperatures. *Geophys. Res. Lett.* **27**, 4101–4.

Wigley, T. M. L., R. Richels, and J. A. Edmonds (1996). Economic and environmental choices in the stabilization of atmospheric CO_2 concentrations. *Nature* **379**, 240–3.

Zwiers, F. W. (1996). Interannual variability and predictability in an ensemble of AMIP climate simulations conducted with the CCC GCM2. *Clim. Dyn.* **12**, 825–47.

3

Energy-balance climate models

GERALD R. NORTH AND MARK J. STEVENS

Department of Atmospheric Sciences, Texas A&M University, TAMU College Station, TX

3.1 Introduction

As discussed in Chapter 1, energy-balance climate models are important tools for studying Earth's climate. Energy-balance climate models (EBCMs) have been used in studies of climate change for more than a quarter of a century. Some papers can be found even before that, but the widespread use of these simple models was made popular by the nearly simultaneous appearance of the papers by Budyko (1968) and Sellers (1969). A review of these and other early works was presented in North *et al.* (1981). The purpose of this chapter is to introduce these models in the modern context, justifying their usefulness in certain situations without hiding their limitations. We also introduce stochastic versions of the models and some recent applications of them.

The EBCMs attempt to model Earth's surface-temperature field $T(r, t)$, where r is a point on the spherical surface and t is the time, by the constraint of energy conservation for individual columns of the Earth–atmosphere system. For each infinitesimal horizontal area element consider an ideal geometric column containing matter extending from roughly the tropopause to just beneath the solid Earth's surface, or to the bottom of the oceanic mixed layer, depending on location. The collection of all such columns covering Earth make up the system. It is assumed that the energetic indices of a column can be unambiguously labeled by the temperature at the surface (the column changes its temperature 'rigidly' from top to bottom). This is the first of a series of idealizations made in formulating the model. There are many fluxes of heat into and out of an individual column (now a box). Each of these fluxes is to be parameterized in terms of the surface temperature and its horizontal derivatives. Once the various fluxes into and out of a column are added

Frontiers of Climate Modeling, eds. J. T. Kiehl and V. Ramanathan.
Published by Cambridge University Press. © Cambridge University Press 2006.

up and equated to the time rate of change of the energy content (sum of enthalpies of slabs in the vertical column) of the column (actually, all columns simultaneously), an equation emerges for the surface-temperature field as a function of position and time. This partial differential equation together with appropriate boundary conditions yields the surface-temperature field as a solution.

The steps in formulating the terms and the resulting equation will be presented briefly in the next section, but first one might ask why set up such a crude model of climate in the face of the progress being made in the construction of coupled ocean–atmosphere general-circulation models of today. First of all, in spite of the almost schematic formulation of the EBCMs they actually do a rather good job of representing the temperature field through the seasonal cycle and even in the fields of natural fluctuations, indicating that the larger scales of the surface-temperature field are insensitive to some of the dynamical details. Second, the models are amenable to a hierarchical formulation from global-average annual-average models to annual-average zonal-average models to two-dimensional seasonal models. This ability of running up and down the hierarchy provides an insight which proves useful. Finally, the models are subject to a certain amount of analytical treatment, allowing the old-fashioned methods of theoretical and statistical physics to be used in understanding the behavior of solutions. Solutions of EBCMs and their properties then form an intuitive framework or a benchmark with which to judge the outcomes of experiments with modern coupled ocean–atmosphere models of the climate system.

3.2 Formulation of EBCMs

In each of the following subsections a component of the energy budget is formulated.

3.2.1 Longwave radiation

Longwave radiation is the radiation energy per unit area per unit time leaving the top of the atmosphere going into space. It is concentrated in the infrared (IR) but the spectrum is complicated by several trace-gas absorbers/emitters in the atmosphere. Also at any one time about half the planet is covered with cloud, and the cloud tops emit roughly as black bodies which are at lower temperatures than the surface. Since the 1970s satellite sensors have been used to estimate the outgoing longwave radiation (Graves *et al.*, 1993). Budyko (1968) introduced a convenient parameterization, Equation (3.1), for the outgoing long wave flux, where T is in

$$F_{\mathrm{LW}}(r, t) = A + BT(r, t) \tag{3.1}$$

°C, A and B are empirical coefficients which can be fitted from satellite data of the outgoing flux $F_{LW}(r, t)$ with the local surface temperature. The range of $T(r, t)$ is provided by its climatology of latitude and seasonal dependence (month averages). The value of B is typically found to be c. 1.90 W m^{-2} °C^{-1}; A is c. 211 W m^{-2} (Graves *et al.*, 1993). The value of B is reasonably well defined in the middle latitudes but is essentially undetermined at low latitudes because of the poor dynamic range of T in the tropics and the fact that clouds are a significant determinant of F_{LW} at this latitude. Locally near the ITCZ there could even be an inverse relationship, since when local surface areas warm, clouds will form leading to less longwave radiation being lost to space from the cloud tops. Nevertheless, we will stick to our simple linear form, keeping in mind this potentially serious limitation.

There can also be a loss of energy flux by exchange with matter below the column usually in consideration. For example, over oceanic areas there is an exchange of heat with the deeper ocean below. This is typically taken to be linear with the difference between the mixed-layer temperature and the temperature of the layer just below it. In simulations one would find this indistinguishable from the simple $BT(r, t)$ term, but with B modified. It may thus be appropriate to use a larger value of B over oceanic areas to account for this phenomenon.

3.2.2 Shortwave radiation

Earth is heated by sunlight which is concentrated mainly in the visible part of the spectrum. The solar constant is the amount of radiation energy per unit time crossing a plane perpendicular to its path at the (annual average) Earth–Sun average distance; it is estimated to be about 1366 W m^{-2}. The amount of this energy impinging at the top of the atmosphere and averaged through the diurnal cycle is $QS(\mu, t)$, where Q is one quarter of the solar constant (because of the ratio of a sphere's area to that of a disk) and $S(\mu, t)$, where μ is the sine of latitude, is derived from elementary celestial mechanics. This function represents the fraction of the sunlight entering the atmosphere perpendicular to a surface element at a particular time of year (for a derivation see North *et al.*, 1981). By definition,

$$\int_0^1 S(\mu, t)\, d\mu = 1 \tag{3.2}$$

The planetary coalbedo is the fraction of the incoming solar beam that is absorbed by the system. In our case we are speaking of the coalbedo associated with the entire geometric column as opposed to say the surface coalbedo. The annual average planetary coalbedo is defined by Equation (3.3), where t is in years, and ϕ is

$$a_p = \frac{1}{4\pi} \int_{-1}^{1} \int_{0}^{2\pi} \int_{0}^{1} a(\mu, \phi, t) S(\mu, t) \, dt \, d\phi \, d\mu \tag{3.3}$$

longitude. For Earth, the planetary coalbedo, a_p, is about 0.70.

The amount of radiation absorbed from sunlight per unit time can now be written as in Equation (3.4). Note that $a(\mathbf{r}, t)$ can have a dependence on position and time

$$F_{SW}(\mathbf{r}, t) = QS(\mu, t) a(\mathbf{r}, t) \tag{3.4}$$

through the seasonal cycle or on other time scales. In particular there is a strong zenith-angle dependence especially over oceans. In fact, the coalbedo also depends upon such features as snowcover, clouds, etc, which can conceivably be parameterized by dependences on the surface-temperature field.

If we average over the planet and through the year, then equate the two fluxes, Equation (3.5), where T_{eq} is the mean annual planetary average temperature, which

$$A + BT_{eq} = Qa_p \tag{3.5}$$

can now be solved as in Equation (3.6). Using the values suggested earlier, we find

$$T_{eq} = \frac{Qa_p - A}{B} \tag{3.6}$$

$T_{eq} \approx 14.8\,°C$, which is reasonably close to the observed value ($\approx 15\,°C$). It might not be inappropriate to adjust the effective value of A downwards to force the global average temperature to agree with the observed value. This kind of fudging might be essential if one is hooking an icecap edge to a particular mean annual isotherm (e.g., Budyko's choice of $-10\,°C$).

3.2.3 Sensitivity

We can now calculate the rate of change of the planetary temperature with respect to a fractional change in solar constant, which we can think of as a control parameter, assuming a_p does not depend on T_{eq}, Equation (3.7). We find that this *static*

$$Q \frac{dT_{eq}}{dQ} = Q \frac{a_p}{B} = \frac{A + BT_{eq}}{B} \tag{3.7}$$

sensitivity is inversely proportional to B, the infrared damping coefficient. We say static sensitivity because in doing the change we wait for equilibrium to re-establish itself after the change in solar constant. The value of B seems to reflect the water-vapor feedback in the atmospheric column, but clouds are also involved. Using the empirical values of A, B, a_p, Q, we find a static sensitivity of $1.26\,°C$ for a 1% increase in solar constant.

If Earth had no atmosphere and its surface behaved as a black body in the infrared, the value of B would exceed 4.61 W m^{-2} $°C^{-1}$. The smaller value of B for the real atmosphere appears to double the sensitivity. One must keep in mind however that the tropics (half Earth's surface) may well be misrepresented by our simple IR formula and it is likely that planetary sensitivity is reduced by tropical effects. Furthermore, in this simple estimate, we used the empirical value of a_p which includes the effect of cloud cover.

If the coalbedo depends on the temperature, we find the relationship shown in Equation (3.8). Hence, dependences of a_p on temperature through cloud, water

$$Q\frac{dT_{eq}}{dQ} = \frac{Qa_p}{B - Q\frac{da_p}{dT}} \tag{3.8}$$

vapor, or snow/ice cover can significantly increase or decrease sensitivity.

A crude way to take into account greenhouse-gas changes is through the parameter A. The dependence on CO_2 concentration is roughly given by Equation (3.9),

$$A(CO_2) = A_0 - 5.35\ln\frac{C}{C_o} \tag{3.9}$$

where C is the CO_2 concentration in ppmv and C_o is the present concentration ($\approx$ 365 ppmv), and the coefficient 5.35 (W m^{-2}) is derived from detailed radiative-transfer calculations (Myhre *et al.*, 1998). We can calculate the sensitivity to doubling CO_2 by calculating ΔT for $C = 2 \times C_o$, Equation (3.10).

$$\Delta T_{2\times CO_2} = \frac{5.35\ln(2)}{B} \approx 1.95\,°C \tag{3.10}$$

Again, the important factor B appears in the denominator. And in the event that the coalbedo depends on T_{eq} we make the replacement $B \to B - \frac{da_p}{dT}$.

3.3 Time dependence

Before taking spatial dependences into account it is interesting to contemplate how the global average temperature responds to imbalances in the heat budget. For our schematic column of matter there is an effective heat capacity $C(r)$. The local magnitude of $C(r)$ will have a rather marked dependence on whether the local surface type is land or sea. For the moment consider an ideal geographically uniform planet in which case $C(r)$ is a constant independent of position. For the planet as a whole we can write Equation (3.11), where the subscript p denotes global average.

$$C\frac{dT_p}{dt} = Qa_p - A - BT_p \tag{3.11}$$

The solution of this simple initial value problem is given by Equation (3.12), where

$$T_p(t) = T_{eq} + (T_p(0) - T_{eq})e^{-t/\tau} \qquad (3.12)$$

the decay time constant is $\tau = C/B$. Again the replacement $B \to B - \frac{da_p}{dT}$ holds if albedo is allowed to depend on T; however, if these dependences are included one must be mindful of the time constants of glacial growth, etc., which might enter, requiring separate governing equations which have to be coupled to the energy-balance equation through the temperature. The important point is that the relaxation time is proportional to the sensitivity ($\tau \propto \frac{1}{B}$): more sensitive climates will have longer relaxation times.

If the system is driven by a sinusoidal forcing of angular frequency ω, the response (take the real part at the end) is given by Equation (3.13); $T'(t)$ is the

$$\frac{dT'}{dt} + \frac{T'}{\tau} = \frac{F_\omega}{C}e^{-i\omega t} \qquad (3.13)$$

departure from the time-independent steady-state solution. The steady-state response is similarly sinusoidal with complex amplitude, Equation (3.14), and the

$$|T_{p\omega}| = \frac{|F_\omega|}{B\sqrt{1 + \omega^2\tau^2}} \qquad (3.14)$$

phase lag (response lagging forcing) $\varphi = \arctan \omega\tau$. Note that even for infinite C the phase lag only reaches $\frac{\pi}{2}$ (quarter cycle). The response amplitude is inversely proportional to the damping coefficient B (larger sensitivity $\to$ larger amplitude if τ is independently known). The amplitude decreases with frequency. Similarly the amplitude of the response is diminished and the phase lag is increased toward its upper limit of $\frac{\pi}{2}$ if τ is larger for a given value of B (maritime effect).

3.3.1 Random forcing

Weather instabilities and other small-scale effects cause temporal disturbances in the local heat balance, especially in mid latitudes. These disturbances cause the temperature field to fluctuate leading to the EBCM's version of natural variability. We introduce these disturbances through a stationary random function of time, $F(t)$. The autocovariance of this function diminishes very rapidly with time lag, being essentially a Dirac delta function, δ; Equation (3.15). This kind of random

$$\langle F(t)F(t')\rangle = \sigma_F^2\delta(t - t') \qquad (3.15)$$

function in time is called *white noise*. The Fourier representation of such a function can be written as in Equation (3.16). The complex Fourier components F_ω are

$$F(t) = \frac{1}{2\pi} \int_{-\infty}^{\infty} F_\omega e^{-i\omega t}\, d\omega \qquad (3.16)$$

random variables, with mean zero, statistically independent of one another at different ω and mean square $\langle |F_\omega|^2 \rangle$, a constant independent of ω. This means the power spectrum of the random forcing ($\propto \langle |F_\omega|^2 \rangle$) is *flat*.

We solved the response problem for a sinusoidal forcing at frequency ω in the last subsection. We can think of the white noise forcing as providing a sinusoidal term on the right-hand side of the energy-balance equation with a random amplitude at each frequency ω; furthermore, the forcings at different frequencies are uncorrelated, a property of stationary time series. The solution for the amplitude of the temperature fluctuation is given by Equation (3.14) with T_ω now a normally distributed random variable. One finds that the power spectrum of the temperature (response) field is given by Equation (3.17). This is the typical *red noise* spectrum which for large ω

$$S_\omega = \langle |T_\omega|^2 \rangle = \frac{\langle |F_\omega|^2 \rangle}{B^2(1 + \omega^2 \tau^2)} \qquad (3.17)$$

goes as ω^{-2}.

Without going into details here one can also find the autocorrelation function, Equation (3.18), for the temperature field.

$$\rho(|t - t'|) = \exp\left(-\frac{|t - t'|}{\tau} \right) \qquad (3.18)$$

In other words, the autocorrelation time (memory) for this process is precisely the relaxation time for the unforced EBCM. The value of τ for an all-land planet is just a month or so, while for a mixed-layer ocean planet (insulated from the deep ocean below) it is of the order of a few years. Note that this dependence of the autocorrelation function on τ permits its possible estimation independent of the knowledge of B and C separately through the study of the time series.

It is interesting to speculate that one might make an estimate of sensitivity, since if τ is known from time-series analysis, one might estimate the response amplitude to some known periodic forcing using Equation (3.14), then infer the sensitivity $\propto \frac{1}{B}$. A good candidate for the external periodic forcing is the approximately 11-year solar cycle (Stevens and North, 1996; North and Stevens, 1998).

3.3.2 Albedo feedback

The icecap feedback mechanism leads to some curious features of global climate. As a simple example, consider the case where the planetary coalbedo increases with global average temperature until all the ice disappears. Similarly as the temperature decreases the coalbedo decreases as the global average temperature decreases until the icecap edge reaches the equator. Such a coalbedo function can lead to more than one solution of Equation (3.19). Typically, there is a solution similar to the

$$A + BT_{eq} = Qa_p(T_{eq}) \qquad (3.19)$$

present climate with another solution at the same value of Q for which the planet is completely iced over. For reasonable dependences of a_p on T_{eq} a decrease of Q of $c.$ 10% can lead to a catastrophic plunge to the ice-covered planet (North *et al.*, 1981; North, 1990; Mengel *et al.*, 1988; Lin and North, 1990). This appears to happen in GCMs as well as EBCMs.

3.4 Adding horizontal dimensions

If we consider an infinitesimal area element on the sphere and its heat budget, we must now take into account the divergence of heat flux leaving the area element due to advection. Also the heat capacity $C(r)$ as well as other macroscopic coefficients might depend on location or phase in the seasonal cycle. The simplest parameterization of heat flux is to take it proportional to the local temperature gradient: $-D(r)\nabla T(r, t)$. The energy balance equation now becomes the partial differential Equation (3.20): Aside from the new term due to advection, there is a term F_{below}

$$C(r)\frac{\partial T}{\partial t} - \nabla \cdot (D(r)\nabla T) + A + BT = QS(x, t)a(r, t; T) + F_{below} + F_{noise}$$
$$(3.20)$$

which is used to take into account fluxes into the oceanic mixed layer from below. This term essentially accounts for the poleward flow of heat in the oceans. It could be of the form $B'(r)(T - T_{below}(r))$ or independent of temperature.

Use of diffusive heat transport always sparks debate. Two assertions of plausibility perhaps help. (1) At the coarsest scales most prevailing winds flow roughly along surface isotherms (i.e., $\mathbf{v} \cdot \nabla T \approx 0$). (2) The differences in time scale between atmospheric eddies (≈ 3 d) and the relaxation time of the surface-temperature field (≈ 1 month) is sufficient for the diffusion or random-walk approximation to hold at least in ensemble average. Some authors have suggested that the diffusion coefficent D should depend on temperature or its gradient, but this introduces a nonlinearity, which though small poses a complication best left out in our philosophical framework. In tuning to the present climate we do allow D to have a dependence on position, diminishing towards the poles.

3.4.1 Length scale

Just as $C/B = \tau$ is a representative time scale for global quantities, it is convenient to explore for a characteristic length scale at low frequency. This static length scale turns out to be $\ell = \sqrt{D/B}$, which can be found in a variety of ways: decay of correlation in space for a randomly forced model, or the decay length of the response to a steady point heat source (Green's function). An interpretation of this length scale is the distance a randomly diffusing thermal anomaly goes during one characteristic decay time. The size of this scale is typically 1000 to 2000 km (for a comparison with data see Hansen and Lebedeff, 1987). It is, of course, dependent on the diffusive mechanism, but it means that features smaller than this length scale will be smeared out quickly. This length scale is curiously close to the Rossby radius in middle latitudes, but their relationship remains obscure.

If we cover Earth with disks of this radius we have only about 64 (=8^2) such 'statistically independent' areas. This means the standard deviation of a local temperature is about eight times that of the global average ($\approx 0.15\,^{\circ}\mathrm{C}$ for annual averages). A gauge located at the center of each of these areas should give a good estimate of the global average temperature; i.e., the standard error of this estimate will be much less than the standard deviation of the planetary temperature. The standard error of a global average estimate based on well distributed point gauges will fall as the square root of the number of gauges, N, until we reach ~ 64 after which the reduction will be at a rate slower than $\propto 1/\sqrt{N}$. The latter occurs because when multiple gauges are located within one correlation radius of one another there is a degree of redundancy in the records.

The diffusive length scale may also play a role in the minimum size of stable icecaps. Dynamic features such as icecaps whose size is determined solely by the temperature field smaller than this length tend to be unstable. Hence, ice sheets tend to be larger than this size or not at all. This has led to conjecture that icecaps such as those on Greenland, Antarctica, or the Laurentide ice sheets might have formed rather suddenly. Actually, only the conditions for growth are sudden – it takes thousands of years to build the ice thickness. On the other hand, ice sheets can disappear rather quickly once the critical size for instability is reached (Lindzen and Farrell, 1977; North, 1984).

3.4.2 Seasonal cycle

Next we wish to show that the seasonal cycle of the surface-temperature field can be represented rather well with such a simple model. In doing so there are two approaches. The first is to demonstrate that the essential physics is captured by the energy balance (for an example, see North *et al.*, 1983). In this approach we keep the number of free parameters in the problem at a minimum. A second approach is to drop the parsimonious parameter issue and tune the phenomenological coefficients in the EBCM to all available data in order to use the model in applications – this is discussed in a later section. Returning to the first approach, for example, we drop the term F_{below} and let $D(r)$ take on a very simple form dependent only on latitude, mildly decreasing toward the poles and symmetric across the equator. The heat capacity takes on one value over land and another over ocean (two orders of magnitude larger), the latter suppressing the seasonal amplitude and increasing the phase lag over the oceans. Moreover, we drop the temperature dependence of the coalbedo, making the problem entirely linear (but to see interesting effects of allowing the snowline to enter, see North *et al.*, 1983; Mengel *et al.*, 1988; Lin and North, 1990; Hyde *et al.*, 1990). In this case the seasonal cycle is composed of a seasonal mean, an annual harmonic, and a semiannual harmonic. This is a very rapidly converging Fourier series, with the semiannual harmonic being everywhere less than *c*. 2 °C, while the annual harmonic peaks at near 30 °C over the interior of the large continents. Similarly the phase lag of the seasonal cycle is only about a month over the continents and near a quarter cycle over the oceans. These fields have been shown to be in qualitatively good agreement with the corresponding ones from the data by North *et al.* (1983) and Hyde *et al.* (1990).

A similar agreement is found for the fields of variance and covariance (from one site to another) of climate fluctuations for month averages as well as for longer time averages (Kim and North, 1991; 1992). In the parsimonious parameter approach we take the noise forcing to be uniform over the globe and white in both space and time. In this way, only one free parameter is introduced in the noise – its overall variance or strength. Because of the land–sea heat capacity variation over the planet, we find large variance over the continents and small variance over the oceans, with a smooth transition joining them which has a length scale of ℓ. The comparisons with data and some general circulation model (GCM) output can be found in a series of papers by Kim, North and colleagues (Kim and North, 1991; 1992; Leung and North, 1991; North *et al.*, 1992). In the comparisons of EBCM solutions for the steady-state seasonal cycle and for the second moment fluctuation statistics it can be seen that there is reasonable basis to believe that the EBCM is capturing the main physics at these space and time scales for the surface-temperature field.

3.4.3 Comparison with GCM simulations

In a series of papers using an early version of the NCAR Community Climate Model an attempt was made to check the EBCM in some controlled situations. To facilitate the comparisons the boundary conditions in the GCM were simplified to a planet called "Terra Blanda." Terra Blanda consists of a planet with no geographical/seasonal features: no topography, all-land, equinox or mean annual forcing, north–south symmetric, no ice-albedo effect, no soil-moisture memory. This set of conditions allowed considerable savings in the computer time necessary to gather essential statistics. For instance, one could treat all longitudes as statistically equivalent as well as all months. By having no ocean, the longest radiation relaxation time is about one month. The EBCM for this configuration is solvable analytically with spherical harmonics (North and Cahalan, 1981; Leung and North, 1991). A control run with the GCM showed that the sequence of relaxation times for the different spatial scales (spherical harmonic degrees) of the GCM solution matched those of the EBCM quite well (Leung and North, 1991). Similarly the spatial-length scales were in good agreement between the models.

In a comprehensive study comparing the two model solutions, it was shown that the power spectrum of the Terra Blanda simulations with the GCM were very similar to those of the EBCM. Finally, the response of the GCM to point and latitudinal-ring heat sources was also in close agreement (North *et al.*, 1992).

3.4.4 Tuned simulations with EBCMs

A reasonable analogy to the above approach to EBCMs is the Bohr model of the atom or the shell model in nuclear physics. We know the model is somewhat schematic and based upon some questionable assumptions. On the other hand, there is clearly a physical basis for the validity of the model. In addition the EBCMs form an extremely useful intuitive introduction and guide to the problem, especially considering the incredible complexity of the real climate system and the GCMs being developed for simulations. The fact that there are only a few free or adjustable parameters in EBCMs makes it hard to cheat by fudging more coefficients than there are actual data to be fitted. This will always be a problem for the large complex models since there is never enough data to constrain all the phenomenological coefficients in the sub-grid scale parameterizations. Scientific testing of models becomes problematic since new data sets that might be used for model testing are very rapidly incorporated in the tuning of the next generation of the model.

The above discussion suggests that EBCMs can be used as a tool in certain applications where only the surface temperature is needed. Also the EBCMs are so cheap to run over long periods they can be used as a laboratory to test such notions as sampling errors when only short segments or a few realizations from GCM simulations are available. It therefore behoves us to relax the criterion of as few adjustable parameters as possible, so that we can achieve the best possible fit to observations in order to use the EBCM in these applications. For this reason in the following tuning exercise we allow some additional dependences in the parameterizations. These are physically motivated, but hardly derived rigorously from a set of fundamental principles. The most direct example is to allow the diffusion coefficient to have different values over land and ocean, to allow the noise forcing to be peaked in the middle latitudes, etc. We also allow the heat from below the mixed layer to have a latitude dependence until the mean annual surface-temperature field is in good agreement with observations. This flux from below is fixed and does not affect climate-change calculations or fluctuation statistics.

Figures 3.1 and 3.2 show some results obtained when the EBCM is tuned to reproduce the current climatology as well as possible given the model's limitations. Figure 3.1a shows the amplitude of the annual harmonic of the surface-temperature field for the tuned model, while Figure 3.1b shows that derived from the observed climatology. The lag of the surface temperature after the solar forcing at the top of the atmosphere (TOA) is shown for both the model (Figure 3.2a) and the observed climatology (Figure 3.2b). Some spatial properties of the statistics of the noise-forced EBCM temperature field are shown in Figures 3.3–3.5. Figure 3.3 shows the one-year lag autocorrelation of the annual mean surface temperature for the model (Figure 3.3a) and observed data (Figure 3.3b). The variance field for the annual mean surface temperature is shown in Figure 3.4. Finally, Figure 3.5 shows the square of the correlation of the local annual mean surface temperature with grid points selected from Asia (Figures 3.5a,b), and the North Atlantic (Figures 3.5c,d). While these figures reveal some of the shortcomings of the EBCM, they also show how well such a simple model can simulate the basic statistical properties of the surface temperature.

3.5 Applications

A rather fruitful area for making use of the EBCM is in the estimation of parameters in the climate system. Since the EBCM can be solved quickly by standard numerical methods, the statistics of the solution fields can be easily obtained. Furthermore, one can establish the sampling errors incurred if only short records or only a few realizations are available.

3.5.1 Paleoclimatology

The ability to modify quickly and run EBCMs to an equilibrium solution make them natural choices for paleoclimate studies. Much of the earlier research using EBCMs was concerned with the investigation of the causes of the ice ages (North *et al.*, 1983; Short *et al.*, 1991; Crowley *et al.*, 1992). This is still a topic of more recent research (Crowley and Kim, 1994), as well as the investigation of climate forcings over the past few hundred years (Crowley and Kim, 1996; 1999). The EBCMs have also been useful for modeling the more distant past (North and Crowley, 1985; Crowley *et al.*, 1986; Crowley and Kim, 1995).

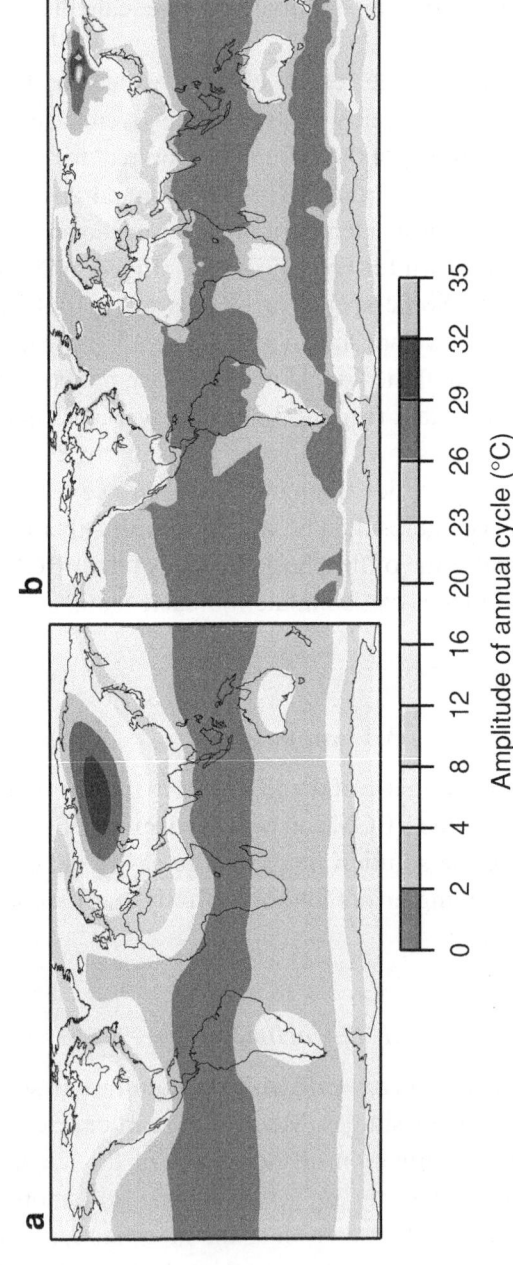

Figure 3.1. Amplitude of annual cycle of surface temperature computed using 40 years of EBCM control-run data (a), and observed climatology (b).

Figures 3.1 and 3.2 are available for download in colour from www.cambridge.org/9780521791328

64

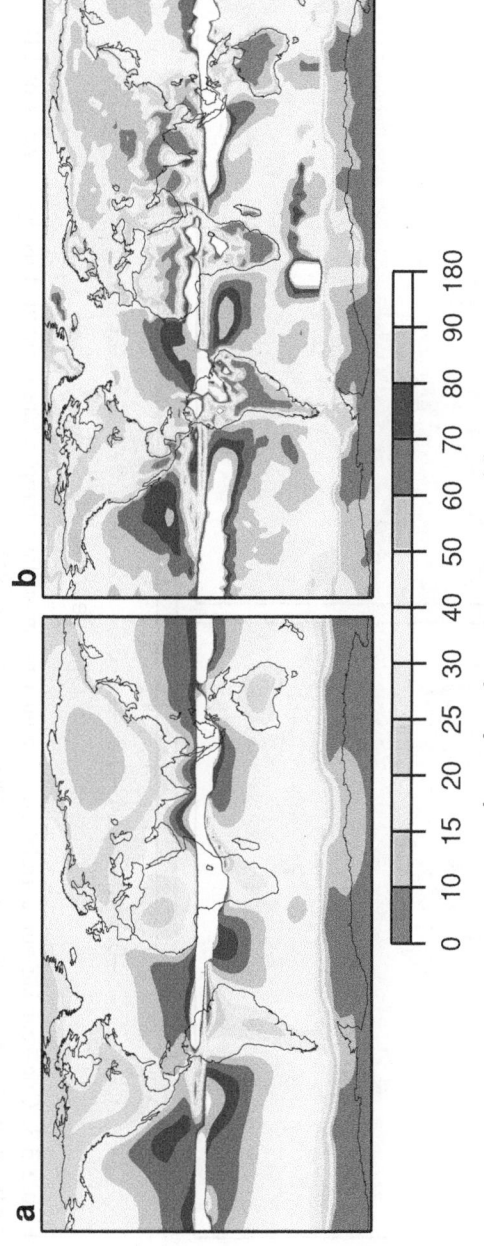

Lag of surface temperature (d)

0 10 15 20 25 30 40 50 60 70 80 90 180

Figure 3.2. Lag of surface temperature after TOA solar forcing computed using 40 years of EBCM control-run data (a), and observed climatology (b).

65

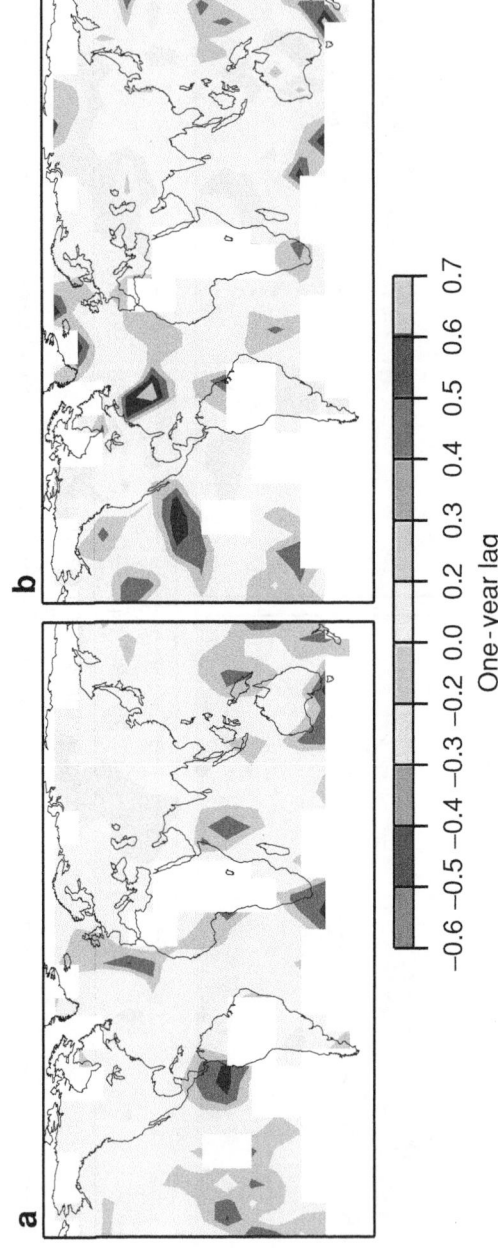

One-year lag

-0.6 -0.5 -0.4 -0.3 -0.2 0.0 0.2 0.3 0.4 0.5 0.6 0.7

Figure 3.3. One-year lag autocorrelation of annual mean surface temperature computed using 40 years of EBCM control-run data (a), and 40 years of detrended observations 1958–97 (b).

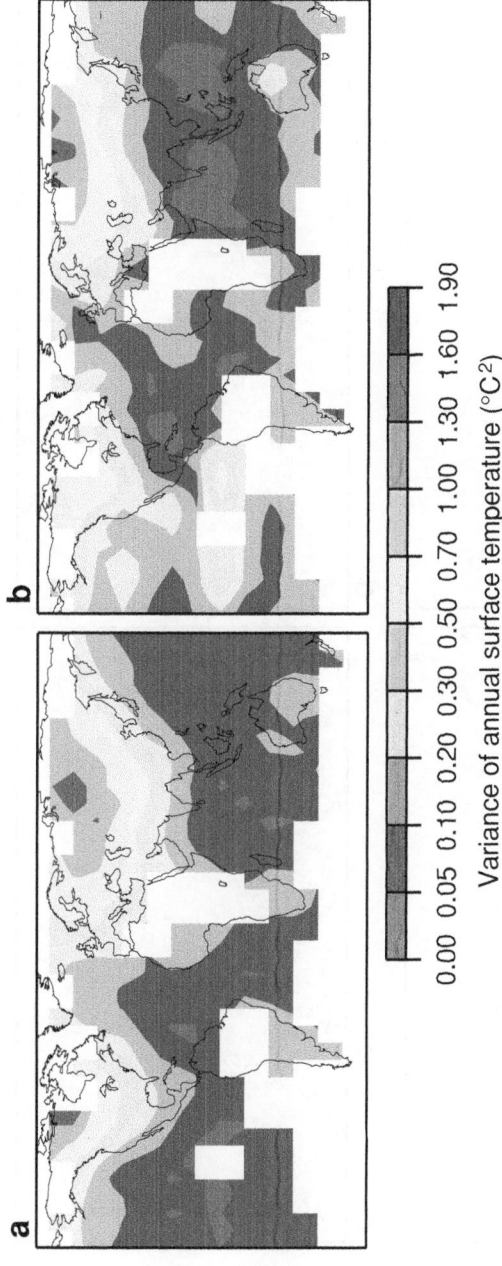

Figure 3.4. Variance of annual mean surface temperature computed using 40 years of EBCM control-run data (a), and 40 years of detrended observations 1958–97 (b).

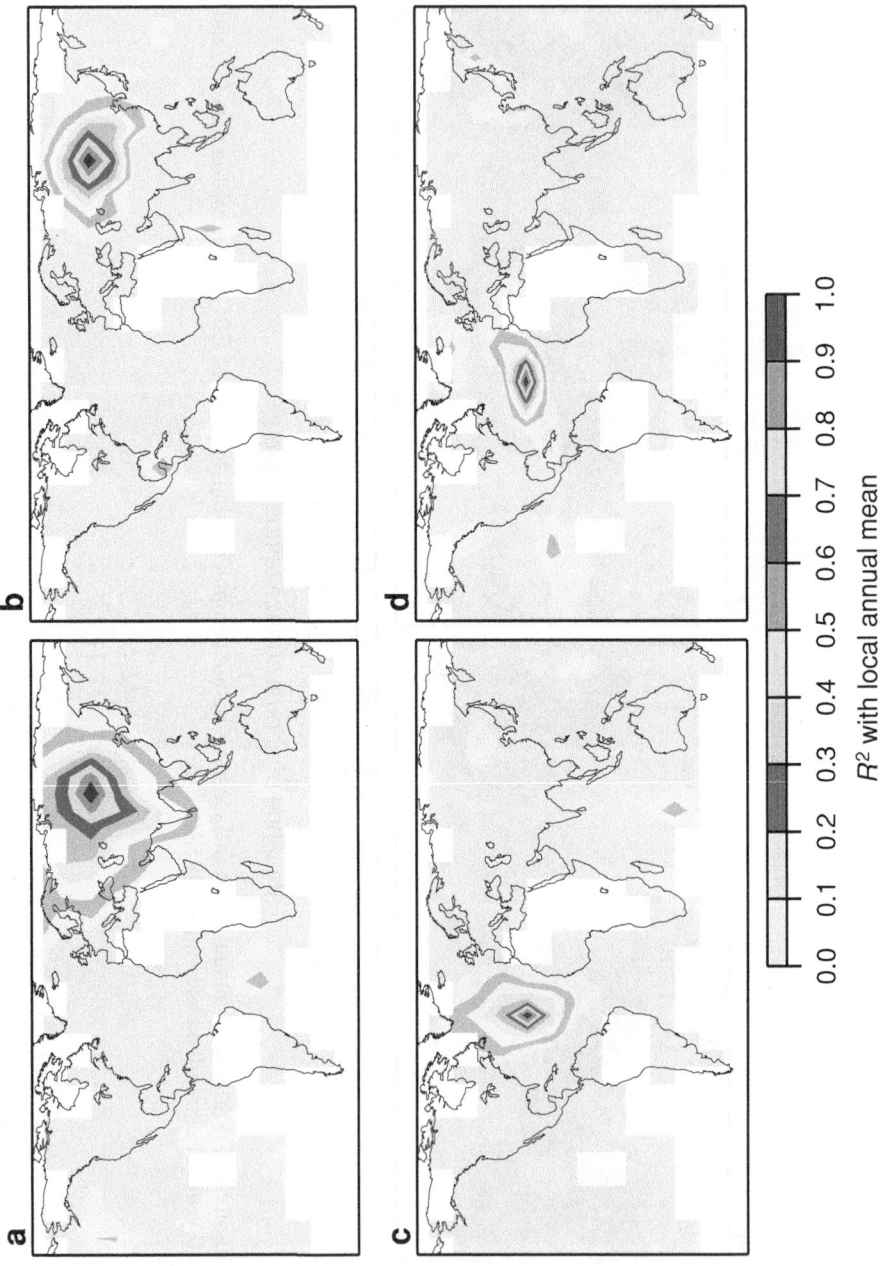

Figure 3.5. Square of the correlation of the local annual mean surface temperature with a grid point in Asia (50° N, 90° E) using 40 years of EBCM control data (a), and detrended observations 1958–97 (b), similarly for a point in the North Atlantic (30° N, 40° W) using EBCM data (c), and observations (d).

R^2 with local annual mean

0.0   0.1   0.2   0.3   0.4   0.5   0.6   0.7   0.8   0.9   1.0

68

3.5.2 Estimation of area averages and other aggregates

A problem of great interest is the errors incurred in estimating a global or hemispheric average from data based on a finite number of point measurements. The errors will depend upon the variance field, the space–time autocorrelation structure and the averaging time. Energy-balance climate models have been used to study this problem and have shown that surprisingly few stations are required to obtain a rather good estimate of the global average temperature (Shen *et al.*, 1994). By good we mean that the sampling error (standard error) is actually much smaller than the natural variability. If the station data are optimally weighted as few as 16 stations are capable of obtaining a satisfactory estimate of the global average. Similar studies have been conducted to find the standard error in estimating spherical harmonic coefficients (Zwiers and Shen, 1997). One can also estimate the spherical harmonic power spectrum (variance of the spherical harmonic coefficients). Because the number of point sites is finite, there will be aliasing biases in the estimates (Li *et al.*, 1997). Li and North (1998) have studied the problem of sampling error and bias when estimating area averages and spherical harmonic amplitudes from polar orbiting satellites.

3.5.3 Signal detection in the climate system

Energy-balance climate models have been used in the study of signal detection in the current climate-change debate (North *et al.*, 1995; North and Kim, 1995; Stevens and North, 1996; North and Stevens, 1998). In this problem one hypothesizes that there are four forced signals in the 100-year record of surface temperatures: greenhouse gases (G), anthropogenic aerosols (A), solar cycle forcing (S), and volcanic aerosols (V). Simulations of the 100 years for each of these forcings is conducted and the EBCM responses form the signals. One then proceeds to use these space–time signal patterns as the regression variables in a multiple regression exercise. The question is, in fitting the observations to these signals, what are the amplitudes of the individual signals, and what are their statistical error properties. For instance, one wishes to know if the null hypothesis that the amplitude G is zero can be rejected at a certain confidence level. One might also ask what the actual amplitude of A is including its confidence interval.

Several GCM groups are working on this problem and one might ask why bother with such a crude tool as the EBCM? The answer is simple, one cannot obtain a 'clean' signal from a GCM, since it is inevitably contaminated with sampling error itself. One must run many realizations of the 100-year forced GCM run for each signal. With the EBCM this is trivial, since one merely turns off the noise forcing to obtain a signal pattern. In fact, one can study the error (bias) incurred from using only a few realizations of the signal runs. In the regression problem it is customary to use long control runs to establish the covariance statistics of the temperature field. Typically these control runs are of 1000-year duration. With the EBCM one can easily make control runs of 10 000 years to establish the sampling error incurred in the shorter control runs. Several studies have now been conducted with the EBCM signals along with a comparative study of natural variability of several GCM control runs. The results for amplitude estimation do not seem to depend very sensitively on the choice of model used in the control runs (North and Stevens, 1998).

3.6 Conclusion

Energy-balance climate models provide a simple introduction to the main physical principles of global climate. The largescale fields including the seasonal cycle and natural fluctuation statistics are rather well simulated by these simple models. Not only are the models easy to understand and provide insight into the various processes involved, they have proven useful in some applications especially those involving statistical sampling error and signal processing studies.

Of course, these models have a number of limitations which must be acknowledged. The models are thermal balance models: they do not provide winds and they seem to only work well near the surface where largescale wave phenomena are strongly damped. They cannot include ocean dynamics beyond the simple upwelling diffusion models (Kim *et al.*, 1992). The EBCM considers the vertical profile of the atmosphere to be rigid not allowing changes in lapse rate or other potential feedbacks which might be important. Clouds are treated only through the macroscopic coefficients A, B, and a. For the most part cloud feedback is ignored. Water-vapor feedback is included empirically in the damping coefficient B, but this is probably only valid for mid latitudes.

References

Budyko, M. I. (1968). The effect of solar radiation variations on the climate of the Earth. *Tellus* **21**, 611–19.

Crowley, T. J. and K.-Y. Kim (1994). Milankovitch forcing of the last interglacial sea level. *Science* **265**, 1566–8.

(1995). Comparison of longterm greenhouse projections with the geologic record. *Geophys. Res. Lett.* **22**, 933–6.

(1996). Comparison of proxy records of climate change and solar forcing. *Geophys. Res. Lett.* **23**, 359–62.

(1999). Modeling the temperature response to forced climate change over the last six centuries. *Geophys. Res. Lett.* **26**, 1901–4.

Crowley, T. J., K.-Y. Kim, J. G. Mengel, and D. A. Short (1992). Modeling 100,000-year climate fluctuations in pre-Pleistocene time series. *Science* **255**, 705–7.

Crowley, T. J., D. A. Short, J. G. Mengel, and G. R. North (1986). Role of seasonality in the evolution of climate during the last 100 million years. *Science* **231**, 579–84.

Graves, C. E., W.-H. Lee, and G. R. North (1993). New parameterizations and sensitivities for simple climate models. *J. Geophys. Res.* **98**, 5025–36.

Hansen, J. and S. Lebedeff (1987). Global trends of measured surface air temperature. *J. Geophys. Res.* **92**, 13345–72.

Hyde, W. T., K.-Y. Kim, T. J. Crowley, and G. R. North (1990). On the relation between polar continentality and climate: Studies with a nonlinear seasonal energy balance model. *J. Geophys. Res.* **95**, 18653–68.

Kim, K.-Y. and G. R. North (1991). Surface temperature fluctuations in a stochastic climate model. *J. Geophys. Res.* **96**, 18573–80.

   (1992). Seasonal cycle and second-moment statistics of a simple coupled climate system. *J. Geophys. Res.* **97**, 20437–48.

Kim, K.-Y., G. R. North, and J. Huang (1992). On the transient response of a simple coupled climate system. *J. Geophys. Res.* **97**, 10069–81.

Leung, L.-Y. and G. R. North (1990). Information theory and climate prediction. *J. Clim.* **3**, 5–14.

   (1991). Atmospheric variability on a zonally symmetric land planet. *J. Clim.* **4**, 753–65.

Li, T.-H. and G. R. North (1998). Sampling errors for meteorological fields observed by low Earth orbiting satellites. *J. Geophys. Res.* **103**, 19595–614.

Li, T.-H., G. R. North, and S. S. Shen (1997). Aliased power of a stochastic temperature field on a sphere. *J. Geophys. Res.* **102**, 4475–86.

Lin, R. Q. and G. R. North (1990). A study of abrupt climate change in a simple nonlinear climate model. *Clim. Dyn.* **4**, 253–61.

Lindzen, R. S. and B. Farrell (1977). Some realistic modifications of simple climate models. *J. Atmos. Sci.* **34**, 1487–1501.

Mengel, J. G., D. A. Short, and G. R. North (1988). Seasonal snowline instability in an energy balance model. *Clim. Dyn.* **2**, 127–31.

Myhre, G., E. J. Highwood, K. P. Shine, and F. Stordal (1998). New estimates of radiative forcing due to well mixed greenhouse gases. *Geophys. Res. Lett.* **25**, 2715–18.

North, G. R. (1984). The small ice cap instability in diffusive climate models. *J. Atmos. Sci.* **41**, 3390–5.

   (1990). Multiple solutions in energy balance climate models. *Paleogeog., Paleoclim., Paleoecol.* **82**, 225–35.

North, G. R. and R. F. Cahalan (1981). Predictability in a solvable stochastic climate model. *J. Atmos. Sci.* **38**, 504–13.

North, G. R. and T. J. Crowley (1985). Application of a seasonal climate model to Cenozoic glaciation. *J. Geol. Soc. London* **142**, 475–82.

North, G. R. and K.-Y. Kim (1995). Detection of forced climate signals. Part II: simulation results. *J. Clim.* **8**, 409–17.

North, G. R. and M. J. Stevens (1998). Detecting climate signals in the surface temperature record. *J. Clim.* **11**, 563–77.

North, G. R., R. F. Cahalan, and J. A. Coakley (1981). Energy balance climate models. *Rev. Geophys. Space Phys.* **19**, 91–121.

North, G. R., J. G. Mengel, and D. A. Short (1983). Simple energy balance model resolving the seasons and the continents: application to the astronomical theory of the ice ages. *J. Geophys. Res.* **88**, 6576–86.

North, G. R., K.-Y. Kim, S. S. Shen, and J. W. Hardin (1995). Detection of forced climate signals. Part I: filter theory. *J. Clim.* **8**, 401–8.

North, G. R., K.-J. Yip, L.-Y. Leung, and R. M. Chervin (1992). Forced and free variations of the surface temperature field in a general circulation model. *J. Clim.* **5**, 227–39.

Sellers, W. D. (1969). A global climatic model based on the energy balance of the Earth–atmosphere system. *J. Appl. Meteor.* **8**, 392–400.

Shen, S. S. P., G. R. North, and K.-Y. Kim (1994). Spectral approach to optimal estimation of the global average temperature. *J. Clim.* **7**, 1999–2007.

Short, D. A., J. G. Mengel, T. J. Crowley, W. T. Hyde, and G. R. North (1991). Filtering of Milankovitch cycles by Earth's geography. *Quat. Res.* **35**, 157–73.

Stevens, M. J. and G. R. North (1996). Detection of the climate response to the solar cycle. *J. Atmos. Sci.* **53**, 2594–608.

Zwiers, F. W. and S. S. Shen (1997). Errors in estimating spherical harmonic coefficients from partially sampled GCM output. *Clim. Dyn.* **13**, 703–16.

4

Intrinsic climatic variability: an essay on modes and mechanisms of oceanic and atmospheric fluid dynamics

JAMES C. MCWILLIAMS

Department of Atmospheric Sciences, UCLA, Los Angeles, CA

4.1 Introduction

A folk adage is "the climate is what you expect; weather is what you get." Taking a somewhat longer and more mechanistically explicit view, we might rephrase this as "the forced response is what you expect; intrinsic variability is what you get."

Transient climatic forcing is defined, somewhat arbitrarily, as the external influences on Earth's climate involving astronomy, geology, evolution, and human activity: changes in the incident radiation from space; the shape of Earth's solid surface; the material composition of air, seawater, and the land surface; the species composition of biota; and anthropogenic land modification and pollutant emission. The other parts of the climatic system thus contribute to its internal dynamics in the atmosphere, ocean, land surface, and sea ice.

The climatic record evinces both forced and intrinsic variability. It is a reasonable view that the longer the time scale of fluctuation, the more likely that the forcing influence dominates. Yet some very simple climatic models – among those that are the subject of this chapter – show that natural variability can occur on all time scales, even well beyond the primary dynamical relaxation or adjustment times (e.g., weeks for the atmospheric troposphere or centuries for an abyssal oceanic basin). It is also a reasonable view that the global mean climate is more strongly influenced by external forcing, while the regional patterns are influenced at least as much if not more by intrinsic variability. Nevertheless, the changes in global mean surface temperature, T, over $\sim 10^5$ y are not larger than those that occur in inter-annual and decadal variations in extra-tropical, winter-mean regional patterns.

Climatic change thus occurs through a combination of forced and intrinsic variability mechanisms. In the present era when anthropogenic forcings are probably

Frontiers of Climate Modeling, eds. J. T. Kiehl and V. Ramanathan.
Published by Cambridge University Press. © Cambridge University Press 2006.

inducing changes that rival those in the paleoclimatic record, it is important to understand how these mechanisms compete and combine, if we are to anticipate the future accurately. However, this understanding is elusive: the empirical record is short and fragmentary; the phenomena of climate are complex; many of the forcing functions are poorly determined, both retrospectively and certainly prospectively; and – our focus here – the dynamical mechanisms of intrinsic variability are subtle.

The conceptual basis for interpreting climatic variability is the use of models of many types, ranging from simple, ad hoc mathematical representations of hypothesized relationships among different variables (sometimes called box models, connoting volume integration, though I prefer to call them analogy models) to general circulation models (GCMs) that, to the degree computationally feasible, embody the fundamental laws of fluid dynamics, material properties, chemical reactions, and radiation. Unavoidably GCMs involve compromises that do – and often should – worry the specialists. However, seemingly inevitably they are becoming the standard models for climatic assessment and prediction, as they already are for weather prediction, because they can strike a best balance among the very many influential processes. Even present GCM solutions simulate many aspects of intrinsic climatic variability in ways that are consistent with empirical evidence to the degree that we know how to make the comparisons. While this kind of success assures us that these models embody relevant ingredients, it does not mean that we comprehend well the mechanisms that produce the answer. Nor should we have confidence that we can yet envision the range of climatic outcomes.

This essay is about models simpler than GCMs that provide paradigms for interpreting intrinsic variability in GCM solutions and nature. Intrinsic variability is generically an instability of the directly forced climatic system. The default paradigm, or null hypothesis, is that intrinsic climatic variability arises primarily from weather fluctuations, sometimes called weather noise. Weather is a synoptic-scale (i.e., over $\sim$ thousands of kilometers and several days) instability of tropospheric, seasonal-mean atmospheric circulation, and it somehow, through fluid-dynamical advective non-linear interactions, induces the rest of the climatic system to produce fluctuations over larger and longer scales. This is an inverse transfer of energy in space and time scales, which is a generically familiar process in homogeneous geostrophic turbulence, but it occurs rather differently and is not yet well understood in the context of climatic dynamics. In this null hypothesis, the land surface, sea ice, and ocean passively respond to weather noise as negative thermal feedbacks, thereby amplifying and lengthening the duration of the fluctuations by providing a greater thermal inertia, hence an increased "memory." Present opinion is that there is considerable validity to this null hypothesis based on the fact that atmospheric GCMs exhibit substantial, structurally realistic, low-frequency, planetary-scale variability even with seasonal-mean solar forcing and surface conditions.

The null hypothesis overlooks all the other positive feedbacks and instabilities known to be present in the climatic system, ranging from ice-albedo feedback on the planetary scale to mesoscale instabilities of the oceanic wind-driven circulation to the many fluid-dynamical instabilities on small scales often lumped into the category of turbulence. However, only if the consequences of these other processes appreciably influence the largescale variability of surface air temperature can they be judged to contribute enough to climatic variability to vitiate the null hypothesis. This null hypothesis is known to be inadequate with respect to El Niño southern oscillation (ENSO) cycle in tropical Pacific sea-surface temperature fluctuations that last several years and are demonstrably related to subsurface and extra-tropical climate. It is probably also inadequate for many other phenomena, although it is indicative of the difficulty of climatic science that unambiguous proofs are mostly lacking. The essential reason is that the fluid-dynamical mechanism whereby weather noise and other instabilities force climatic variability is not yet well enough understood that we can confidently identify their influences in competition with other mechanisms.

How does one identify exceptions to the null hypothesis? The most convincing means is demonstrating causality from, e.g., oceanic fluctuations to atmospheric fluctuations on climatic scales. When the underlying causes of both fluctuations are not well understood, then one is forced to resort to simple statistical measures of single links in the causal chain, such as a time-lag correlation with the ocean leading the atmosphere or a prevalent correlation between, say, oceanic surface temperature and upward surface heat flux (e.g., the empirical analyses of Czaja and Frankignoul, 1999; 2002, for the North Atlantic oscillation, NAO). In addition, in models one can demonstrate causality by including or eliminating particular oceanic processes not part of the null hypothesis (e.g., Sutton and Mathieu, 2002, showing the atmospheric impact in the North Atlantic of including oceanic lateral heat flux convergences in a model).

This essay is intended as a didactic survey of the present understanding of various dynamical mechanisms for intrinsic variability. It is not meant as a review article, which would require a much lengthier discussion and bibliography. It is being written at a time when ideas are very much in ferment, hence presumably also in transit. As yet there is little agreement among scientists about the essential dynamical ingredients in climatic variability. However, if we are to achieve this understanding, we must work through the types of models discussed below, even if we can anticipate that history will judge them to be somewhat incomplete and naive.

4.2 The empirical record

To set the stage for considering mechanisms, we first consider several examples from the empirical record of climatic variability. The totality of the record is extremely difficult to interpret, mainly because the long perspective demanded by the phenomena greatly exceeds the ~ 100-year period in which instruments have been widely available to measure temperature, pressure, winds, etc. Instead much of our empirical knowledge comes from indirect inferences from geologically preserved materials such as glacial ice, sediments, etc., which are fraught with potential ambiguity in their interpretation.

Figure 4.1 shows some of the indirect evidence over the last $\sim 10^5$ y. In this broad sweep, we see the transition from a glacial era to the current Holocene era, where temperature changes are global in extent. This type of variation is most likely a forced response to changes in the seasonal insolation, sometimes called Milankovitch cycles, although fluid dynamical models of this response are still lacking because of the length of time integration involved. In detail, though, the evidence shows variability on periods of $\sim 10^3$–10^4 y that does not correspond to any obvious astronomical forcing and is not globally uniform; hence we infer that it may be evidence of intrinsic climatic variability. This variability is most evident in the North Atlantic region (though correlations exist with other regions too) in the rather abrupt changes in oxygen isotopic ratio (a surrogate for temperature) and sea-ice-rafted debris. These changes have been called Dansgaard–Oeschger cycles and Heinrich events during the glacial period, and Figure 4.1 shows that millennial variability persists through the Holocene as well. The oceanic thermohaline circulation is a significant mode for both global and North Atlantic heat transport, it is dynamically delicate, and it has millennial adjustment times if not also instabilities (Section 4.3.1); therefore, it may be implicated in this millennial intrinsic variability in association with changes in land ice and the global hydrological cycle. The modeling case for this is perhaps best developed for the Younger-Dryas interruption of the transition from glacial to Holocene eras (at age $\approx 11 \times 10^3$ y in Figure 4.1), interpreted as a cessation and recovery of the thermohaline circulation following large freshwater runoff from North America (Saravanan and McWilliams, 1995; Manabe and Stouffer, 1997).

Within the modern era of more direct atmospheric and oceanic measurements, a general warming trend has occurred in the global mean surface air temperature. It has recently accelerated probably in response to increased human emissions of radiatively active (i.e., greenhouse) gases. There also is evident inter-annual and decadal variability, both in the global mean T and, with larger T amplitude, in regional patterns. The GCM modeling studies with reconstructed anthropogenic and natural climatic forcing histories indicate that much of the recent global T

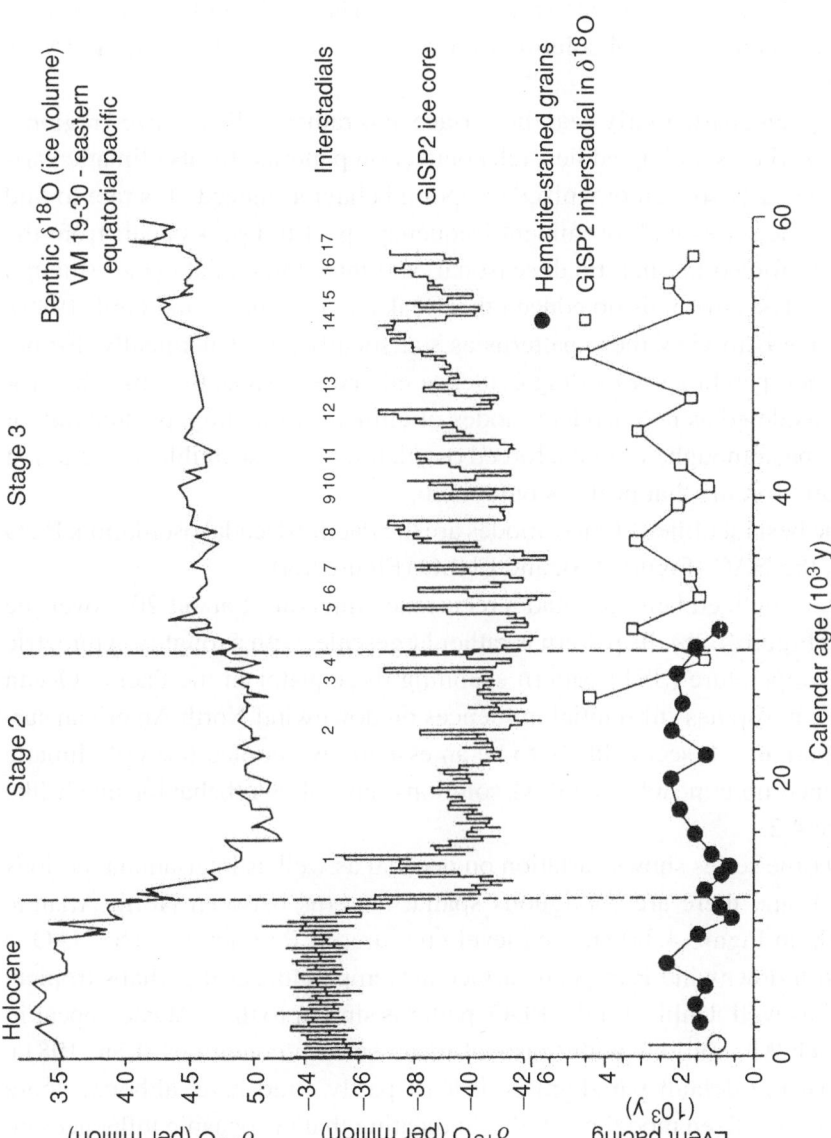

Figure 4.1. Time series over the past ~ 10^5 years of oxygen isotope anomaly, $\delta^{18}O$ (a surrogate for T), from a Greenland ice core (GISP2) and an eastern equatorial Pacific sediment core (VM 19–30). Also shown are the time intervals between peaks in hematite grain concentration in sediment cores in the North Atlantic Ocean west of Greenland over the past 0.3×10^5 y (solid dots) and the warm events (interstadials) in the ice-core record before that (open squares). Adapted from Bond *et al.*, 1997.

variability is a forced response (Delworth and Knutson, 2000; Dai *et al.*, 2001; also see Chapters 1 and 2 in this book), but the regional variations in the modern era are almost certainly more the manifestations of intrinsic climatic variability. Furthermore, we should not be complacent that the short instrumental record has shown us the full range of intrinsic variability; e.g., Hall and Stouffer (2001) show an example in a coupled GCM solution of a very rare regional fluctuation of very large amplitude.

The atmosphere, particularly near the surface, has rather well organized regional spatial patterns (i.e., standing eddies, teleconnection patterns) for its climatic variability but not nearly so well organized temporal behavior. Indeed, it is rare to find atmospheric time series with prominent frequency spectrum peaks apart from the astronomically forced diurnal, tidal, seasonal, and inter-millennial cycles (though specialized analysis methods do educe other peaks; e.g., Vautard and Ghil, 1989). Present practice is to view these patterns as statistically and dynamically distinct from each other, pending compelling evidence otherwise. Thus, they may be provisionally considered as independent modes of climatic variability, presumably of an intrinsic type, although, as with a forced oscillator, also susceptible to excitation by any climatic forcing that projects onto them.

Among the best identified climate modes are the Pacific decadal oscillation, PDO (Figure 4.2), the NAO (Figure 4.3), and ENSO (Figure 4.4).

The PDO, as defined here, has had a recurrence interval of about 20 y over the past century (Figure 4.2b). Its pattern is rather largescale, with a roughly symmetric sea-surface temperature (SST) pattern spanning the equator in the Pacific Ocean (Figure 4.2a); it also has substantial influences on downwind North American surface air temperature. It seems likely to be an essentially coupled form of climatic variability since no atmospheric GCM solutions have shown behavior much like that in Figure 4.2.

The NAO time series shows variation on decadal as well as inter-annual periods (Figure 4.3b), and there are contiguous spatial patterns between North Atlantic SST (a tripole in Figure 4.3a) and sea-level pressure, SLP (a dipole). The NAO is correlated with downwind European surface air temperature and perhaps tropical Atlantic SST as well. Unlike for the PDO, patterns similar to the NAO do appear in atmospheric GCM solutions with seasonal-mean surface conditions (Lau, 1981), indicating that the default paradigm is at least partly valid here, although there are a variety of idealized modeling studies indicating that the oceanic influences go beyond this (e.g., Selton *et al.*, 1999; Saravanan *et al.*, 2000; Watanabe and Kimoto, 2000b; Czaja and Marshall, 2001).

El Niño southern oscillation shows episodic warming of the eastern and central equatorial SST (Figure 4.4a). Its time series (Figure 4.4b) shows a broad spectrum peak between about 1.5 and 7 y, though more clearly so in oceanic SST than

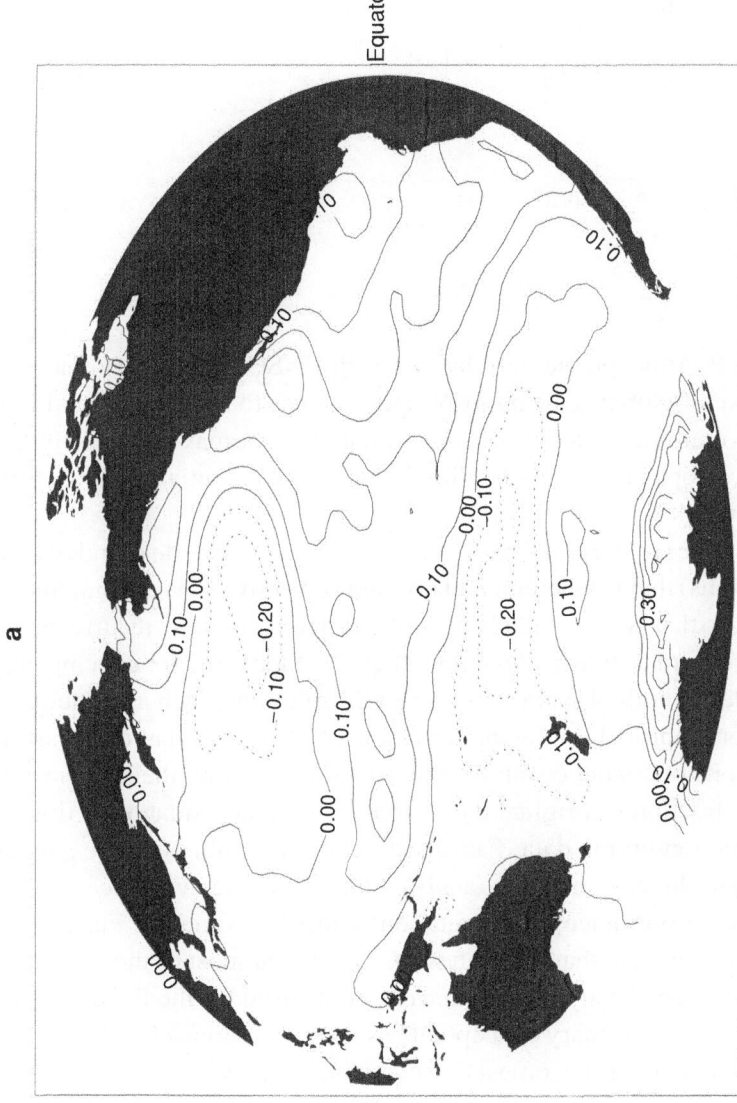

Figure 4.2. (a) Pattern of SST change (K) in the Pacific Ocean for a composite of three transition events (1978–1982 minus 1971–1975, 1959–1963 minus 1952–1956, 1926–1930 minus 1919–1923). (b) Time series of the projection of the spatial pattern in (a) onto Pacific SST anomaly with a five-year running average. (Adapted from Chao *et al.*, 2000.)

b

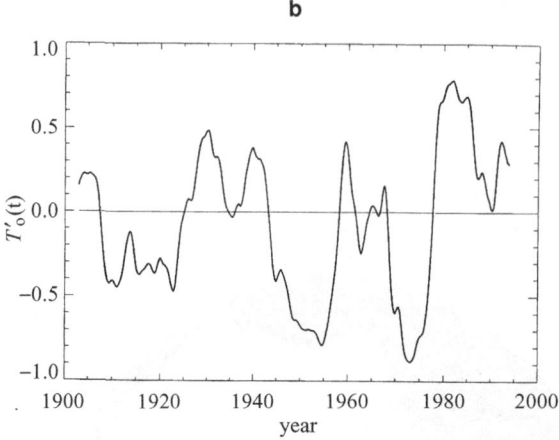

Figure 4.2. *(cont.)*

atmospheric SLP. Atmospheric correlations with ENSO have a very wide spatial extent, indicative of global-scale atmospheric response to tropical thermal forcing, and this leads to remote oceanic surface correlations that may not causally participate in ENSO. Tropical air–sea coupling is, however, essential to the occurrence of ENSO (Section 4.3).

All of the analyses in Figures 4.2–4.4 are made by a composite or index methodology (see captions) that is inherently ad hoc and subjective. The extenuating virtue is its simplicity, so that no one is likely to doubt the reality of these recurrent regional patterns. A more formal statistical methodology is widely used for educing climatic variability modes in both observations and GCM solutions, such as principal components, PCs, or empirical orthogonal functions, EOFs. Principal components are the eigenfunctions of spatial covariance matrices, ordered by their eigenvalues in the represented fraction of original data variance, with accompanying time series from their projection on the data. Faced with comprehending complex signals in multivariate, multidimensional data, analysts find PCs attractive since they are by construction the most efficient representation of the data variance. Furthermore, it is the common experience that only a few PCs represent most of the variance for a given climatic data set. Many variants have been devised for the PC methodology, and they now provide a primary conceptual basis for describing climatic variability, which to a large extent has become a discourse on patterns.

The efficient representation of variance by PCs does not imply that an individual PC is a dynamically coherent mode. In part this is because the PC methodology enforces spatial and temporal orthogonality between the modes, which dynamics does not. More importantly, PC patterns are often non-robust with respect to different physical variables analyzed, the choices of variance norm and dataset, and

a

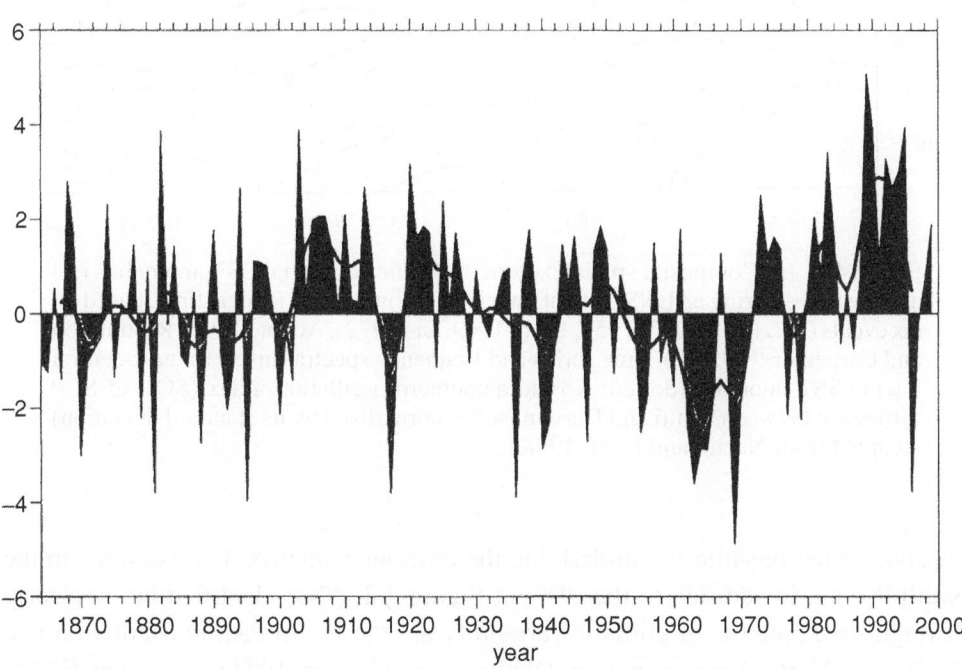

b

Figure 4.3. (a) The 1960–1995 regression patterns of winter SLP (dark contours and H/L labels; contour interval = 1 mb) and SST (light contours and hatch marks [vertical = positive; random = negative]; contour interval = 0.1 K) anomalies (Deser, personal communication), based on (b) the NAO index of normalized SLP difference between Lisbon and Iceland (solid line = five-year running average; Hurrell, 1995).

a

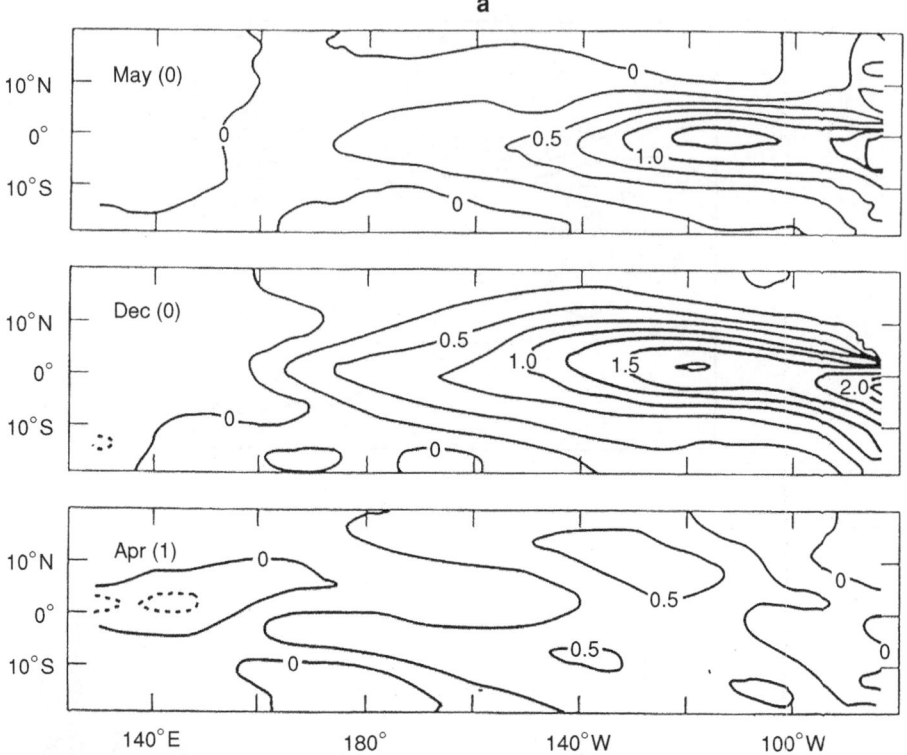

Figure 4.4. (a) Composite spatial pattern of Pacific equatorial SST anomalies (K) at three times during an ENSO event, labeled as month(year) in local time, based on six events (i.e., 1951, 1953, 1957, 1965, 1969, and 1972). Adapted from Rasmusson and Carpenter, 1982. (b) Time series and frequency spectra (inset) for east-central Pacific SST anomaly (dotted; K) and a southern oscillation index (SOI) of SLP difference between Tahiti and Darwin (solid; normalized by its standard deviation) Adapted from Neelin and Latif, 1998.

the space–time pre-filtering underlying the covariance matrix. For example, in the North Pacific, in addition to the PDO in Figure 4.2, other identified but unclearly distinguished patterns of climatic variability are the North Pacific oscillation and the Pacific–North America pattern (Wallace and Gutzler, 1981), as well as ENSO itself. Another example is the Arctic oscillation (Thompson and Wallace, 1998), whose spatial pattern overlaps and whose time series is highly correlated with the NAO in Figure 4.3 (Deser, 2000). From one analysis to the next, the distinctions among statistically reduced patterns frequently become blurred. This poses a considerable dilemma for climate theory. How can a dynamical mechanism be meaningfully tested against pattern analyses whose uncertainties across different analyses are largely unquantified? What is the right way to ask whether two patterns

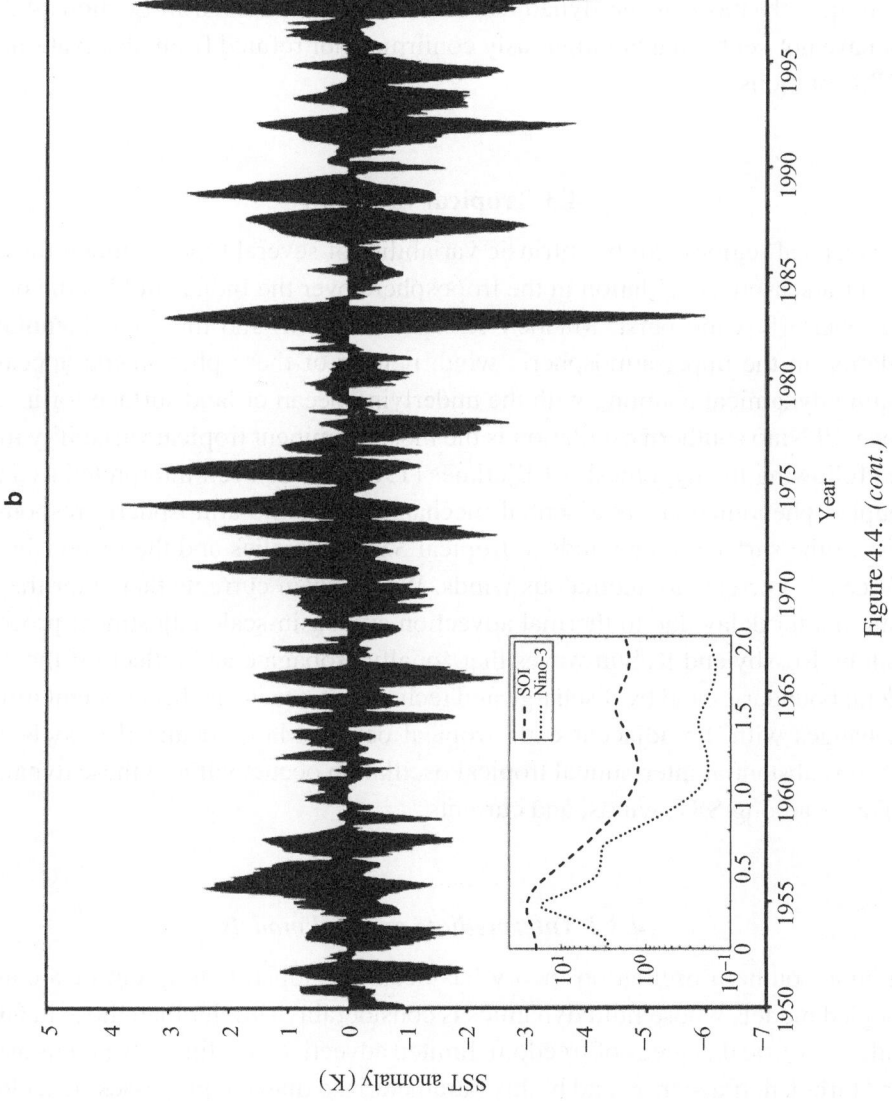

Figure 4.4. (*cont.*)

83

are similar enough to claim consistency between a model solution and nature? A traditional comparison methodology is the pattern correlation coefficient – but what framework should be used for assessing the significance of a partial correlation? Partial correlations among variables and locations – the underlying basis for PC analysis – cannot logically be identified with dynamical causality or its absence. Consequently most of the dynamical mechanisms described in Sections 4.3 and 4.4 have not yet been unambiguously confirmed nor refuted from observations and GCM solutions.

4.3 Tropical variability

The tropical regions exhibit intrinsic variability of several types. Among these are the intra-seasonal oscillation in the troposphere over the Indian and Pacific oceans (i.e., westerly wind burst, Madden–Julian oscillation) and the quasi-biennial oscillation in the upper-atmospheric wind; neither of these phenomena appears to require dynamical coupling with the underlying ocean or land surface for its existence. El Niño southern oscillation is the most prominent tropical variability mode, and following the hypothesis of Bjerknes (1969), it has been interpreted as a truly coupled phenomenon. Its essential mechanisms are the atmospheric responsiveness of the surface trade winds to tropical SST anomalies and the responsiveness of oceanic currents to anomalous winds. Wind-driven currents then alter the SST pattern after delay due to thermal advection and basin-scale adjustment processes both by Rossby and Kelvin waves that zonally propagate and reflect off the continental boundaries and by discharge and recharge of equatorial heat content through exchanges with the adjacent extra-tropical ocean. Many idealized models show that no substantial inter-annual tropical oscillation occurs without these dynamical linkages among SST, winds, and currents.

4.3.1 Intermediate coupled models

El Niño southern oscillation theory has been developed mainly out of a class of coupled models whose fluid dynamics is considerably simpler than those in GCMs with few vertical degrees of freedom, limited advective non-linearity and its associated turbulent transience, and highly parameterized diabatic processes (i.e., clouds, radiation, and precipitation). The prototypical model of this class by Cane and Zebiak (1985) is for a single oceanic basin and contains three elements:

• a linear, horizontal, upper-ocean, anomaly dynamics for zonal equatorial waves and quasi-steady currents with a viscous, Coriolis balance against anomalous surface wind stress, τ';

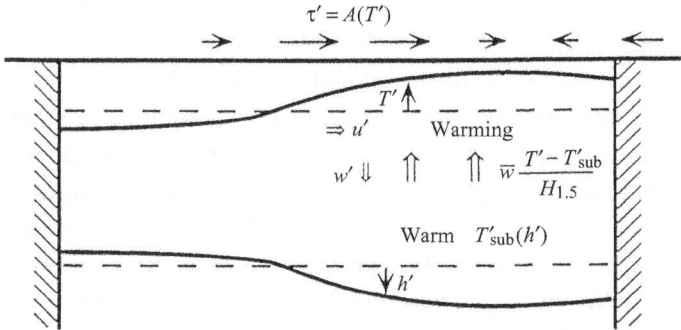

Figure 4.5. Schematic of warming mechanisms that amplify and maintain eastern and central Pacific equatorial SST anomalies during an ENSO warm phase. u' is the ocean zonal current anomaly; w' is the ocean vertical current anomaly; $H_{1.5}$ is the ocean surface layer thickness. The converse applies during a cold phase. Wind-induced convergence elevates the sea level and deepens the thermocline. Anomalous zonal advection from the mean warmer western region and mean upwelling from the anomalously warm subsurface both contribute to the warming. Not depicted are the oceanic dynamics that supply the memory needed for oscillation. From Neelin *et al.*, 1998.

- a two-dimensional (2D) SST anomaly, T'_o, evolution equation with vertical upwelling and horizontal advection in the presence of seasonal-mean thermal gradients, $\nabla \overline{T}$, and damping by air–sea heat flux;
- a non-local, diagnostic (i.e., slaving) relationship between surface wind stress and SST anomaly.

The latter is written symbolically as in Equation (4.1), where $\mathcal{A}$ is a non-local functional

$$\tau' = \mathcal{A}[T'_o; x, y] \tag{4.1}$$

of T'_o over the entire basin. We denote (east, north, up) coordinates by (x, y, z). The paradigm for Equation (4.1) is the steady-state, tropical, tropospheric response to anomalous heating as embodied in the linear viscous model of Gill (1980), though in some formulations it has also been specified from empirical regression relations. It is taken to be an instantaneous relation, presuming a rapid atmospheric adjustment to SST compared to the climatic evolution of interest.

Figure 4.5 is a schematic drawing of the mechanisms for SST warming in ENSO, as embodied in intermediate coupled models.

The great success of such intermediate coupled models is that they exhibit self-sustained oscillations with periods in the ENSO range, given model parameters and spatial fields for $\nabla \overline{T}$ and $\mathcal{A}$ that are not physically implausible. They have provided targets for extensive asymptotic theoretical analyses (reviewed in Neelin *et al.*, 1998) and motivated some even simpler analogy models (Section 4.3.2). They even contain regimes with temporally irregular oscillations (as observed) arising from wholly deterministic dynamics through coupling with the seasonal-mean cycle (Jin *et al.*, 1994; Tziperman *et al.*, 1995), although irregularity can also be induced through stochastic excitation (Kleeman and Moore, 1997).

The essential mechanisms of this mode of intrinsic climatic variability clearly depart from the null hypothesis, with no essential role for weather-noise forcing (Section 4.1), although it is plausible that weather variability may accentuate the irregularity of ENSO. Of course, such intermediate models are too idealized to be expected to compete quantitatively with GCM simulations and forecasts (though they have done so for the time being while GCM developments ensue). Nor have they yet provided much insight into the decadal variations of ENSO (e.g., the eastern extremum in Figure 4.2a), most likely because their dynamics is limited to only shallow tropical mechanisms (see Section 4.4.2). However, a diagnostic analysis of how ENSO properties change in an intermediate model due to changes in $\nabla \overline{T}$ and $\mathcal{A}$ (Federov and Philander, 2000) may offer at least a partial interpretation. Another partial interpretation is that Indo–Pacific coupling increases the decadal variability of the inter-annual ENSO mode in the Pacific (Yu *et al.*, 2001).

4.3.2 Delayed and recharge oscillators

Fluid dynamics cannot rigorously be represented by low-order dynamical systems except under rare circumstances near marginal instability where few spatial modes are active. Nevertheless, such systems can be more readily comprehensible than fluid models of intrinsic climatic variability and as such are quite useful in providing a quantitative conceptual language by analogy (see also Section 4.4). The solutions of analogy models are metaphors for climatic variability, to be judged by the verisimilitude of their behavior. Many low-order models have been built around the Bjerknes hypothesis (e.g., McWilliams and Gent, 1978). Two such models are widely considered as particularly useful, the delayed oscillator (Suarez and Schopf, 1988; Battisti and Hirst, 1989) and the recharge oscillator (Jin, 1997a).

A delayed oscillator equation for the eastern equatorial SST anomaly, $T = T_0'(x_E)$, has the form of Equation (4.2), where k is the positive (i.e., unstable)

$$\dot{T} = kT - eT^3 - \alpha T(t - \delta), \qquad k, e, \alpha, \delta > 0 \qquad (4.2)$$

air–sea feedback rate of the Bjerknes hypothesis that supports growth of fluctuations, e is the negative feedback strength usually associated with the non-linear dependence of subsurface temperature on thermocline depth (NB $T_{sub}'(h')$ in Figure 4.5), δ is a time delay associated with a basin-scale adjustment of the thermocline depth by zonal transit and reflection of Kelvin and Rossby waves, and α represents the memory efficiency of the delay. Both e and α represent negative feedbacks that largely control the finite-amplitude equilibration and oscillation period of the cycle. There are periodic solutions of Equation (4.2) in the ENSO range for physically plausible choices of the constants.

A recharge oscillator equation for the eastern equatorial SST anomaly, $T = T'_o(x_E)$, and western equatorial thermocline depth anomaly, $h = h'(x_W)$, has the form shown in Equations (4.3) and (4.4), with all constants positive. Here k and e

$$\dot{T} = kT - e(h/b + T)^3 + \gamma h \qquad (4.3)$$
$$\dot{h} = -rh - abT \qquad (4.4)$$

have similar meanings as in Equation (4.2), γ represents upwelling advection (a positive feedback), r is the basin-wide wave-adjustment rate (a negative feedback), b represents wind response to SST, a represents the current response to wind, abT; is the recharging rate of equatorial heat content, due to mass exchange with the extra-tropical ocean by Ekman advection and leakage by reflecting waves at the zonal boundaries (including Indonesian throughflow), which largely controls the oscillation period. This model too has periodic solutions in the ENSO range for plausible choices of the constants.

Neither oscillator model exhibits irregular cycles, although this can be achieved by adding stochastic excitation (weather noise). Both models emphasize the importance of unstable air–sea coupling, as is also implicit in Equation (4.1). These models neglect surface heat and water fluxes as important causes of SST change in ENSO (unlike extra-tropical variability; Section 4.4), although they do provide a negative feedback (*i.e.*, anomaly damping) and they are implicit in the mechanism for τ' being slaved to T'_o. The delayed oscillator emphasizes the zonal wave adjustment of the thermocline depth, while the relaxation oscillator also emphasizes the importance of changes in total equatorial heat content (as originally argued by Wyrtki, 1975, and Zebiak, 1989). Assessing these behaviors by ENSO observations is a subtle task. Nevertheless, a recent coupled GCM solution (Figure 4.6) shows a remarkably simple PC pattern of relationships among equatorial wind, SST, and heat content; in this pattern the volume integral of the latter shows substantial changes (implying extra-tropical exchange) as a precursor to eastern and central SST changes, as embodied in the recharge oscillator model.

4.4 Extra-tropical variability

Some parts of extra-tropical climatic fluctuations are a remote response to tropical ENSO fluctuations, conveyed by atmospheric mechanisms (e.g., rainfall anomalies in the southeast and southwest US). Other parts, such as sub-monthly blocking patterns and intra-seasonal oscillations associated with the flow of westerly winds

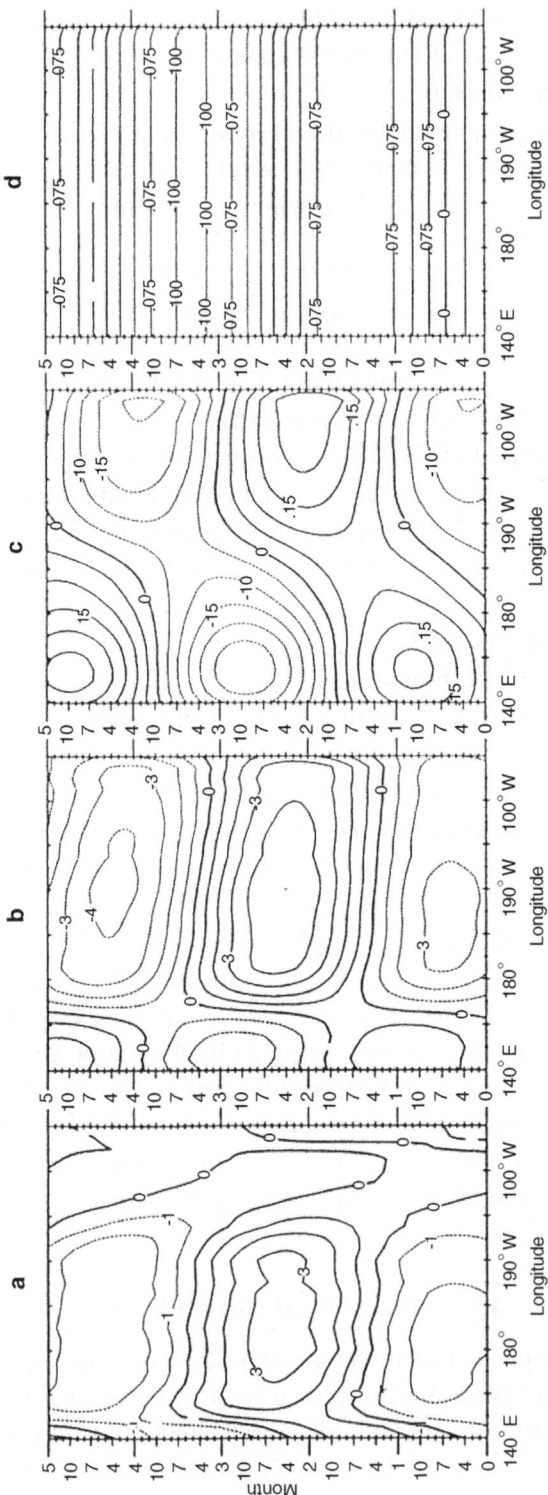

Figure 4.6. Principal-component structure along the equator for the ENSO cycle in a coupled GCM with only a Pacific-basin ocean: (a) $\tau'^{(x)}$, (b) T'_o, (c) depth-averaged oceanic T' anomaly, (d) zonal mean of c. Contour intervals are 10^{-2} N m^{-2} for a, 0.1 K for b, and 0.05 K for c and d. From Yu and Mechoso, 2001.

over mountain ranges, only involve internal atmospheric extra-tropical dynamics without significant oceanic coupling. These phenomena will not be discussed here.

There is considerable largescale, low-frequency, extra-tropical, intrinsic variability that does not fit into these categories. An apt illustration comes from a long-duration solution for a simplified atmospheric GCM with either seasonal-mean or time-invariant insolation and uniform surface conditions (James and James, 1989). It shows considerable low-frequency variability – extending at least through periods of many decades with local frequency spectra that are approximately independent of ω (i.e., white in shape) for small ω, below $\mathcal{O}(0.1)$ d^{-1} – somehow generated by internal non-linear (turbulent) fluid dynamics. On the largest scales much of the low-frequency variability is associated with the zonally averaged westerly wind jet waxing and waning in strength and wandering in its meridional position; its frequency spectrum also becomes white in shape though only for ω smaller than about 1 y^{-1}. This low-frequency behavior is familiar from many other atmospheric GCM calculations with specified surface conditions having only seasonal-mean variation, although very long atmospheric integrations are relatively rare. With spatially non-uniform surface conditions, much of the low-frequency variability occurs in standing eddies qualitatively similar to the observed patterns (e.g., the NAO in Figure 4.3). With increasing altitude there is a shift from more energetic near-surface, standing-eddy patterns to more energetic zonal-jet variability near the tropopause. This low-frequency variability is not linked to any intrinsic dynamical time scale in the atmosphere; consequently, with some poetic license, it is sometimes referred to as a low-frequency noise process or, more aptly, a low-frequency dynamical response to weather-noise forcing (Section 4.1).[1]

Held (1983) argues that the low-frequency dynamics of standing eddies is primarily a steady linear response to specified orography, zonal mean wind, $\overline{u}_a(y, z)$, where a = atmosphere, and y and z are latitude; altitude coordinates diabatic heating distribution; and transient-eddy (weather) flux divergence; i.e., excluded is any significant transient or non-linear dynamics of the low-frequency standing eddies themselves. This, of course, is not a complete theory since it does not explain where the variability in $\overline{u}_a$, heating, and eddy flux comes from; no doubt these are products of a large-scale turbulent dynamics – of the atmosphere at least if not the larger climatic system – that is not yet well understood. One can formally write an equation for the low-pass advective dynamics of, say, T as in Equation (4.5), where T is

$$\frac{\partial \langle T \rangle}{\partial t} = -\langle \mathbf{u} \rangle \cdot \nabla \langle T \rangle - \mathcal{T} + \dots \tag{4.5}$$

---

[1] It seems useful to distinguish the terminology of "climate noise" from the null hypothesis discussed here. The former is the statistical sampling concept (Leith, 1976; Madden, 1976) that long-time averages of measurements exhibit variability even in the absence of any long-time correlations in the process that generates them.

defined by Equation (4.6). Here $\langle \cdot \rangle$ denotes a low-pass (climate) filter in space and

$$\mathcal{T} = \langle \mathbf{u} \cdot \nabla T \rangle - \langle \mathbf{u} \rangle \cdot \nabla \langle T \rangle \tag{4.6}$$

time, and the dots indicate non-advective dynamical contributions. Even if the dynamics of low-frequency variability does not have important low-frequency advection, i.e., the first right-hand-side term in Equation (4.5), it still does have an important advective influence through $\mathcal{T}$, which is the inverse transfer associated with higher-frequency variability. The term $\mathcal{T}$ may be loosely described as the weather-noise forcing, or even explicitly represented as such in a stochastic analogy model, but this is not its fundamental nature. The true distributions and mechanisms of $\mathcal{T}$ are well worth more attention.

   These results provide a rational basis for skepticism that coupling with land-surface, sea-ice, and ocean dynamics is essential for extra-tropical intrinsic climatic variability. On the other hand, these latter dynamics do have longer intrinsic time scales than the atmosphere, so that there is the possibility that they may organize the low-frequency variability beyond simply a noise response. To a considerable degree the determination of the appropriate balance between these alternatives – noise vs. organization – is still unaccomplished.

4.4.1 Analogy models

The delayed and recharge oscillator models (Section 4.3.2) provide analogies to fluid dynamical behavior, whose designs are guided by physical hypotheses but whose derivations from first principles are obscure. An analogy model for atmospheric low-frequency variability is a simple stochastic ordinary differential equation (ODE) for a low-frequency atmospheric temperature anomaly, $T = T_a'$,

$$\dot{T} = WN - \alpha T, \tag{4.7}$$

where WN is a white-noise forcing representing weather events (i.e., weather noise) and α^{-1} is an extra-tropical dynamical relaxation time towards the climatic mean state, whose magnitude is (10) days. The frequency spectrum for T from Equation (4.7) is quite simple (Figure 4.7a): it is monotonically decreasing (i.e., red in shape), $\propto \omega^{-2}$, at frequencies higher than the relaxation rate and white at all lower frequencies, roughly as observed for the atmosphere. At all frequencies its amplitude is proportional to that of the noise forcing.

   The most evident weakness of Equation (4.7) compared to observations is that its transition frequency from red to white is too high for some variables, especially those associated with largescale patterns. Hasselmann (1976) proposes a generalization with local thermal coupling to the underlying ocean, see Equations (4.8)

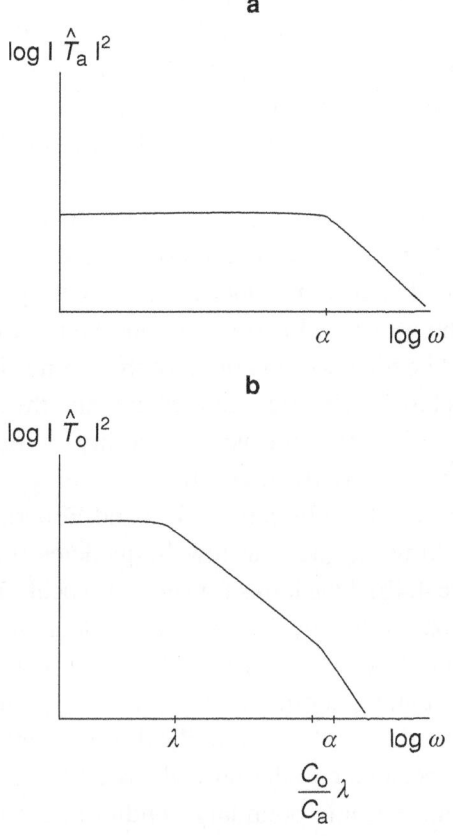

Figure 4.7. Frequency spectra for (a) T_a' in the atmospheric stochastic model given by Equation (4.7) and (b) T_o' in the coupled stochastic model given by Equations (4.8) and (4.9).

and (4.9), where T_o' is an oceanic surface temperature anomaly, c_a and c_o are

$$\dot{T}_a' = WN - \alpha T_a' + \lambda \frac{c_o}{c_a} \left[T_o' - T_a' \right] \qquad (4.8)$$

$$\dot{T}_o' = \lambda \left[T_a' - T_o' \right], \qquad (4.9)$$

atmospheric and oceanic heat capacities (with $c_o \gg c_a$ and $c_o \propto h$, the upper-ocean depth for active vertical mixing), and λ^{-1} is a thermal relaxation time between ocean and atmosphere, involving stratospheric infrared radiative equilibration and c_o, whose magnitude is $\mathcal{O}(1)$ y. The feedback in Equation (4.9) due to coupling is entirely negative: it acts to damp the atmospheric fluctuations instigated by weather noise. The spectrum for T_a' from Equation (4.9) is only modestly different due to coupling: it is red in shape and $\propto \omega^{-2}$ for $\omega \gg \alpha, \lambda \frac{c_o}{c_a}$; white in shape for $\omega \ll \lambda$; and more complex but not strongly varying for intermediate frequencies. The spectrum for T_o', however, is red in shape for all frequencies greater than λ and even redder than T_a' for $\omega \gg \alpha$, $\lambda \frac{c_o}{c_a}$ (Figure 4.7b). Thus, the thermal coupling of the atmosphere with the much larger thermal inertia of the ocean causes a red spectrum shape over a much broader range of frequencies for some climatic variables (here T_o'). Note also that Equations (4.8) and (4.9) yield a different behavior for $T_a'(t)$ than it would if only the first equation were used with a slowly varying T_o' independently specified, because the coupling term acts to diminish the surface flux by making the T' values in both media evolve towards each other. This demonstrates a fallacy in the widespread practice of using atmospheric GCMs with specified SST fields, even when specified from the observed history: without the thermal damping feedback, surface-heat fluxes will be too large and often of the wrong sign if the ocean cannot co-evolve with the stochastic (chaotic) atmosphere, and the atmospheric evolution will be too predictable (i.e., the intrinsic variability is too small; Barsugli and Battisti, 1998; Bretherton and Battisti, 2000).

The frequency spectra in both Equation (4.7) and Equations (4.8) and (4.9) are wholly broadband, lacking any peak at low-frequencies (unlike, e.g., the ENSO SST spectrum in Figure 4.4b). One looks for spectrum peaks in hopes of diagnosing the presence of organized dynamical modes. Saravanan and McWilliams (1998) extend the seven by model Equations (4.8) and (4.9) to include in one dimension a preferred atmospheric spatial pattern with scale L (interpreted as the projection of weather noise onto a standing eddy pattern like those in Figures 4.2a and 4.3a) and advection by the mean oceanic circulation with speed V, see Equations (4.10) and (4.11), with an upstream oceanic boundary condition, $T_o'(0, t) = 0$ when $V > 0$.

$$\dot{T}_a' = W N \cdot \sin\left[\frac{2\pi y}{L}\right] - \alpha T_a' + \lambda \frac{c_o}{c_a}\left[T_o' - T_a'\right] \qquad (4.10)$$

$$\dot{T}_o' = \lambda\left[T_a' - T_o'\right] - V \frac{\partial T_o'}{\partial y}, \qquad (4.11)$$

This model has no organized dynamical modes of climatic variability; i.e., in the absence of forcing, all fluctuations decay to zero while propagating in y. Nevertheless, its solutions exhibit a spectrum peak in T_o' near $\omega = t_{adv}^{-1} = L/V$ whenever the parameter, Γ, shown in Equation (4.12) is large enough (Figure 4.8);

$$\Gamma = \left[\frac{1}{\lambda} + \frac{c_o}{c_a \alpha}\right] \cdot \frac{V}{L} \qquad (4.12)$$

Γ measures the competition between oceanic advection and thermal damping; small values occur for relatively shallow oceanic thermal anomalies and slow mean circulation, while large values occur for deeper anomalies and faster circulation. It expresses the likelihood that an oceanic thermal anomaly generated in one location will

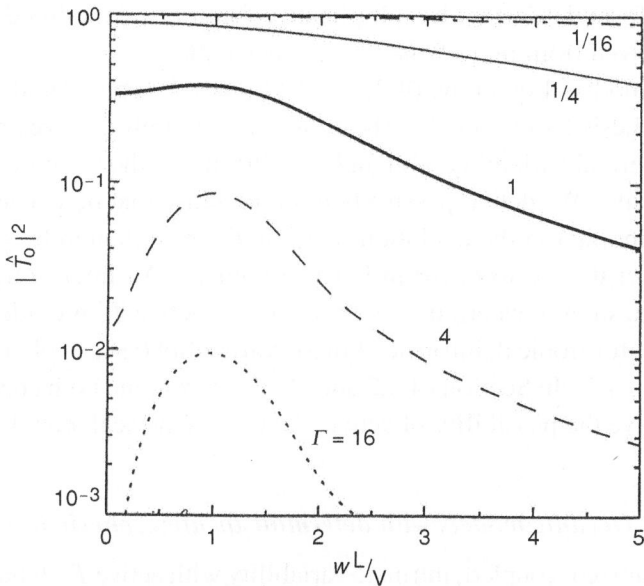

Figure 4.8. The T_o' frequency spectra for the coupled stochastic model given by Equations (4.10) and (4.11) with a preferred atmospheric pattern and mean oceanic circulation, as a function of Γ defined in (4.12).

be moved to another location before releasing its heat back to the atmosphere (i.e., the thermal coupling is non-local, though the sense of feedback remains negative).

Analogy models are useful for illustrating other climatic possibilities. Leith (1975) uses a conservative, stochastic model to demonstrate a statistical–mechanical, fluctuation–dissipation relation between the mean response to external forcing of a dynamical system and the behavior of its unforced intrinsic variability. Lorenz (1990) uses a forced-dissipative, deterministic, chaotic dynamical system with multiple statistical-equilibrium states under steady forcing – an analogy model for the extra-tropical atmospheric zonal jet and its eddies – to show how enhanced low-frequency variability occurs by transitions between different equilibria when the forcing is slowly varying (i.e., extra inter-annual variability with seasonal-mean forcing).

Most stochastic analogy climatic models, like those above, specify the low-frequency effects of high-frequency variability as an ad hoc additive noise forcing without any modification of the form of the deterministic dynamics, linear in Equations (4.7) to (4.11) but non-linear in general. However, as shown by Majda *et al.* (1999) with examples derived from several low-order dynamical systems with both high- and low-frequency elements, these simple attributes are not general. If analogy models are to become more serious tools for climatic variability, then more sophisticated formulations with multiplicative forcing and modified non-linearity, more fully derived from fluid dynamics, are desirable.

All of the models Equations (4.7) to (4.11) may be considered as variants of the null hypothesis (Section 4.1). The source of variability is weather noise, and the oceanic thermal variability, although significant for the evolution of T_a', is dynamically passive. We define passive here in the sense that no oceanic circulation anomaly is important in the evolution of T_a' or T_o', even though by geostrophy we certainly expect $u_o' \neq 0$ to occur in balance with T_o'. An interesting, and largely unresolved, question is where does a dynamically active T_o' provide an important feedback for extra-tropical, intrinsic climatic variability (as it so obviously does for ENSO; Section 4.3). In Sections 4.4.2 and 4.4.3 we examine both analogy and fluid models that have the possibility of active, delayed, non-local, coupled feedback.

4.4.2 Oceanic models with deterministic atmospheric feedback

Modeling organized, coupled, intrinsic variability with active T_o' dynamics involves the mechanisms and consequences of oceanic circulation. Some simple ingredients for the latter are Ekman transport and pumping, Equations (4.13) and (4.14), where

$$\int_{-h_{ek}}^{0} u_{ek}(x, y, z)dz = \frac{\tau(x, y)}{f\rho_o} \tag{4.13}$$

$$w_{ek}(x, y) = \text{curl}\left[\frac{\tau(x, y)}{f\rho_o}\right], \tag{4.14}$$

h_{ek} is the surface boundary-layer depth ($\mathcal{O}(50)$ m), $f(y)$ is the Coriolis frequency, ρ_o is the oceanic density, u_{ek} is the ageostrophic horizontal velocity within the boundary layer, τ is the surface wind stress vector, and w_{ek} is the vertical velocity at the interior edge of the boundary layer that forces an interior geostrophic horizontal circulation by planetary vortex stretching. As a consequence of Equations (4.13) and (4.14), the depth-integrated, total meridional transport is shown in Equation (4.15), where

$$\int_{-H}^{0} v(x, y, z)dz = \frac{\text{curl}\left[\tau\right]}{\beta\rho_o} \tag{4.15}$$

$H(x, y)$ is the full oceanic depth, v is the meridional velocity, and $\beta = df/dy$; this is usually called the Sverdrup transport relation. Many simplifying assumptions lie behind Equations (4.14) and (4.15) that certainly are not universally true for large-scale oceanic dynamics, even as realized in an oceanic GCM configured for climatic problems; nevertheless, they are often invoked in the context of climatic variability. These relations explain, at least qualitatively, both the mean and the low-frequency variability of the sub-tropical and sub-polar, wind-driven gyres and of the equatorial upwelling, as a consequence of the extra-tropical westerly and tropical easterly winds, τ.

If we further assume that the geostrophic, horizontal current Equation (4.16),

$$\mathbf{u}_g \equiv \frac{1}{f}\hat{z} \times \nabla\phi \tag{4.16}$$

(with ϕ the geopotential function, equal to dynamic pressure divided by ρ_o, and $\hat{z}$ the unit vertical vector), is confined to an upper-ocean thermocline layer, with mean thickness h_u, then its approximate (i.e., low-frequency, largescale, away from boundaries and strong currents) evolution Equation (4.7) is obtained, where C is

$$\frac{\partial\phi}{\partial t} - C\frac{\partial\phi}{\partial x} = -g'\text{curl}\left[\frac{\tau}{f\rho_o}\right] \tag{4.17}$$

the long, baroclinic Rossby wave speed and g' is the reduced gravitational acceleration ($= g\Delta\rho_o/\rho_o$, where $\Delta\rho_o$ is the density difference across the thermocline), with $C = \sqrt{g'h_u}/f$. Note that Equation (4.17) is consistent in its steady state with the Sverdrup relation, Equation (4.15), for a total meridional transport equal to $h_{ek}v_{ek} + h_u v_g$. In its transient behavior it implies a westward-progressing adjustment towards steady-state Sverdrup balance on a Rossby-wave, zonal-transit time, $t_{rw} = L/C$ (e.g., ≈ 10 y for $C = 0.02$ m s^{-1} and $L = 6000$ km). Figure 4.9 shows that Equation (4.17) aptly captures the mid-ocean dynamic height response to decadal wind fluctuations in the sub-tropical North Atlantic; this demonstrates its relevance for climatic-variability studies. Similarly, the latitudinal position of the Gulf Stream covaries in phase with the NAO, in a way that is at least qualitatively consistent with the Sverdrup circulation anomaly balance and associated T_o advection (Taylor and Stephens, 1998; Taylor and Gangopadhyay, 2001). Thus, atmospheric climatic wind fluctuations do force anomalous oceanic currents.[2]

Together with some form of western-boundary closure for absorption/dissipation of Rossby waves and zonally integrated transport balance, the preceding relations are representative of the extra-tropical, wind-driven oceanic dynamics in many simple models for climatic variability. In a way similar to tropical intermediate coupled models (Section 4.3.1), the calculated circulation may be combined with a T_o' equation, with Ekman and geostrophic advection of mean and anomalous ∇T_o, and with

---

[2] A somewhat different suggestion is that oceanic basin modes may be quasi-resonantly excited by atmospheric wind fluctuations, with at least the possibility of consequent climatic feedbacks. The usual view is that such modes are dense in wavenumber and frequency space, hence unlikely to be well distinguished in their response to broad-band forcing. However, Cessi and Primeau (2001) show that an effect of weak lateral diffusion is to make a few basin modes with decadal periods become well separated from the others in a shallow-water model linearized about a state of rest, hence susceptible to climatic excitation (Cessi and Paparella, 2001). These special modes behave as a combination of largescale, westward, Rossby wave propagation and boundary (and potentially equatorial) Kelvin wave propagation connecting arrivals at the western boundary to subsequent reemissions at the eastern boundary.

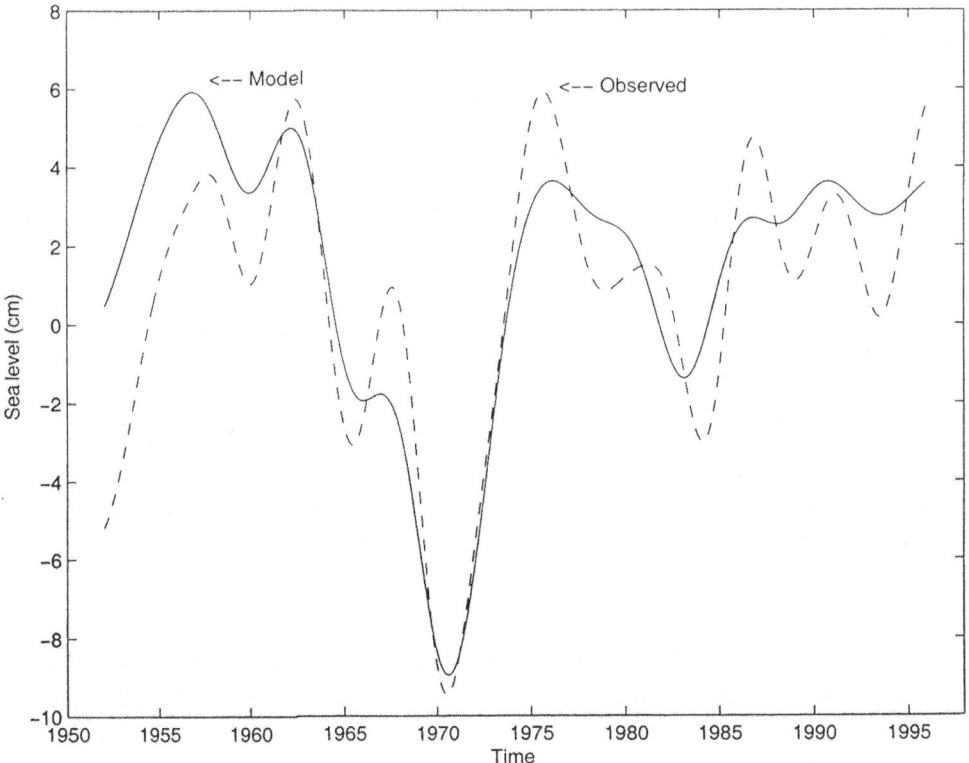

Figure 4.9. Low-frequency variability of the surface dynamic height (i.e., sea level; ϕ'/g; cm) near Bermuda: measured by hydrography (dashed) and calculated from the circulation model, Equation (4.13), with measured wind anomalies, $\tau'(x, y)$ (solid). These curves have been low-pass filtered with a cut-off frequency of 0.3 y^{-1}. Updated from Sturges and Hong, 1995.

some type of diagnostic or stochastic atmospheric model, even one as simple as a posited deterministic feedback relation with the functional form of Equation (4.1), albeit with a different functional $\mathcal{A}$ appropriate to extra-tropical air–sea coupling. Models of this class lack any mechanism for generating organized intrinsic climatic variability apart from that introduced either by the coupling relation or by stochastically represented weather noise.

A long-standing conceptual barrier for models of this type is the lack of an extra-tropical air–sea coupling hypothesis as compelling as the Bjerknes (1969) hypothesis for the tropics. To illustrate possibilities, Liu (1993) assumes for Equation (4.1) a local linear relation between atmospheric streamfunction, ψ_{a}, and SST, Equation (4.18), with $\gamma > 0$ and $A = A^{r} + i A^{im}$ a complex constant. With

$$\psi_{a}' = AT_{o}', \qquad \tau' = -\gamma \hat{\mathbf{z}} \times \psi_{a}', \tag{4.18}$$

Equation (4.18) in an intermediate coupled model as described in the preceding paragraph (also including an evaporative cooling term $\propto \tau$ in the T_o equation), at least three types of (unstable) positive feedback mechanisms exist in the absence of oceanic boundary influences, depending of the sign of A:

- an upwelling mode with $f \cdot A^r > 0$, where $T_0' > 0$ induces a cyclonic atmospheric circulation anomaly $\Rightarrow w_{ek} < 0$ by Equation (4.14) $\Rightarrow$ SST heating by downwelling advection;
- a SST-Sverdrup mode with $\partial \overline{T}_o / \partial y \cdot A^r < 0$, where $T_0' > 0$ induces an anti-cyclonic atmospheric circulation anomaly $\Rightarrow$ poleward Sverdrup transport and geostrophic flow by Equations (4.15) and (4.17) $\Rightarrow$ SST heating by poleward advection away from the warmer tropics;
- a SST-evaporation mode with $\overline{\tau}^{(x)} \cdot A^{im} < 0$, where $T_0' > 0$ induces a phase-shifted atmospheric pressure anomaly $\Rightarrow$ a locally diminished westerly geostrophic wind $\Rightarrow$ SST heating by reduced evaporation.

On the other hand, reversal of the sign of A in each of these mechanisms implies a negative air–sea feedback and decaying anomalies. Also, since positive feedback requires different signs of A for these different mechanisms, there can be destructive interference among them in combination. Thus, while the possibility of unstable coupled modes exists in the extra-tropics, the often asked, but still essentially unanswered question of what anomalous atmosphere wind occurs in response to T_0' (i.e., magnitude and spatial phase relationship) remains a serious obstacle to drawing conclusions about such mechanisms.

4.4.2.1 Oceanic gyres and Rossby waves

The circulation dynamics in Equation (4.17) gives the possibility of an organized climatic oscillation on the decadal time scale of oceanic-basin adjustment to anomalous wind forcing, t_{rw}. With stochastic (weather-noise) forcing whose spectrum shape is white, the oceanic-response spectrum has a broad, weak peak around t_{rw}^{-1} (Frankignoul *et al.*, 1997; Jin, 1997b). With the addition of a T_0' equation and the further assumption of a diagnostic feedback relation like Equation (4.1) – whether based on an ad hoc hypothesis (Munnich *et al.*, 1998); based on a simple, steady-state dynamical model for atmospheric response to heating (Jin, 1997b; Goodman and Marshall, 1999; Cessi, 2000); or based on a PC regression analysis of atmospheric GCM solutions (Weng and Neelin, 1998; Xu *et al.*, 1998; Watanabe and Kimoto, 2000a), even combined with a stochastic, multiplicative-noise component (Neelin and Weng, 1999) – then a coupled decadal oscillation can occur in either the absence or presence of weather-noise forcing. In some of these models, the mean SST front across the boundary between the sub-tropical and sub-polar gyres is particularly important (e.g., the Kuroshio and Gulf Stream extension currents). Thermal advection by a wind-driven v_0' where $\partial \overline{T}_0 / \partial y$ is large generates a large T_0' in the middle of the atmospheric storm track where the air–sea heat flux is large. This mechanism is particularly effective if the anomalous Sverdrup transport gyre is meridionally displaced across the mean gyre boundary (i.e. an "inter-gyre gyre," as Marshall et al. [2000] argue occurs for NAO wind anomalies). The term T_0' may also be advected by the mean circulation, $\bar{u}_0$. This contributes another potentially important time scale, $t_{adv} = L/\bar{V}$, as in the Equations (4.10) and (4.11) analogy model, to the coupled oscillation period (Latif and Barnett, 1994; Xu et al., 1998). As with equatorial wave adjustments (Section 4.3.2), the effects of the extra-tropical, Rossby-wave adjustment process can be modeled as a delayed oscillator with a delay time $\sim \mathcal{O}(t_{rw})$ (Gallego and Cessi, 2000; Marshall et al., 2000). Phase-locking of oscillations in the North Atlantic and Pacific oceans might also occur, with the zonal-mean, extra-tropical atmospheric anomaly as the conduit of inter-basin coupling (Gallego and Cessi, 2001).

4.4.2.2 Antarctic circumpolar wave

In the Antarctic circumpolar region, a largescale anomaly pattern in SST, SLP, surface wind, sea-ice extent, and sea level (i.e., $\eta' = \phi'/g$) has been observed propagating eastward with a speed of $\mathcal{O}(0.1)$ m s^{-1} and an oscillation period of $\mathcal{O}(4)$ y (White and Peterson, 1996; Jacobs and Mitchell, 1996). The zonal phase relationship between curl$[\tau']$ and ϕ' is consistent with Equation (4.13) augmented by a term representing mean advection by the Antarctic circumpolar current, $\bar{u}_0 \cdot \partial\phi'/\partial x$ (Jacobs and Mitchell, 1996). In several different intermediate coupled models – each of which has a different prescription for a steady-state atmospheric response to SST anomalies plus oceanic u_0' and T_0' balances, like those described in Section 4.4.2.1 but with the addition of zonal advection by the mean circumpolar current, $\bar{u}_0$ – differing conclusions are reached about whether positive air–sea feedbacks will sustain an organized circumpolar-wave oscillation (i.e., YEA: Qiu and Jin, 1997; White et al., 1998; Baines and Cai, 2000; Colin de Verdiere and Blanc, 2001; NAY: Goodman and Marshall, 1999). If the air–sea feedbacks induce only decaying climatic anomalies, then the fluctuations require an external excitation for their occurrence; both weather noise and an atmospheric teleconnection with ENSO (White and Peterson, 1996; Cai and Baines, 2001) are possibilities. In solutions from simplified GCMs, both Weisse et al. (1999) and Haarsma et al. (2000) find an inter-annual spectrum peak in oceanic quantities in the circumpolar region without significant positive air–sea feedbacks; in the latter study the essential mechanism for this peak is diagnosed as due to $\bar{u}_0 \partial T_0'/\partial x$ advection, as in the analogy model, Equations (4.10) and (4.11). From this perspective the Antarctic circumpolar wave is essentially a forced oceanic response consistent with the null hypothesis for climate.

4.4.2.3 Sub-tropical overturning cell

The PDO pattern (Figure 4.2a) spans both the tropics and extra-tropics in the Pacific. What mechanisms might effect this? The null-hypothesis answer would be that it is a planetary-scale atmospheric standing eddy pattern, forced by weather noise. However, since this pattern involves the tropics with its coupled ENSO dynamics (Section 4.3), this seems somewhat implausible as a complete explanation. Another answer is that it the Pacific decadal oscillation is some kind of low-frequency analog of ENSO with the essential air–sea coupling confined to the tropics and an atmospheric teleconnection response that forces extra-tropical T_o' with a purely passive SST dynamics, as in the Equations (4.8) and (4.9) analogy model. As yet no purely tropical, single-basin, coupled model has been identified that exhibits organized decadal oscillations.

A demonstration that tropical/extra-tropical oceanic coupling can be important comes from oceanic GCM solutions where initial extra-tropical T_o' differences alter the subsequent tropical T_o' evolution (Lysne et al., 1997). Gu and Philander (1997) argue that subduction and mean advection of oceanic thermal anomalies, $\bar{u}_o \cdot \nabla T_o'$, by the sub-tropical meridional overturning cell (i.e., the circulation implied by Equation (4.14) with sub-tropical sinking and equatorial upwelling) could generate an equatorial T_o' after a delay time $t_{adv}\mathcal{O}(10)$ y, and they illustrate the possibility with a simple analogy model. However, empirical and GCM solution analyses have not as yet demonstrated a coherent migration of thermal anomalies along the whole sub-surface circuit of the sub-tropical cell.

In an intermediate coupled model of Kleeman et al. (1999) with tropical/sub-tropical coupling, a slightly irregular decadal oscillation occurs (in addition to inter-annual fluctuations resembling ENSO, as in Section 4.3.1). It is comprised of an upper-ocean dynamics for the response of Ekman, geostrophic, and equatorial zonal currents to wind fluctuations; a SST equation; a diagnostic atmospheric feedback relation, like Equation (4.1), from an empirical PC analysis of SST and wind covariance over the whole Pacific (Figure 4.10a); and a latent-heat surface flux feedback relation utilizing the diagnostic wind anomaly. The latter ingredient is declared to be essential for the decadal oscillation, and its basis is a concatenation of correlations (i.e., $T_o' \rightarrow v_a' \rightarrow Q'$, where v_a' is the atmospheric wind fluctuation, and Q' is the air–sea heat flux fluctuation) that may misrepresent causality. (This point is made by Frankignoul [1999] using a simple analogy model that is an extension of Equation (4.7).) The spatial pattern of the model's decadal oscillation involves both the tropics and extra-tropics (Figure 4.10b). Note that it has both similarities and differences with the inter-decadal PDO (Figure 4.2a) and the inter-annual ENSO (Figure 4.4a) patterns. This illustrates the ambiguity in characterizing climatic variability by its spatial patterns. This tropical/sub-tropical decadal mode involves equatorial T_0' changes primarily by anomalous local vertical advection, $w' \cdot \partial \overline{T} / \partial z$, where the equatorial w' occurs in response to extra-tropical wind forcing of the sub-tropical cell, without coherent subsurface T' traversing the cell circuit. Thus t_{adv} for the sub-tropical cell is not a controlling time scale for the oscillation, but t_{rw} may be. The primary positive feedback mechanism implicit in the pattern in Figure 4.10a occurs in the eastern sub-tropical gyre, and it is of the SST-evaporation type.

Thus, organized low-frequency extra-tropical oscillations of several types can occur through coupling of oceanic and atmospheric dynamics that have no internal mechanisms for intrinsic variability without coupling. An essential ingredient for this behavior is a positive feedback mechanism implicit in the deterministic atmospheric response to T_0', as expressed in Equation (4.1). But the actual strengths and mechanisms behind these deterministic responses are significantly uncertain in nature. Models with only deterministic feedback relations are inadequate for assessing how their mechanisms compete with the intrinsic variability; the origins of the latter are not fundamentally dependent on air–sea coupling, either due to atmospheric weather, as in the null hypothesis, or due to other instability modes for the general circulation; see Section 4.3.

4.4.3 Idealized fluid models with intrinsic variability

As discussed at the beginning of Section 4.4, the atmospheric general circulation is not known to have important low-frequency instabilities to provide a basis for organized variability. The oceanic circulation has at least two types of instability modes that might lead to organized intrinsic climatic variability through coupling. These modes are for the thermohaline circulation, either by itself or through interactions with sea ice, and for the strong currents in the wind-driven circulation. The primary climatic questions in this context are:

- How responsive is the extra-tropical atmosphere to such intrinsic variability in the oceanic circulation and sea ice?
- How does this type of atmospheric response compete with the intrinsic variability due to the inverse spectral transfer of weather noise (i.e., the null hypothesis)?

In the null hypothesis the important feedbacks are predominantly negative, involving a passive local T_0' dynamics, whereas intrinsic variability modes in the ocean and sea ice will have an active T_0' dynamics with delayed, non-local feedbacks that could be positive. This latter behavior is not necessarily inconsistent with variability arising from deterministic positive feedback relations (Section 4.4.2), but as yet their synthesis has not been examined.

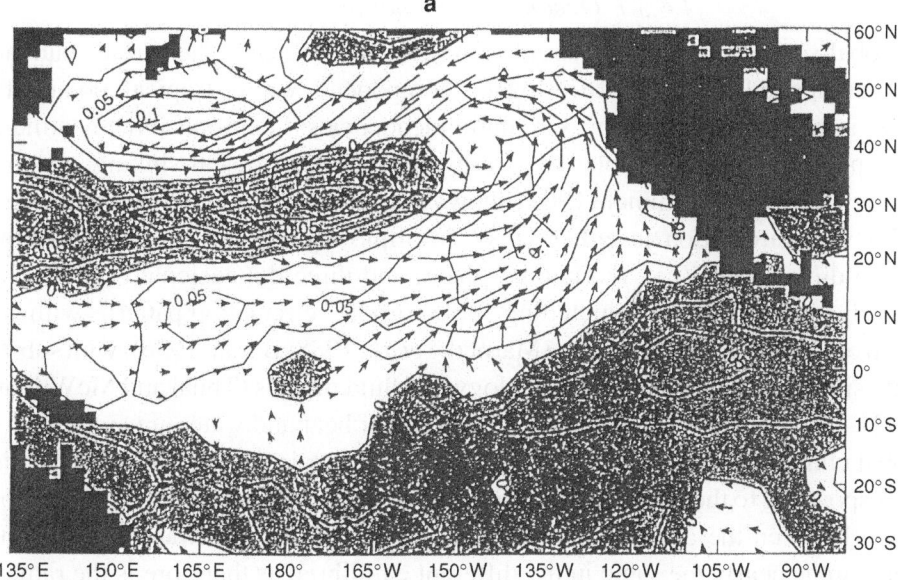

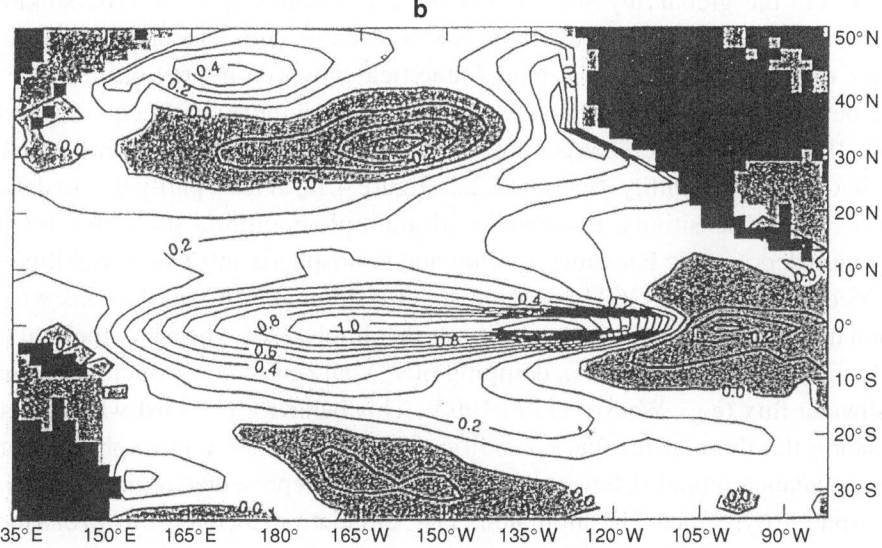

Figure 4.10. (a) Patterns T_0' and τ' associated with an empirical PC analysis, used as the diagnostic feedback Equation (4.1) in an intermediate coupled model; (b) T_0' pattern for the decadal oscillation in the model solution in phase with the stress pattern in a (K per standard deviation). Regions with $T_0' < 0$ are shaded. Adapted from Kleeman *et al.*, 1999.

4.4.3.1 Oceanic thermohaline circulation

A global-scale, thermohaline circulation (THC) is forced by the surface buoyancy flux, comprised of heat and freshwater components whose tropical–polar differences imply opposing oceanic circulation tendencies; i.e., tropical net heating implies polar sinking, and polar net freshening implies tropical sinking. A simple analogy model of Stommel (1961) shows that this permits multiple stable steady states with circulation in either sense. Multiple equilibria have been confirmed in two-dimensional (i.e., meridional-plane) and three-dimensional oceanic GCM fluid-dynamical solutions, including "pole-to-pole" circulation patterns with sinking in only one polar hemisphere (Marotzke *et al.*, 1988; Bryan, 1986), with substantially similar regime diagrams in analogy and fluid models (Thual and McWilliams, 1992). The meridional transports by the THC of heat and water are significant for global climatic balance. Multiple climatic equilibria also occur in coupled models, corresponding to the different THC equilibria, with a substantial degree of compensation between the oceanic and atmospheric meridional heat transports such that their sum is nearly the same in the different equilibria, as therefore is the radiative balance at the top of the atmosphere (Saravanan and McWilliams, 1995). In the present era the global abyssal water mass is fed primarily by a THC sinking of $\sim 20 \times 10^6 \, \mathrm{m^3 \, s^{-1}}$ of seawater in the North Atlantic, with a secondary deep-water source of $\sim 5 \times 10^6 \, \mathrm{m^3 \, s^{-1}}$ around Antarctica; however, in previous eras the balance between these source regions has probably been more nearly equal (Broecker *et al.*, 1999). Also, THC changes are likely important contributors to millennial intrinsic climatic variability (Section 4.2.2; Figure 4.1), at least partly due to delicacy and near-intransitivity associated with multiple equilibria subjected to slowly varying paleoclimatic buoyancy forcing and atmospheric intrinsic variability.

Uncoupled oceanic GCM solutions exhibit decadal THC oscillations with so-called mixed surface-boundary conditions that imply a local negative thermal feedback with the atmosphere (i.e., damping of T_o') but no feedback fluctuations in the freshwater flux (e.g., Weaver *et al.*, 1993). This behavior is robust with respect to replacing the thermal feedback condition with a horizontal atmospheric thermal energy-balance model (Chen and Ghil, 1996) as a representative or more general atmospheric dynamics. The anomalous THC circulation, $\mathbf{u}'$, has the form of a meridional overturning cell spanning the sub-tropical and sub-polar gyres closest to the site of mean THC polar sinking. Upper-ocean T_o' and S_o' anomalies are generated by $\mathbf{u}_o'$ advection of $\nabla \overline{T}_o$ and $\nabla \overline{S}_o$, and, once generated, they move mainly with the mean THC and wind gyres. The period of the fluctuation is too short for significant deep T' and S' anomalies to traverse the lower branch of the overturning cell. The THC oscillation period is $\mathcal{O}(t_{\mathrm{adv}})$, though the relevant advective pathway is still in dispute (and may not be universal across different models or THC regimes in nature).

Decadal THC oscillations also occur in models with flux boundary conditions for both T and S, and they have been shown to occur as a linear instability of the mean THC circulation (Colin de Verdiere and Huck, 1999). However, comparisons with idealized coupled solutions indicate that THC fluctuation behavior with mixed boundary conditions is much more realistic than with flux conditions (Saravanan et al., 2000), and the THC oscillations more often occur as global bifurcations to isolated limit cycles, rather than linear instabilities (a preliminary result of research jointly with J. Molemaker).

To investigate the importance of THC fluctuations in climate requires a different type of model than previously discussed. Its oceanic dynamics must encompass the THC and wind gyres, and this most commonly is done with an oceanic GCM with sufficiently large eddy diffusivities that mesoscale instabilities do not occur (but see Section 4.4.3.3). The atmospheric dynamics must encompass the thermal and hydrological cycles and variability of the extra-tropical zonal-mean and standing-eddy circulations. Given the latter it is easy also to calculate explicitly the weather fluctuations, thereby encompassing the intrinsic climatic variability mechanism of the null hypothesis. This is a much greater degree of completeness than in the atmospheric component of intermediate coupled models in Sections 4.3.1 and 4.4.2 – i.e., a deterministic wind-stress feedback relation, Equation (4.1), plus possibly a stochastic forcing – and this choice is compelled by ignorance about how to formulate an adequately complete generalization of Equation (4.1). Coupled models of this type might be called idealized or simplified GCMs, insofar as they contain less elaborate forms for radiation, precipitation, boundary-layer, land-surface, and sea-ice parameterizations; idealized continental, orographic, and topographic shapes; dynamical approximations, such as quasi- or planetary-geostrophy; and reduced spatial resolution for computational efficiency (cf. James and James, 1989). Their weakness compared to simpler models is the complexity of their solution behavior, but their strength relative to GCMs is their exclusion of extraneous mechanisms and geographical specificity (à la Ockham's razor).

An idealized GCM for investigating intrinsic climatic variability mechanisms associated both with the null hypothesis and with the THC is comprised of a longitudinal-sector ocean (with either meridional-plane [Saravanan and McWilliams, 1995; 1997] or fully three-dimensional dynamics [Saravanan et al., 2000]), a flat land surface, and no sea ice. In coupled solutions with the three-dimensional active ocean,[3] the low-level atmospheric circulation has mean

---

[3] This solution also has a passive oceanic sector with a specified latitudinally varying SST, located in the opposite hemisphere (see Figure 4.11). The motivation for choosing this configuration is to allow both the total meridional oceanic heat transport and the magnitude of the THC overturning streamfunction, $\overline{\Phi}$, to be about as observed in the present climate. This is done by making the active basin about as wide as the Atlantic and having an implied meridional heat flux (about half the total) associated by surface heat flux over the passive basin about as wide as the Pacific.

extra-tropical westerly winds and standing eddies with sub-tropical high and sub-polar low pressure cells because of the ocean basin (Figure 4.11a); and the oceanic meridional overturning circulation is primarily pole-to-pole with northern hemispheric sinking but also has shallow Ekman cells associated with the wind-driven gyres (Figure 4.11b). When the oceanic and atmospheric components are calculated separately (using surface boundary conditions taken from the mean state of a coupled solution), each exhibits low-frequency intrinsic variability – broad-band fluctuations in the zonal jet and standing eddies in the atmosphere, and somewhat irregular decadal THC oscillations in the ocean – as expected from other idealized GCM discussed above.

The same phenomena are evident in a coupled solution. The dominant PC for the overturning circulation, Φ'_o, shows a pattern of changing intensity of the mean THC in the sub-tropical and sub-polar regions of the sinking hemisphere (Figure 4.12a), and the spectrum of its associated time series, though broadly red in shape, does have a significant decadal peak (Figure 4.12b). The co-varying SST patterns (Figure 4.13a) have the largest anomalies generated near the gyre boundary by anomalous THC advection, $v'_o \partial \overline{T}_o / \partial y$, and they spread northwestward along the extension current and atmospheric storm track at a rate $\mathcal{O}(t_{adv}^{-1})$ due to the mean circulation. The co-varying T'_o and upward surface heat flux, Q', are in phase with each other (Figure 4.13b), indicating an oceanic forcing of the atmosphere and an active SST dynamical mechanism. The two leading PCs for upper-level atmospheric circulation, ψ'_a, have the structure primarily of meridional displacements of the westerly jets (Figures 4.14a,b). Their simultaneously correlated Q' and T'_o patterns are largest in the western extra-tropical part of the basin (Figures 4.14c–f), where $\overline{Q} > 0$ is largest and weather cyclogenesis is most frequent. For the dominant PC 1, the ψ'_a-correlated T'_o (Figure 4.14e) has a broad zonal scale with opposite signs in the sub-tropical and sub-polar gyres. Its pattern has some similarity with the upper-ocean thermal anomaly associated with Φ'_o (not shown). It is of opposite sign to the ψ'_a-correlated Q' (Figure 4.14c), indicating an atmospheric forcing of the ocean, as in the null hypothesis. Thus, we interpret the first PC 1 as an intrinsic atmospheric variability mode that acts to force passive oceanic THC fluctuations by a stochastic spatial resonance (i.e., coincidence of spatial anomaly patterns for the intrinsic modes of the uncoupled ocean and atmosphere; Saravanan and McWilliams, 1997). This atmospheric forcing contributes to the oceanic THC spectrum peak (Figure 4.12b) without any implication of a significant non-local feedback on the atmosphere; this type of behavior is also shown in the analogy model of Griffies and Tziperman (1995). In contrast, for the subordinate PC 2, the ψ'_a-correlated T'_o pattern (Figure 4.14f) strongly resembles the Φ'_o-correlated pattern (cf., Figure 4.13a), and it has the same sign as Q' (Figure 14.14d) at least in the western region where T'_o is large. We interpret PC 2 as the result of a positive, non-local feedback on

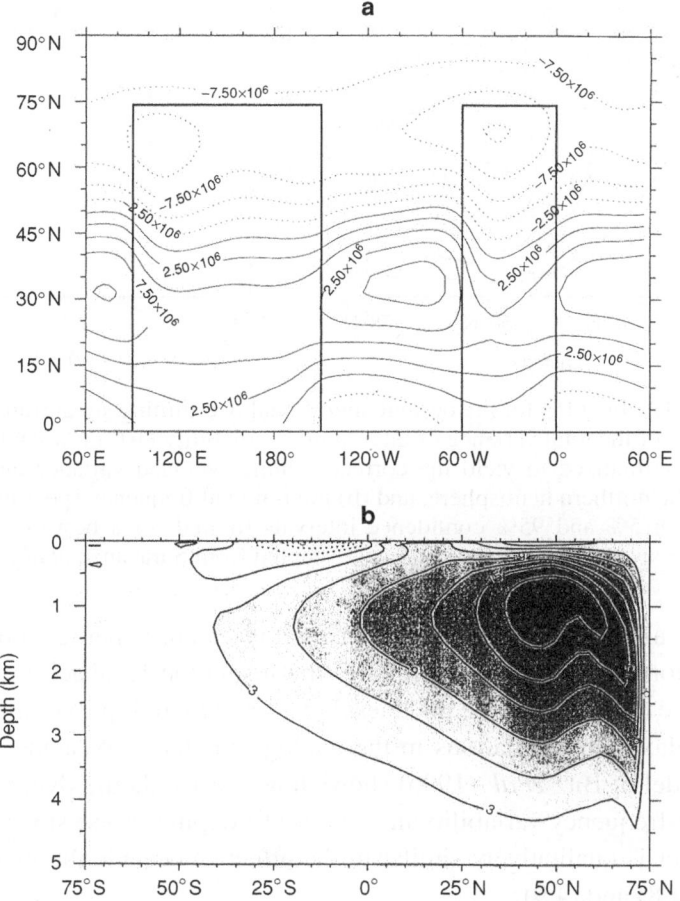

Figure 4.11. Mean atmospheric and oceanic circulations for an idealized coupled model: (a) horizontal wind streamfunction, ψ_a, at 750 mb in the northern hemisphere ($m^2 s^{-1}$), with the oceanic basin boundaries marked by the heavy lines (active basin to the right) and (b) meridional overturning streamfunction, $\overline{\Phi}_o$, in the active oceanic basin [$10^6 m^3 s^{-1}$]. Adapted from Saravanan et al., 2000.

the atmosphere. Note that this organized intrinsic climatic variability mechanism is not as important as the mechanism that embodies the null hypothesis. Furthermore, both of these atmospheric modes have no significant low-frequency peaks in their frequency spectra (Figures 4.14g,h), whose shape is close to white (as in the simple stochastic analogy models of Section 4.4.1).

4.4.3.2 Sea ice

Sea ice is an important ingredient in climatic dynamics, and it is well known to induce very sensitive behavior in climatic GCMs. The ice-albedo feedback

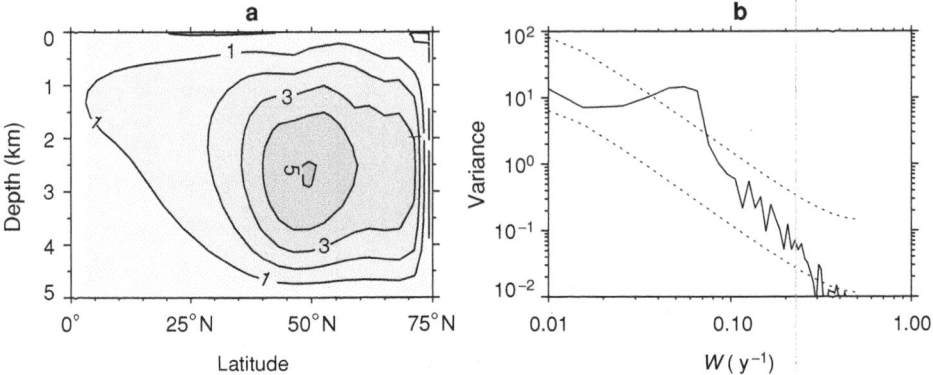

Figure 4.12. First PC for the oceanic meridional overturning streamfunction, Φ_o, with 49% of the total variance of inter-annual variability (10^6 m³ s⁻¹): (a) spatial pattern, normalized to yield the correct volume-averaged variance and plotted only in the northern hemisphere, and (b) inter-annual frequency spectrum (solid), along with 5% and 95% confidence intervals (dotted) of a best-fit, first-order autoregressive process (i.e., red-noise). Adapted from Saravanan *et al.*, 2000.

mechanism[4] allows multiple equilibria in energy-balance analogy models (see the article by North in this book). Similarly the insulation feedback mechanism between surface heat flux and ice thickness[5] gives rise to multiple equilibria and even sustained relaxation oscillations in the analogy model of Welander (1977). The analogy model of Bitz *et al.* (1996) shows how sea-ice thermodynamic processes amplify low-frequency variability in response to weather-noise stochastic forcing. This behavior is qualitatively similar to the effect of oceanic thermal coupling in Equations (4.8) and (4.9).

The interaction of sea ice with oceanic circulation provides additional mechanisms for climatic variability. In a brine–THC feedback cycle,[6] the slowest process is advection by the THC circulation anomaly, $v'_o \cdot \partial \bar{T}_o / \partial y$, over a meridional distance, L_y, between adjacent ice-free and ice-covered oceanic regions. Yang and Neelin (1993; 1997) report solutions showing a sustained oscillation due to this feedback cycle with a period of $t_{adv} \sim \mathcal{O}(10)$ y for modest transport anomaly magnitude ($\approx 10\%$ of the mean THC) and anomaly advection distance $L_y \approx 10^3$ km. Another mechanism for intrinsic variability is due to the seasonal cycle of winter sea-ice freezing and summer melting (hence $S'_o > 0$ and < 0, respectively) that, in

---

[4] The presence of sea ice increases the surface albedo compared to its absence ⇒ reduces solar absorption ⇒ cools the surface ⇒ increases the ice amount.

[5] A larger flux for a given $T_a - T_o$ with thinner ice, and especially so without ice present ⇒ greater oceanic cooling ⇒ increasing ice thickness ⇒ decreasing surface flux.

[6] A cold polar surface $T'_o < 0$ ⇒ more sea ice by freezing ⇒ polar surface $S'_o > 0$ by brine rejection ⇒ polar surface $\rho'_o > 0$ ⇒ stronger THC ⇒ polar warming by increased meridional advection ⇒ warm polar surface $T'_o < 0$, etc.

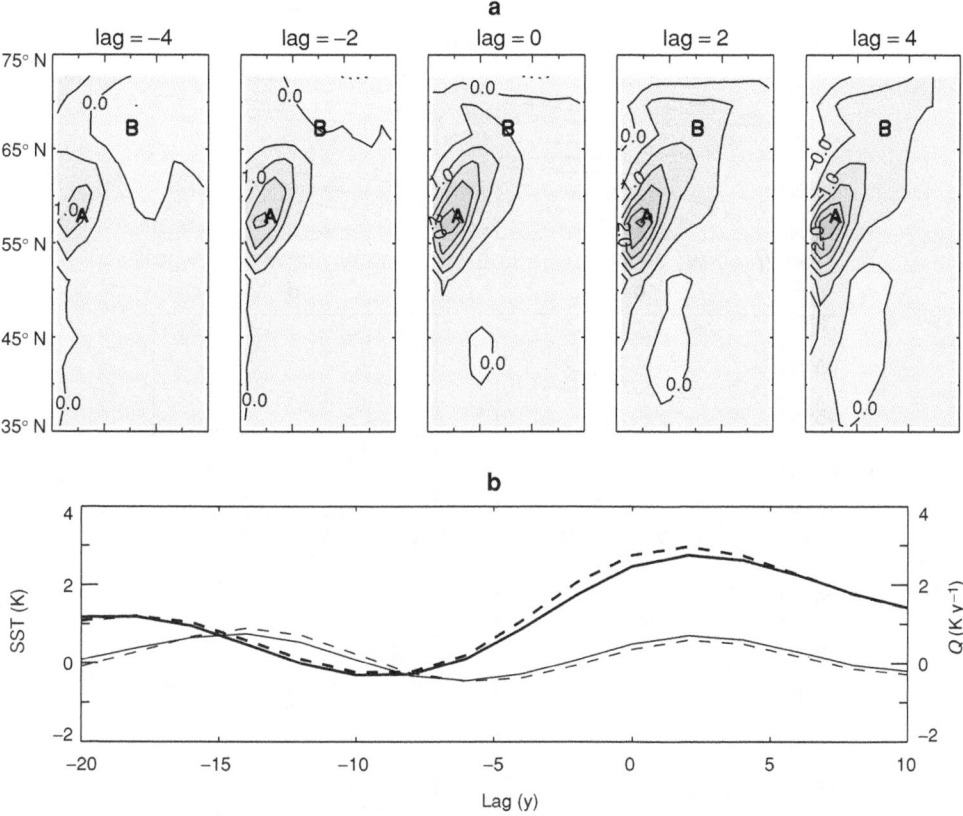

Figure 4.13. Regressions against the time series of the oceanic Φ_o PC in Figure 4.12: (a) SST anomaly, $T_o'(\lambda, \phi)$ (K), at several time lags from $t_L = -4$ to $+4$ y and (b) $T_o'(t_L)$ (solid) (K) and upward surface heat flux, $Q'(t_L)$ (dashed) (K y^{-1}, assuming heat is exchanged with oceanic layer of thickness $h = 100$ m), at the locations A (thick lines) and B (light lines) marked in panel a. The positive correlations in b indicate oceanic forcing of the atmosphere throughout the cycle. Adapted from Saravanan *et al.*, 2000.

combination with deeper oceanic convective mixing in winter, causes a net freshening of the upper ocean in an annual average. This generates both THC circulation and T_o and S_o advection anomalies, as in the feedback loop above, causing a decadal oscillation under circumstances where it is absent with only time-mean surface forcing (Yang and Huang, 1996). Finally, the thermal insulation feedback also yields a decadal oscillation when coupled with the THC, where surface heat flux anomalies are more due to variations in the extent of ice coverage than in its thickness (Zhang *et al.*, 1995; cf., Welander, 1977). As yet, the questions of how these sea-ice modes of variability couple with intrinsic instability of the THC (Section 4.4.3.1) and feedback onto the atmosphere are open.

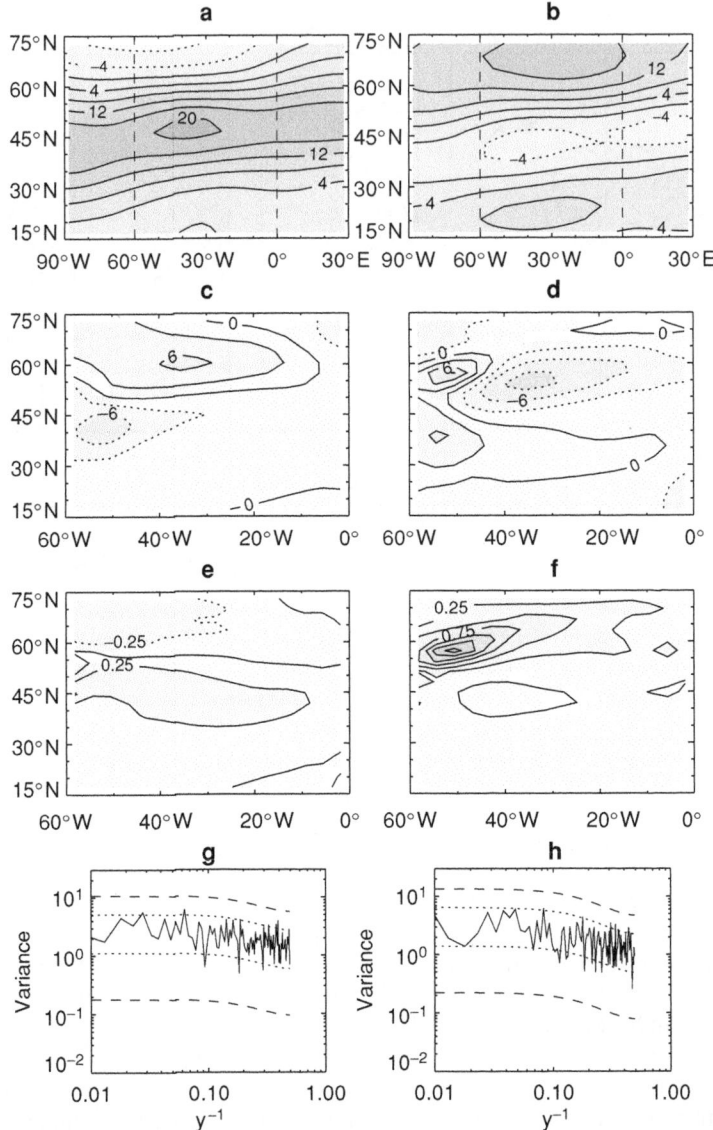

Figure 4.14. First two atmospheric PCs for horizontal streamfunction, ψ_a', at 250 mb 40% (left column) and 20% (right column) of the total inter-annual variability, respectively: (a), (b) Φ_0 patterns in the northern hemisphere near the active oceanic basin (10^5 m^2 s^{-1}); simultaneous regression patterns for (c), (d) upward surface heat flux, $Q'(\lambda, \phi)$, K y^{-1}, assuming heat is exchanged with an oceanic layer of thickness $h = 100$ m; (e), (f) $T_o'(\lambda, \phi)$, in K, in the northern part of the active oceanic basin; and (g), (h) inter-annual ψ_a' frequency spectra, along with 5% and 95% confidence intervals of a best-fit, first-order autoregressive process (i.e., red noise). Adapted from Saravanan *et al.*, 2000.

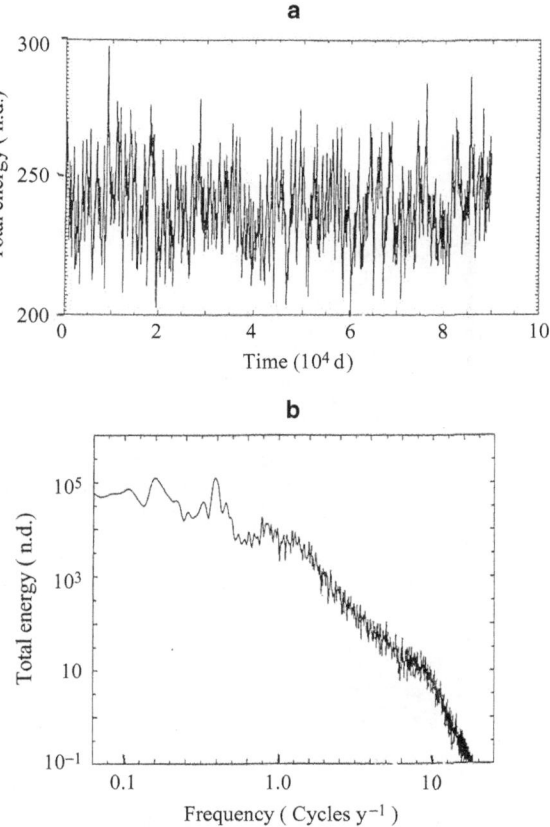

Figure 4.15. Volume-integrated energy in an idealized oceanic gyre solution with steady wind forcing at moderately high Reynolds number (i.e., with a horizontal eddy viscosity of $\nu = 400$ m^2 s^{-1}): (a) time series and (b) frequency spectrum. Adapted from Berloff and McWilliams, 1999.

4.4.3.3 Oceanic wind gyres

The oceanic wind gyres have strong, upper-ocean, western-boundary currents $\mathcal{O}(1)$ m s^{-1} that close the meridional mass transport balance with the interior Sverdrup transport Equation (4.15) and then separate to flow eastward into the interior as extension currents. These mean currents are unstable to mesoscale eddies that are geographically quite widespread, have horizontal scales $\mathcal{O}(10^2)$ km, and develop comparable velocity magnitudes. Eddy momentum and buoyancy fluxes in turn reshape the mean currents into recirculation gyres with stronger-than-Sverdrup westward return flows flanking the extension currents and extending into the abyss.

Mesoscale eddies are usually viewed as too small and evanescent to interact directly with the atmospheric general circulation. This is one reason – computational efficiency is the other – why most oceanic GCMs configured for climatic

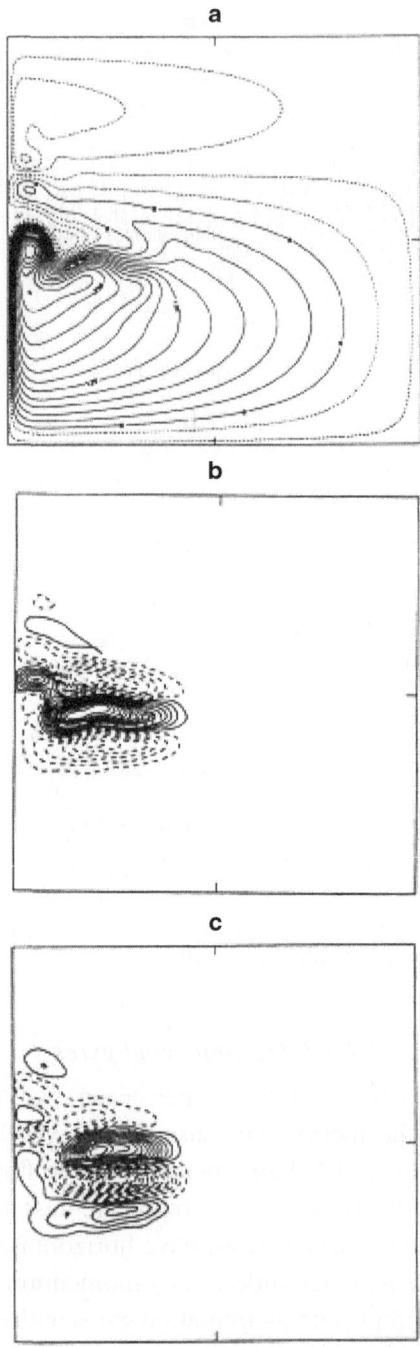

Figure 4.16. Upper-ocean streamfunction patterns, $\psi(x, y)$ in an idealized oceanic gyre solution with steady wind forcing at moderately high Reynolds number: (a) time-mean and (b), (c) first two PC patterns of low-frequency variability, representing 29% and 19% of the low-pass (i.e., $\omega < 0.6$ cycles y^{-1}) variance, respectively. (Adapted from Berloff and McWilliams, 1999.)

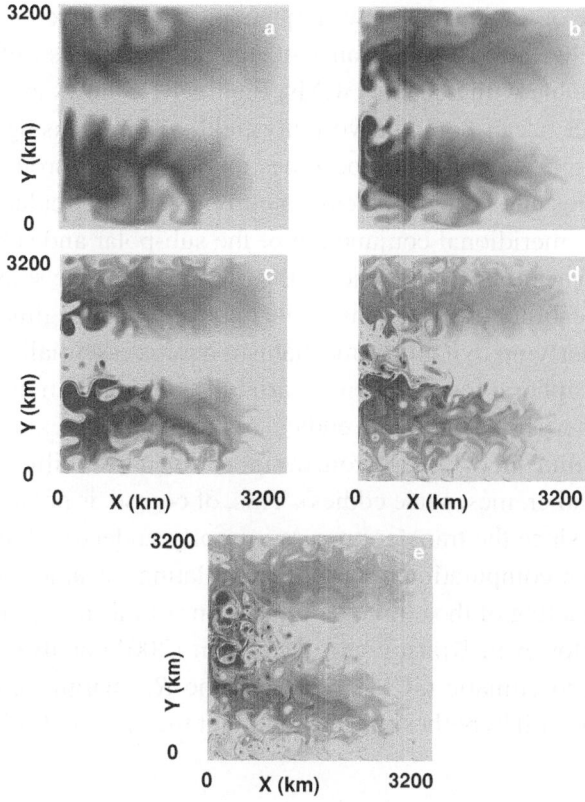

Figure 4.17. Instantaneous, upper-ocean potential vorticity, $q(x, y)$ in idealized oceanic gyre solutions with steady wind forcing. The Reynolds number (Re) increases from (a)–(e) in steps of a factor of 4. From Siegel *et al.*, 2001.

studies have coarse spatial resolution and parameterizations that suppress mesoscale instability. However, Figure 4.15 shows that even with steady wind forcing the solutions of idealized oceanic gyre models exhibit a broad-band, low-frequency, largescale intrinsic variability when the parameterization Reynolds number (Re) is high enough (e.g., Jiang *et al.*, 1995; McCalpin and Haidvogel, 1996; Spall, 1996; Berloff and McWilliams, 1999), in addition to what is forced by low-frequency wind variability (cf., Figure 4.9). The associated spatial patterns are reconfigurations of the extension currents and recirculation gyres as meridional displacements and changes in the intensity and zonal extent (Figure 4.16). Thus, the potential for feedback onto the atmospheric circulation is plausible since the spatial scale of the fluctuation pattern is moderately large and its location is near the gyre boundary (where $\nabla \overline{T}_o$ is large $\Rightarrow$ active SST fluctuations, induced by $\mathbf{u}'_o \cdot \nabla \overline{T}_o$, are large $\Rightarrow$ Q' is large; cf. Section 4.4.2.1).

The magnitude of the intrinsic low-frequency variability increases in gyre models with the Re value, at least over a range of values that extends well beyond what is presently computable with oceanic GCMs. Figure 4.17 shows instantaneous snapshots of the upper-ocean potential vorticity field with increasing Re, illustrating both how the mesoscale variability becomes much more vigorous and coherent in its vortex structures and how the extension currents and recirculation gyres evolve in shape (NB, the meridional conjunction of the sub-polar and sub-tropical extension currents into a single eastward jet at the highest Re value shown). Since there are no known instability modes in this scale and frequency regime, it is very plausible that the underlying dynamical mechanism here is essentially the same as that for intrinsic atmospheric low-frequency variability under the null hypothesis: preferred spatial patterns (i.e., standing eddies, recirculation gyres) are excited by the turbulent inverse transfer of energy from the more direct instabilities of the mean circulation (i.e., weather, mesoscale eddies). This, of course, is not a very satisfactory characterization, since the transfer process is poorly understood in both contexts.

Given the large computational cost for calculating oceanic gyres at high Re, little coupled modeling of the climatic implications of intrinsic gyre variability has been done yet. However, Kratsov and Robertson (2001) analyzed solutions of a particular idealized climatic GCM on at least the Re margin of this regime and concluded that the null hypothesis is the dominant mechanism for intrinsic climatic variations.

4.5 Discussion

The Earth's climate has intrinsic variability on many scales. An undoubtedly important mechanism, declared as the null hypothesis, is the extra-tropical, atmospheric inverse transfer of weather variability, accompanied by negative feedbacks and slow adjustments through coupling with the underlying land surface, sea ice, and ocean. However, other mechanisms for more organized low-frequency variability may importantly amend the null hypothesis. This is certainly true for inter-annual ENSO variability; it seems likely to be true for THC and sea-ice variability, both over decades and millennia, at least to some degree; and it may be true for oceanic gyre and circumpolar current inter-annual and decadal variability as well. Hall and Manabe (1997) analyze various empirical records and argue that the null hypothesis is evidently incomplete in the tropics; in the Antarctic circumpolar region; near sea-ice margins; in the sub-polar North Atlantic (where THC fluctuations are largest); and in the western extra-tropics near gyre boundaries, extension currents, and recirculation gyres; nevertheless, this leaves roughly half the oceanic surface where the null hypothesis is not contradicted. The western portion of the boundary between sub-tropical and sub-polar gyres, where $\nabla \overline{T}_o$ is large, is a key location for active SST dynamics and non-local extra-tropical atmospheric feedback in several transient mechanisms (i.e., Rossby-wave adjustment, Sverdrup balance, THC, and mesoscale-eddy inverse transfer). Nevertheless, the dynamical mechanisms of intrinsic climatic variability are still very much in debate.

The scientific study of intrinsic climatic variability is difficult and hard to unscramble: the phenomena are complex and chaotic; the empirical record is brief and fragmentary; the products of statistical analyses are subtle to interpret; realistically configured climatic GCMs are not credibly comprehensive in representing various potentially important mechanisms; stochastic analogy models illustrate potential mechanisms but are difficult to justify fluid-dynamically; deterministic feedback relations that represent the atmospheric response to changing surface conditions have uncertain validity; and idealized fluid models demonstrate the phenomena of intrinsic variability but do so with partially mysterious inverse transfer mechanisms. Nevertheless, progress has been made and more is foreseeable along all these avenues. Since climatic variability is a scientific problem truly on the human scale, its vigorous pursuit is certain, by whatever mixture of plodding along or leaping ahead we may achieve.

Acknowledgments

This essay is adapted from a lecture presented at the symposium honoring the career of Robert Cess. I thank Jeffrey Kiehl and V. Ramanathan for organizing this symposium and inviting me; David Neelin and R. Saravanan for discussions about climatic models; and Yi Chao and Clara Deser for assistance in preparing unpublished parts of Figures 4.2 and 4.3. Research support was provided by the National Science Foundation through grant OCE 96-33681, the NCAR, and by the National Oceanic and Atmospheric Administration (NOAA) through grant NA86GP0361.

References

Baines, P. G. and W. Cai (2000). Analysis of an interactive instability mechanism for the Antarctic Circumpolar wave. *J. Clim.* **13**, 1831–44.

Barsugli, J. J. and D. Battisti (1998). The basic effects of atmosphere–ocean thermal coupling on mid-latitude variability. *J. Atmos. Sci.* **55**, 477–93.

Battisti, D. S. and A. C. Hirst (1989). Interannual variability in the tropical atmosphere–ocean system: influence of the basic state, ocean geometry, and nonlinearity. *J. Atmos. Sci.* **46**, 1687–1712.

Berloff, P. and J. C. McWilliams (1999). Large-scale, low-frequency variability in wind-driven ocean gyres. *J. Phys. Oceanogr.* **29**, 1925–49.

Bitz, C. M., D. S. Battisti, R. E. Moritz, and J. E. Beesley (1996). Low-frequency variability in the Arctic atmosphere, sea ice, and upper-ocean climate system. *J. Clim.* **9**, 394–408.

Bjerknes, J. (1969). Atmospheric teleconnections from the equatorial Pacific. *Mon. Wea. Rev.* **97**, 163–72.

Bond, G., W. Showers, M. Cheseby *et al.* (1997). A pervasive millennial-scale cycle in Holocene and Glacial climates. *Science* **278**, 1257–66.

Bretherton, C. S. and D. B. Battisti (2000). An interpretation of the results from atmospheric general circulation models forced by the time history of the observed sea surface temperature distribution. *Geophys. Res. Lett.* **27**, 767–70.

Broecker, W. S., S. Sutherland, and T.-H. Peng (1999). A possible 20th-century slowdown of southern ocean deep water formation. *Science* **286**, 1132–35.

Bryan, F. (1986). High-latitude salinity effects and interhemispheric thermohaline circulations. *Nature* **323**, 301–4.

Cai, W. and P. G. Baines (2001). Forcing of the Antarctic Circumpolar wave by El Niño–Southern Oscillation teleconnections. *J. Geophys. Res.* **106**, 9019–38.

Cane, M. and S. Zebiak (1985). A theory for El Niño and the Southern Oscillation. *Science* **228**, 1084–7.

Cessi, P. (2000). Thermal feedback on wind stress as a contributing cause of climate variability. *J. Clim.* **13**, 232–44.

Cessi, P. and F. Paparella (2001). Excitation of basin modes by ocean–atmosphere coupling. *Geophys. Res. Lett.* **28**, 2437–40.

Cessi, P. and F. Primeau (2001). Dissipative selection of low-frequency modes in a reduced-gravity basin. *J. Phys. Oceanogr.* **31**, 127–37.

Chao, Y., M. Ghil, and J. C. McWilliams (2000). Pacific interdecadal variability in this century's sea surface temperatures. *Geophys. Res. Lett.* **27**, 2261–4.

Chen, F. and M. Ghil (1996). Interdecadal variability in a hybrid coupled ocean–atmosphere model. *J. Phys. Oceanogr.* **26**, 1561–78.

Colin de Verdiere, A. and M. L. Blanc (2001). Thermal resonance of the atmosphere to SST anomalies. Implications for the Antarctic Circumpolar wave. *Tellus* **53**, 403–24.

Colin de Verdiere, A. and T. Huck (1999). Baroclinic instability: an oceanic wavemaker for interdecadal variability. *J. Phys. Oceanogr.* **27**, 823–910.

Czaja, A. and C. Franignoul (1999). Influence of the North Atlantic SST on the atmospheric circulation. *Geophys. Res. Lett.* **26**, 2969–72.

  (2002). Observed impact of Atlantic SST anomalies on the North Atlantic Oscillation. *J. Clim.* **15**, 606–23.

Czaja, A. and J. Marshall (2001). Observations of atmosphere–ocean coupling in the North Atlantic. *Quart. J. Roy. Meteor. Soc.* **127**, 1893–1916.

Dai, A., T. M. L. Wigley, B. A. Boville, J. T. Kiehl, and L. E. Buja (2001). Climates of the twentieth and twenty-first centuries simulated by the NCAR Climate System Model. *J. Clim.* **14**, 485–519.

Delworth, T. L. and T. R. Knutson (2000). Simulation of early 20th century global warming. *Science* **287**, 2246–50.

Deser, C. (2000). On the teleconnectivity of the "Arctic Oscillation". *Geophys. Res. Lett.* **27**, 779–82.

Federov, A. and S. G. H. Philander (2000). Is El Niño changing? *Science* **288**, 1997–2002.

Frankignoul, C. (1999). A cautionary note on the use of statistical atmospheric models in the middle latitudes: comments on "Decadal variability in the North Pacific as simulated by a hybrid coupled model." *J. Clim.* **12**, 1871–2.

Frankignoul, C., P. Muller, and E. Zorita (1997). A simple model of decadal response of the ocean to stochastic forcing. *J. Phys. Oceanogr.* **27**, 1533–46.

Gallego, B. and P. Cessi (2000). Exchange of heat and momentum between the atmosphere and the ocean: a minimal model of decadal oscillations. *Clim. Dyn.* **16**, 479–89.

(2001). Decadal variability of two oceans and an atmosphere. *J. Clim.* **14**, 2815–32.

Gill, A. (1980). Some simple solutions for heat-induced tropical circulation. *Quart. J. Roy. Meteor. Soc.* **106**, 447–62.

Goodman, J. and J. Marshall (1999). A model of decadal middle-latitude atmosphere–ocean coupled modes. *J. Clim.* **12**, 621–41.

Griffies, S. M. and E. Tziperman (1995). A linear thermohaline oscillator driven by stochastic atmospheric forcing. *J. Clim.* **8**, 2440–53.

Gu, D. and S. G. H. Philander (1997). Interdecadal climate fluctuations that depend on exchanges between the tropics and extratropics. *Science* **275**, 805–7.

Haarsma, R. J., F. M. Selton, and J. D. Opsteegh (2000). On the mechanism of the Antarctic circumpolar wave. *J. Clim.* **13**, 1461–80.

Hall, A. and S. Manabe (1997). Can local linear stochastic theory explain sea surface temperature and salinity variability? *Climate Dynamics* **13**, 167–80.

Hall, A., and R. Stouffer (2001). An extreme climate event in a coupled ocean–atmosphere simulation without external forcing. *Nature* **409**, 171–4.

Hasselmann, K. (1976). Stochastic climate models. *Tellus* **28**, 473–85.

Held, I. M. (1983). Stationary and quasi-stationary eddies in the extratropical troposphere: theory. In *Large-scale Dynamical Processes in the Atmosphere*, eds., R. P. Pierce and B. J. Hoskins, New York, NY, Academic Press, pp. 127–68.

Jacobs, G. A. and J. L. Mitchell (1996). Ocean circulation variations associated with the Circumpolar wave. *Geophys. Res. Lett.* **23**, 2947–50.

James, I. N. and P. M. James (1989). Ultra-low-frequency variability in a simple atmospheric circulation model. *Nature* **342**, 53–5.

Jiang, S., F.-F. Jin, and M. Ghil (1995). Multiple equilibria, periodic, and aperiodic solutions in a wind-driven, double-gyre, shallow-water model. *J. Phys. Oceanogr.* **25**, 764–86.

Jin, F.-F. (1997a). An equatorial ocean recharge paradigm for ENSO. Part I: conceptual model. *J. Atmos. Sci.* **54**, 811–29.

  (1997b). A theory of interdecadal climate variability of the North Pacific ocean–atmosphere system. *J. Clim.* **10**, 1821–35.

Jin, F.-F., D. Neelin, and M. Ghil (1994). ENSO on the devil's staircase. *Science* **264**, 70–2.

Kleeman, R. and A. M. Moore (1997). A theory for the limitation of ENSO predictability due to stochastic atmospheric transients. *J. Atmos. Sci.* **54**, 753–67.

Kleeman, R., J. P. McCreary, and B. A. Klinger (1999). A mechanism for generating ENSO decadal variability. *Geophys. Res. Lett.* **26**, 1743–6.

Kravtsov, S. V. and A. W. Robertson (2002). Midlatitude ocean–atmosphere interaction in an idealized coupled model. *Clim. Dyn.* **19**, 693–711.

Latif, M. and T. P. Barnett (1994). Causes of decadal climate variability over the North Pacific/North American sector. *Science* **266**, 634–7.

Lau, N.-C. (1981). A diagnostic study of recurrent meteorological anomalies appearing in a 15-year simulation with a GFDL general circulation model. *Mon. Wea. Rev.* **11**, 2287–311.

Leith, C. E. (1975). Climate response and fluctuation dissipation. *J. Atmos. Sci.* **32**, 2022–6.

  (1976). The standard error of time-average estimates of climate means. *J. Appl. Meteor.* **12**, 1066–9.

Lorenz, E. (1990). Can chaos and intransitivity lead to interannual variability? *Tellus* **42A**, 378–89.

Lysne, J., P. Chang, and B. Giese (1997). Impact of the extratropical Pacific on equatorial variability. *Geophys. Res. Lett.* **24**, 2589–92.

Madden, R. A. (1976). Estimates of the naturally occurring variability of time-averaged sea-level pressure. *Mon. Wea. Rev.* **104**, 942–52.

Majda, A. J., I. Timofeyev, and E. Vanden Eijnden (1999). Models for stochastic climate prediction. *Proc. Nat. Acad. Sci.* **96**, 14687–91.

Manabe, S. and R. J. Stouffer (1997). Coupled ocean–atmosphere model response to freshwater input: comparison to Younger Dryas event. *Paleooceanography* **12**, 321–36.

Marotzke, J., P. Welander, and J. Willebrand (1988). Instability and multiple steady states in a meridional-plane model of the thermohaline circulation. *Tellus* **40A**, 162–72.

Marshall, J., H. Johnson, and J. Goodman (2000). A study of the interaction of the North Atlantic Oscillation with ocean circulation. *J. Clim.* **14**, 1399–1421.

McCalpin, J. and D. B. Haidvogel (1996). Phenomenology of the low-frequency variability in a reduced-gravity, quasigeostrophic double-gyre model. *J. Phys. Oceanogr.* **26**, 739–52.

McWilliams, J. C. and P. R. Gent (1978). A coupled air–sea model for the tropical Pacific. *J. Atmos. Sci.* **35**, 962–89 (also: **36**, 181).

Munnich, M., M. Latif, S. Venske, and E. Maier-Reimer (1998). Decadal oscillations in a simple coupled model. *J. Clim.* **11**, 3309–19.

Neelin, J. D. and M. Latif (1998). El Niño dynamics. *Physics Today,* 32–6.

Neelin, J. D. and W. Weng (1999). Analytic prototypes of ocean–atmosphere interaction at midlatitudes. Part I: coupled feedbacks as a sea surface temperature dependent stochastic process. *J. Clim.* **12**, 697–721.

Neelin, J. D., D. S. Battisti, A. C. Hirst *et al.* (1998). ENSO theory. *J. Geophys. Res.* **103**, 14261–90.

Qiu, B. and F.-F. Jin (1997). Arctic circumpolar waves: an indication of ocean–atmosphere coupling in the extratropics. *Geophys. Res. Lett.* **24**, 2585–8.

Rasmusson, E. M. and T. H. Carpenter (1982). Variations in tropical sea surface temperature and surface wind fields associated with the Southern Oscillation/El Niño. *Mon. Wea. Rev.* **110**, 354–84.

Saravanan, R. and J. C. McWilliams (1995). Multiple equilibria, natural variability, and climate transitions in an idealized ocean–atmosphere model. *J. Clim.* **8**, 2296–323.

(1997). Stochasticity and spatial resonance in interdecadal climate fluctuations. *J. Clim.* **10**, 2299–320.

(1998). Advective ocean–atmosphere interaction: an analytical stochastic model with implications for decadal variability. *J. Clim.* **11**, 165–88.

Saravanan, R., G. Danabasoglu, S. C. Doney, and J. C. McWilliams (2000). Decadal variability and predictability in the midlatitude ocean–atmosphere system. *J. Clim.* **13**, 1073–97.

Selton, F. M., R. J. Haarsma, J. D. Opsteegh (1999). On the mechanisms of North Atlantic decadal variability. *J. Clim.* **12**, 1956–73.

Siegel, A., J. B. Weiss, J. Toomre, J. C. McWilliams, P. S. Berloff, and I. Yavneh (2001). Eddies and vortices in ocean basin dynamics. *Geophys. Res. Lett.* **28**, 3183–6.

Spall, M. A. (1996). Dynamics of the Gulf Stream/Deep Western Boundary Current crossover. Part II: low-frequency internal oscillations. *J. Phys. Oceanogr.* **26**, 2169–82.

Stommel, H. (1961). Thermohaline convection with two stable regimes of flow. *Tellus* **113**, 224–8.

Sturges, W. and B. G. Hong (1995). Wind forcing of the Atlantic thermocline along 32°N at low frequencies. *J. Phys. Oceanogr.* **25**, 1706–15.

Suarez, M. J. and P. S. Schopf (1988). A delayed action oscillator for ENSO. *J. Atmos. Sci.* **45**, 3283–7.

Sutton, R. and P.-P. Mathieu (2002). Response of the atmosphere–ocean mixed layer system to anomalous heat flux convergence. *Quart. J. Roy. Meteor. Soc.* **128**, 1259.

Taylor, A. H. and A. Gangopadhyay (2001). A simple model of interannual displacements of the Gulf Stream. *J. Geophys. Res.* **106**, 13849–60.

Taylor, A. H. and J. A. Stephens (1998). The North Atlantic Oscillation and the latitude of the Gulf Stream. *Tellus* **50A**, 134–42.

Thompson, D. J. W. and J. M. Wallace (1998). The Arctic Oscillation signature in wintertime geopotential height and temperature fields. *Geophys. Res. Lett.* **25**, 1297–300.

Thual, O. and J. C. McWilliams (1992). The catastrophe structure of thermohaline convection in a two-dimensional fluid model and a comparison with low-order box models. *Geophys. Astrophys. Fluid Dyn.* **64**, 67–95.

Tziperman, E., M. A. Cane, and S. E. Zebiak (1995). Irregularity and locking to the seasonal cycle in an ENSO prediction model as explained by the quasi-periodicity route to chaos. *J. Atmos. Sci.* **52**, 293–306.

Vautard, R. and M. Ghil (1989). Singular spectrum analysis in nonlinear dynamics, with applications to paleoclimate time series. *Physica D* **35**, 395–424.

Wallace, J. M. and D. S. Gutzler (1981). Teleconnections in the geopotential height field during Northern Hemisphere winter. *Mon. Wea. Rev.* **109**, 784–812.

Watanabe, M. and M. Kimoto (2000a). Behavior of midlatitude decadal oscillations in a simple atmosphere–ocean model. *J. Meteor. Soc. Jpn.* **78**, 441–60.

(2000b). Atmosphere–ocean coupling in the North Atlantic: a positive feedback. *Quart. J. Roy. Meteor. Soc.* **126**, 3343–69.

Weaver, A. J., J. Marotzke, P. Cummins, and E. Sarachik (1993). Stability and variability of the thermohaline circulation. *J. Phys. Oceanogr.* **23**, 39–60.

Weisse, R., U. Mikolaijewicz, A. Sterl, and S. S. Drijfhout (1999). Stochastically forced variability in the Antarctic Circumpolar Current. *J. Geophys. Res.* **104**, 11049–64.

Welander, P. (1977). Thermal oscillations in a fluid heated from below and cooled to freezing from above. *Dyn. Atmos. Oceans* **1**, 215–23.

Weng, W. and J. D. Neelin (1998). On the role of ocean–atmosphere interaction in midlatitude interdecadal variability. *Geophys. Res. Lett.* **25**, 167–70.

White, W. B. and R. G. Peterson (1996). An Antarctic circumpolar wave in surface pressure, wind, temperature, and sea-ice extent. *Nature* **380**, 699–702.

White, W. B., S.-C. Chen, and R. G. Peterson (1998). The Antarctic circumpolar wave: a beta effect in ocean–atmosphere coupling over the Southern Ocean. *J. Phys. Oceanogr.* **28**, 2345–62.

Wyrtki, K. (1975). El Niño – the dynamic response of the equatorial Pacific Ocean to atmospheric forcing. *J. Phys. Oceanogr.* **5**, 572–84.

Xu, W., T. P. Barnett, and M. Latif (1998). Decadal variability in the North Pacific as simulated by a hybrid coupled model. *J. Clim.* **11**, 297–312.

Yang, J. and R. X. Huang (1996). Decadal oscillations driven by the annual cycle in a zonally averaged coupled ocean–ice model. *Geophys. Res. Lett.* **23**, 269–72.

Yang, J. and J. D. Neelin (1993). Sea-ice interaction with the thermohaline circulation. *Geophys. Res. Lett.* **20**, 217–20.

(1997). Decadal variability in coupled sea-ice–thermohaline circulation systems. *J. Clim.* **10**, 3059–76.

Yu, J.-Y. and C. R. Mechoso (2001). A coupled atmosphere–ocean GCM study of the
    ENSO cycle. *J. Clim.* **14**, 2329–50.
Yu, J.-Y., C. R. Mechoso, A. Arakawa, and J. C. McWilliams (2002). Impacts of the
    Indian oceans on the ENSO cycle. *Geophys. Res. Lett.* **29**, 461–4.
Zebiak, S. E. (1989). Ocean heat content variability and El Niño cycles. *J. Phys.
    Oceanogr.* **19**, 475–86.
Zhang, S., C. A. Lin, and R. J. Greatbatch (1995). A decadal oscillation due to the
    coupling between an ocean circulation model and a thermodynamic sea-ice model.
    J. Marine Res. **53**, 79–106.

5

The radiative forcing due to clouds and water vapor

V. RAMANATHAN AND ANAND INAMDAR

Center for Atmospheric Sciences, Scripps Institution of Oceanography, University of California, San Diego, CA

5.1 Introduction

As the previous chapters have noted, the climate system is forced by a number of factors, e.g., solar impact, the greenhouse effect, etc. For the greenhouse effect, clouds, water vapor, and CO_2 are of the utmost importance. The emergence of computers as a viable scientific tool in the 1960s in conjunction with the availability of spectroscopic data enabled us to treat the numerous complexities of infrared-radiative transfer in the atmosphere. While such calculations set the stage for estimating accurately (decades later in the 1990s) the radiative forcing due to greenhouse gases and clouds, they did not yield the necessary insights into the physics of the problem nor did they yield any explanation of the relevant phenomenon. Such insights needed physically based analytic approaches to the problem. It is in this arena that Dr. Robert Cess excelled and provided the community with important insights into numerous radiative processes in the atmosphere of Earth and other planets including Mars, Venus, Jupiter, and Saturn. A few examples that are relevant to the main theme of this chapter are given below.

Within the lower atmosphere of many planets (first 10 km of Earth; 5 km for Mars; and 60 km for Venus) the greenhouse effect is dominated by pressure-broadened vibration–rotational lines (e.g., CO_2 and CH_4) or pure rotational lines (H_2O) of polyatomic gases. Typically, the absorption and emission of radiation occurs in discrete bands with thousands of rotational lines within each band. Even with modern day supercomputers it is impossible to estimate the radiative transfer due to all of these lines and bands through the atmosphere for the entire planet. Thus a three-dimensional characterization of the radiative heating rates from equator to pole using the line-by-line approach is impractical. What is normally done is to use

Frontiers of Climate Modeling, eds. J. T. Kiehl and V. Ramanathan.
Published by Cambridge University Press. © Cambridge University Press 2006.

so-called band models that approximate the effects of the thousands of lines with an equivalent line. Numerous scientists contributed to this important development (see Goody, 1964 for the details). Cess pioneered the development and, more importantly, the use of such band models (Cess and Tiwari, 1972) for understanding the thermal structure of planetary atmospheres (Cess, 1971; Cess and Khetan, 1973) in radiative as well as in radiative-convective equilibrium. For a CO_2-dominant atmosphere (such as Mars and Venus), he gave the first analytical solutions for the integro-differential equation and obtained the vertical thermal structure of Mars and Venus (Cess and Ramanathan, 1972; Cess, 1982). This solution revealed many important insights that evaded most numerical solutions. It showed that under radiative-convective equilibrium, the temperature should be continuous across the tropopause, provided the line center is optically thick even if the opacity in the line wings is small. This is always the case for CO_2 in Mars and Venus and for H_2O and CO_2 on Earth. The line-wing opacity, however, determines the location of the tropopause and the lower stratospheric temperature, for two reasons: First, the line center occupies only a minor fraction of the total band width, while the wings occupy more than 90% of the total band width; and second, the line-center optical depth is almost independent of pressure (i.e., altitude) (for a uniform mixed gas such as CO_2), whereas the line-wing optical depth scales as square of the pressure. As a result, the line-wing opacity decreases rapidly with altitude. Thus the vertical variation of the line-wing optical depth dominates the vertical gradient of fluxes and the temperature.

Using these techniques developed for planets, Cess (1974; 1976) studied the greenhouse effect of H_2O in Earth's atmosphere. He elucidated the role of clouds in regulating the water-vapor greenhouse effect. For example, Cess showed that fixing cloud altitude, as opposed to fixing cloud temperature, had a large effect on climate sensitivity to CO_2 doubling. He was one of the earliest to recognize the fundamental implication of the radiative-convective equilibrium assumption, i.e., the global temperature change can be determined by the radiative forcing at the top of the atmosphere, TOA (Cess, 1975). He brought his analytical skills in radiative transfer to guide the development of the Earth Radiation Budget Experiment (ERBE), one of NASA's successful climate missions. This experiment yielded a global and regional perspective of the radiative forcing due to clouds (e.g. Ramanathan *et al.*, 1989; Harrison *et al.*, 1990), and the greenhouse effect of the atmosphere (Raval and Ramanathan 1989), and settled many of the debates regarding the role of clouds in climate. One primary reason for the major role of ERBE is the fact that it undertook one of the most exhaustive and careful calibration and instrument-characterization efforts to date under the guidance of Cess along with ERBE project scientist Bruce Barkstrom. Employing a remarkably insightful and simple analytical procedure, Cess *et al.* (1989) used the ERBE cloud-forcing and clear-sky greenhouse effect for validating the performance of over 10 general circulation models (GCMs) developed by various institutions around the world. This study conducted in the early 1990s (Cess *et al.*, 1990) revealed cloud feedback to be the major source of model differences in climate sensitivity and established cloud feedback as a major focus of research.

In this chapter, we use ERBE results to summarize our understanding of the radiative forcing due to water vapor and clouds. We begin with a historical perspective.

5.2 A historical perspective

5.2.1 Evolution of the greenhouse theory

The origins of the concept of the atmospheric greenhouse effect can be traced back to Jeen-Baptiste Joseph Fourier (1827). He realized, almost two centuries ago, that the atmosphere is relatively transparent to solar radiation, but highly absorbent to the terrestrial radiation, thus helping to maintain a higher temperature for Earth's surface. A few decades later, John Tyndall (1861), using results from his detailed laboratory experiments, deduced that water vapor is the dominant gaseous absorber of infrared radiation.

However, it was S. Arrhenius (1896) who laid the formal foundation linking atmospheric gases to climate change. His main goal was to estimate the surface-temperature increase due to an increase in CO_2. To this end, he developed a detailed and quantitative model for the radiation budget of the atmosphere and surface. Chamberlin (1899) also should get a major credit for this development. Arrhenius recognized the importance of water vapor in determining the sensitivity of climate to external forcing such as increase in CO_2 and solar insolation. A simple explanation for the water-vapor feedback among the early studies of climate sensitivity was the fact that the relative humidity of the atmosphere is invariant to climate change. As Earth warmed, the saturation vapor pressure (e_s) would increase exponentially with temperature according to the Clausius–Clapeyron relation, and the elevated (e_s) would (if relative humidity remains the same) enhance the water-vapor concentration, further amplifying the greenhouse effect. Although it is well known that atmospheric circulation plays a big role, a satisfactory answer as to why the relative humidity in the atmosphere is conserved is still elusive. Möller (1963) used the assumption of constant relative humidity and obtained a surprisingly large sensitivity for the surface temperature. The flaws of the surface-balance approach followed by Möller and others preceding him were illustrated by Syukuro Manabe and his collaborators. In a series of studies (Manabe and Strickler, 1964; Manabe and Wetherald, 1967), they employed a one-dimensional radiative-convective model using a global mean atmosphere, which included both radiative and convective heat exchanges between Earth's surface and the atmosphere. Their study clearly illustrated that the radiative energy balance of the surface–atmosphere system is the fundamental quantity that governs global mean surface temperature. This finding, perhaps, provided much of the motivation for satellite-radiation budget studies, measuring the radiative fluxes at the top of the atmosphere from which we can deduce the radiative forcing of the surface–atmosphere system.

5.2.2 Development of quantum theory and spectroscopy

John Tyndall measured the heat absorption by gases (CO_2 and H_2O) through carefully designed experiments in the laboratory. Based on the results of these experiments, he concluded in his Bakerian lecture (Tyndall, 1861, p. 273–285), that the chief influence on terrestrial rays is exercised by the aqueous vapor, every variation of which must produce a climate change. His laboratory measurements were extended and supplemented by direct observations of atmospheric transmission by others during the following 30 years; the most important and reliable of such atmospheric data were taken by Samuel P. Langley (1834-1906). Langley (1884; 1889) designed a high precision thermal detector, the bolometer, and recorded numerous observations of the lunar and solar spectra. He recorded both the broadband radiance and the spectral radiance in about 20 spectral bands between 0.9 and 30 μm. Langley's data enabled Arrhenius to attempt his pioneering calculation. The wavelength-dependence of the line absorption and the complexity of the radiative transfer in an inhomogeneous atmosphere were not included by Arrhenius, for these physics awaited the discovery of quantum mechanics. The birth of the quantum mechanical theory in the early twentieth century heralded the beginning of the theoretical and experimental spectroscopy, eventually leading to the availability of improved spectroscopic data (Goody, 1964).

Quantum theory postulated that infrared absorption and emission of thermal radiation result from molecular transitions involving both vibrational and rotational states that are quantized. Transitions between the lower-energy rotational states give rise to lines and those between the higher vibrational states to vibrational bands accompanied by numerous rotational lines within each band. High-resolution spectroscopic data reveal thousands of monochromatic absorption lines in each of the absorption bands. However, in practice, radiation emitted is never monochromatic, but instead spectral lines of finite widths are observed. Michelson (1895) initiated the theory of line broadening based on strong encounters in the sense of kinetic theory. This broadening of the spectral lines is due to: (1) natural causes (loss of energy in emission), (2) pressure broadening occurring due to molecular collisions, and (3) Doppler effect resulting from the difference in thermal velocities of atoms and molecules. In the upper atmosphere, the Doppler broadening prevails in combination with pressure broadening, while the latter dominates in the lower atmosphere. For the larger pressures typical of the lower atmosphere, the most important line-broadening mechanism is pressure broadening (e.g., Goody, 1964), and the variation of the spectral absorption coefficient is expressed by the Lorentz profile. The spread of the line absorption with frequency is determined by the line-shape factor, which depends upon assumptions regarding the mode of broadening. Laboratory measurements of total band and line absorption, complemented by theoretical models of the strengths of molecular transitions, have enabled the creation of a large spectroscopic database (Rothman *et al.*, 1987) that is constantly being refined and improved.

It is possible to model the longwave radiative fluxes in the atmosphere by directly evaluating the contribution of the individual lines of atmospheric absorption bands. The earliest attempts to derive the infrared radiation fluxes in the atmosphere are discussed in the monograph by Elsasser (1942), who developed a simple graphical integration method to derive the total absorption coefficient at a given frequency. He assumed that each individual spectral line has a Lorentz shape and that the total absorption coefficient at a given frequency can be represented by the sum of absorption coefficients from an infinite series of equally spaced lines. Elsasser can also be credited with introducing the pressure dependence of the absorption coefficient for inhomogeneous atmospheric paths by appropriately scaling the optical mass. The inherent assumptions in Elsasser's pressure-scaling method were considered inadequate in a later study by Kaplan (1959). He described a more accurate method for the calculation of infrared fluxes and flux divergence by properly accounting for the temperature dependence of the absorption coefficient, and by applying the now-famous Curtis–Godson approximation (Goody, 1964) for the treatment of the inhomogeneous paths.

Although detailed spectroscopic information on line widths and intensities were available, early attempts at deriving the theoretical spectra without any kind of approximations were hampered by the non-availability of sufficient computing power and data-storage capabilities. With the advent of digital computers, methods that directly integrated the line spectra gained popularity. The first such attempt was made by Hitschfeld and Houghton (1961), who integrated the flux equation directly for the 9.6 μm ozone band between 9.5 and 32.5 km. They did not use a spectral model and their treatment of the pressure effects was exact. A similar study by Drayson (1967) for the 15 μm CO_2 band extended the approach to include non-homogeneous atmospheric paths by performing analytical integration over pressure. Use of the exact line-by-line techniques became more widespread with the availability of documented spectral-line data (McClatchey *et al.*, 1973) and the advent of high-speed digital computers.

However, the enormous number of spectral lines and the complexity of the inhomogeneous atmosphere still demanded copious amounts of computer time and resources and precluded such a task being undertaken for climate-model applications. However, they have been used as benchmark references for other simpler models (as discussed in the next subsection). The approximate approaches for treating absorption over finite spectral intervals involve band models. An illustration of how spectroscopic information is incorporated into the formulation of a band model along with a discussion of different types of models has been given by Cess and Tiwari (1972). Several different band models have also been discussed by Goody (1964), and applications to flux calculations have been numerous. Kiehl and Ramanathan (1983) compared the results of various band-model parameterizations to measurements and showed that the calculated and measured frequency-integrated absorptions agreed to within the experimental accuracy of 5 to 10%.

The inclusion of the quantum-mechanical details also proved to be important. For example it was shown (Augustsson and Ramanathan, 1977) that tens of weak isotopic and hot bands of CO_2 in the 8 to 20 μm region played a dominant role in the greenhouse effect for CO_2 concentrations exceeding present-day values by factor of 10 or larger (values that are considered to be reasonable for Earth's Archean atmosphere of two to four billion years ago), see Chapter 13.

In summary, with the state-of-the-art spectroscopy, the complexity of modeling the absorption and emission in the atmosphere of gases consisting of thousands of spectral lines has ceased to be a source of uncertainty. However, the same is not true for the far wings of distant spectral lines where the so-called "continuum absorption" prevails. More discussion on this appears in the following section.

5.2.3 Radiation modeling

The radiation-flux calculation algorithms used in earlier climate-sensitivity experiments were simple and represented an approximation to the exact solutions of the radiative-transfer equation. The models differed appreciably from each other. However the availability of better spectroscopic data contributed to the development of detailed radiation models that could generate the vertical distribution of fluxes in the atmosphere. These model simulations far exceeded our ability to make such observations. As a result, radiation models used in climate studies were developed in isolation. Symptomatic of this process is the treatment of clouds as flat plates with no inhomogeneities. Differences in model approaches together with an insufficient knowledge of the distribution of the radiatively active species in the atmosphere (water vapor, clouds, and aerosols) caused significant differences between the results of these models.

The perception of possibly significant uncertainties in the simulation of fundamental radiative processes affecting the proposed climate-change mechanisms, led
the World Climate Research Program (WCRP) to sponsor an international effort
called the Intercomparison of Radiation Codes in Climate Models (ICRCCM) with
the goal of evaluating and improving solar and longwave radiative computations
in climate models (Luther *et al.*, 1988). The ICRCCM made extensive use of the
most detailed models and performed explicit integration over each of the tens of
thousands of spectral lines in the thermal region of the spectrum. Figure 5.1 shows
an example of the radiance spectrum measured by a High-resolution Interferometer
Sounder (HIS) radiometer flying at 20 km altitude (Revercomb *et al.*, 1988). The
curve clearly highlights the important spectral regimes for absorption by H_2O, CO_2,
and O_3.

While the spectra from these detailed models are believed to be accurate to better
than 1%, uncertainties still remain due to the background absorption in the far wings
of distant spectral lines and especially in the "continuum" region of the atmospheric
window between 8 and 12 μm. The cumulative effect of strong distant lines, although not important for gases like CO_2, are critical for water vapor, especially in
regions of weak absorption. One of the conclusions of the ICRCCM study was that
in the absence of a widely accepted theory, the H_2O continuum absorption masked
the observed differences among the simpler models. The continuum absorption,
which varies as the square of the water-vapor density, is an important mode of
cooling in the lower troposphere (Roberts *et al.*, 1976). The most widely followed
continuum formulation (Roberts *et al.*, 1976) ascribes most of the continuum absorption to the *e*-type (i.e., dependent solely on the water-vapor partial pressure).
However, Clough and his collaborators (Clough *et al.*, 1989; Clough *et al.*, 1992)
concluded, based on theoretical models of line shape and careful examination of
the experimental data of Burch, that continuum is important at all frequencies. The
p-type contribution (foreign-broadening) to the continuum caused the largest differences with Robert's formulation at 200 mb in the troposphere (Clough *et al.*,
1992). Incidentally, this latter study also brought into focus the prominence of the
far-infrared region of the longwave beyond 15 μm. This far infrared consists of
the pure rotation bands of H_2O, the dominant mechanism for radiative cooling to
space in the mid to upper troposphere (altitudes with pressure levels less than about
500 mb).

Another important outcome of the ICRCCM exercise was the call for an organized effort simultaneously to measure the spectral radiance along with the
information on the atmospheric state, which, in early 1990s, culminated in the
DOE-sponsored Atmospheric Radiation Measurement project at the Southern Grid
Plains site in Oklahoma. The measured radiance spectra display (Figure 5.1) as
much spectral detail as those produced from the line-by-line models themselves.

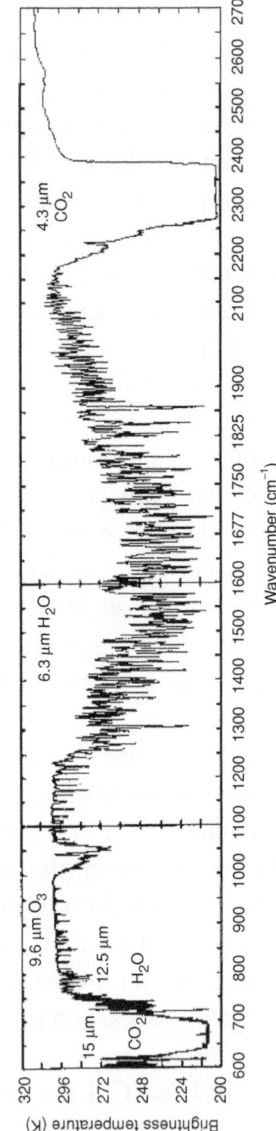

Figure 5.1. An example of a typical Earth-emitted radiance spectrum measured by the HIS radiometer (see text).

5.2.4 Beginning of the satellite era

With the advent of the satellite era in the early 1960s, it was possible to measure the radiation budget at the top of the atmosphere. Radiation-budget instruments measure the outgoing longwave radiation (OLR), incoming solar radiation, and reflected solar radiation. It required nearly two decades of evolution of satellite instrumentation before we could obtain a comprehensive database of radiative energy budget at the top of the atmosphere (Barkstrom, 1984). With the launch of the scanning radiometer as part of ERBE aboard the NOAA satellites, it was possible for the first time to get an observational assessment of the atmospheric greenhouse effect (Raval and Ramanathan, 1989), and cloud-radiative forcing (Ramanathan *et al.*, 1989; Harrison *et al.*, 1990). Another advantage provided by the scanning ERBE instrument was a simultaneous twin view of the Earth (paraphrasing Cess' oft-quoted description of ERBE): one with clouds and one without clouds. This approach provided the data needed to understand two fundamental issues in radiative forcing of climate: cloud forcing and atmospheric greenhouse effect.

5.3 Cloud-radiative forcing

Until the advent of satellite radiation-budget experiments, model studies were the main source of insights into the global radiative effects of clouds (Manabe and Wetherald, 1967; Schneider, 1972). These models suggested that clouds have a large net cooling effect on the planet. Satellite radiation budget instruments measure the OLR, incoming solar radiation, and reflected solar radiation. Numerous studies have attempted to estimate the net radiative effect of clouds with satellite radiation-budget data collected during the 1970s and early 1980s (see summary in Hartmann *et al.*, 1986). The accuracy of these studies with regards to the global and regional effects of clouds has been diminished by the limitation of the data available to define clear-sky and cloudy-sky radiation fields.

In the ERBE approach the TOA radiation budget can be written as: $H = S(1 - A) - F$; where H is the net heating of the surface-atmosphere column, A is the albedo and F is the OLR to space. We can express A and F as the weighted sum of clear-sky value and overcast-sky value as follows: $A = A_c(1 - f) + fA_o$ and $F = F_c(1 - f) + fF_o$ where f is the cloud fraction and the subscripts c and o denote clear-sky and overcast-sky values. Substituting these definitions in H, we obtain Equation (5.1), where C_l and C_s are the longwave and shortwave cloud

$$H = \{S(1 - A_c) - F_c\} + C_l + C_s \tag{5.1}$$

forcing and, as described below, are given by the difference in the radiation budget of the clear- and cloudy-sky values (Ramanathan *et al.*, 1989). In the present context, the term "cloudy" refers to a mix of clear and overcast skies. Satellites mostly observe cloudy-sky values mixed with infrequent clear-sky values, and it takes great care and validated algorithms to infer clear-sky values. The clear-sky OLR also yields the atmospheric greenhouse effect, provided the longwave emission from the surface is known. These quantities are defined first.

Greenhouse effect of the atmosphere (G_a). This follows the classical definition of the greenhouse effect, which is the effect of the atmosphere in reducing the longwave cooling to space. The term G_a is defined as $G_a = E - F_c$ where F_c is the measured OLR for clear skies and E is the emission from the surface, which for a black body equals σT_s^4 with $\sigma = 5.67 \ 10^{-8}$ W m^{-2} K^{-4} and T_s is the surface temperature.

Longwave cloud forcing (C_l). Clouds reduce the OLR further and enhance the greenhouse effect. This enhancement of the greenhouse effect by clouds can be obtained by letting: $C_l = F_c - F$, where F is OLR for average cloudy conditions (mixed clear and cloudy skies). In general, C_l is positive.

Total greenhouse effect (G). The combined effect of cloud and the atmosphere on the greenhouse effect is: $G = G_a + C_l = E - F$.

Short wave cloud forcing (C_s). Clouds also enhance the albedo (reflectivity) of the planet, when compared with cloudless skies, thereby decreasing the planetary solar heating. This effect, referred as the cloud shortwave forcing, C_s, is estimated by differencing the solar energy absorbed by a cloudy region from that over just the clear-sky portion of that region, i.e., $C_s = S(1 - A) - S_c$, where S is the solar insolation at TOA, A is the column albedo, and S_c is the clear-sky solar absorption; C_s is negative, indicating a cooling effect.

Net cloud forcing (C). The net effect of clouds is given by the sum of the longwave and the shortwave cloud forcing, i.e., $C = C_l + C_s$. If C is negative, then clouds have a net cooling effect on the surface–atmosphere column.

With the above definitions, the radiation budget of the system can be expressed as in Equation (5.2):

$$H = \{S(1 - A_c) - E\} + G_a + C_l + C_s \tag{5.2}$$

In summary, we have observable definitions for determining the effect of clouds and the atmosphere. The results here are based on the five-year scanner data collected by ERBE between 1985 and 1990.

5.3.1 Radiation budget of the twin Earth

The OLR for clear and cloudy skies is shown in Figure 5.2. Clear skies reveal a general decrease in emission from tropical to polar regions, except for the following important exceptions: the maximum OLR is in the dry sub-tropical

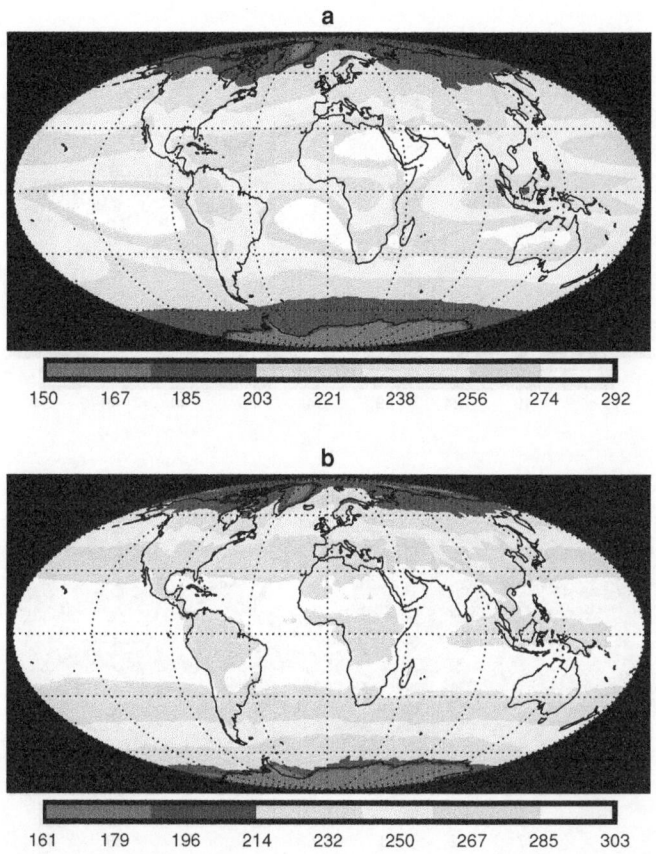

Figure 5.2. Earth Radiation Budget Experiment data averaged over 1985 to 1989 for (a) cloudy-sky (average of clear and cloudy skies) OLR and (b) clear-sky OLR (in W m^{-2}).

subsidence regimes, a minimum in the humid inter-tropical convergence zone (ITCZ) and western Pacific warm-pool regions. As expected, cloudy skies reveal significant east–west and north–south variations. The difference between the cloudy-sky OLR (Figure 5.2a) and clear-sky fluxes (Figure 5.2b) is the longwave cloud forcing.

The albedos are shown in Figure 5.3. It is important to note that these are calibrated broad-band (0.3 to 4 μm) albedos (as opposed to the narrow band visible albedos from GOES or NOAA weather satellites). Darkest are the equatorial oceans and tropical forests; brightest are the deserts and snow-covered polar regions. With clouds, the regional differences practically disappear. It is intriguing to note that, with clouds, the very cloudy regions of the equatorial rain forests and the Indo-Pacific warm pool are just as bright as the nearly cloudless major desert regions in northern Africa, Australia, and others in the world.

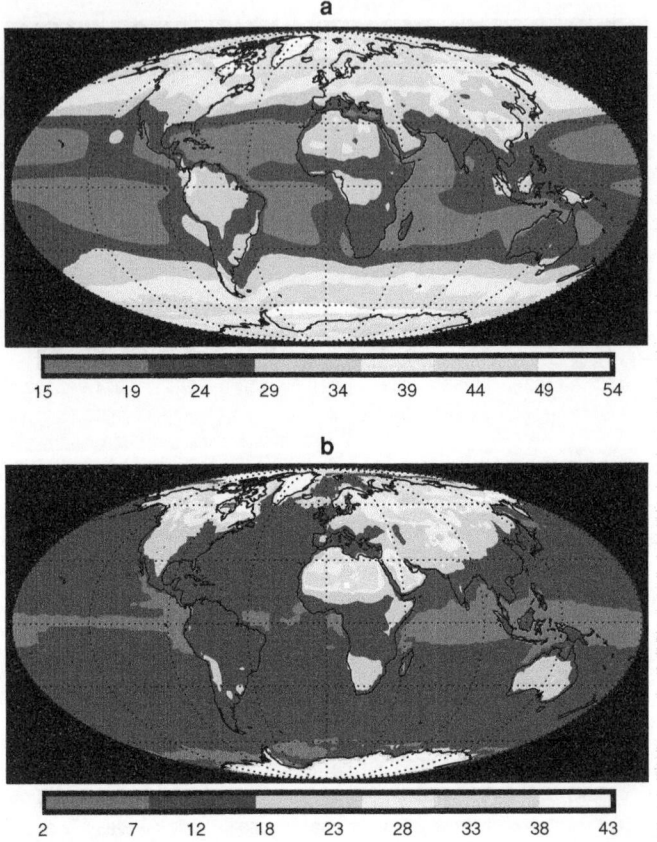

Figure 5.3. Earth Radiation Budget Experiment data (%) averaged over 1985 to 1989 for (a) cloudy-sky albedo (average of clear and overcast skies) and (b) clear-sky albedo.

5.3.2 Global cloud forcing: do clouds heat or cool the planet?

The five-year global mean energy budgets for clear and cloudy regions are illustrated in Figure 5.4. Clouds reduce the absorbed solar radiation by 48 W m^{-2}($C_s =$ -48 W m^{-2}) while enhancing the greenhouse effect by 30 W m^{-2} ($C_1 = 30$ W m^{-2}), and therefore clouds cool the global surface–atmosphere system by 18 W m^{-2}($C =$ -18 W m^{-2}) on average. The mean value of C is several times the 4 W m^{-2} heating expected from doubling of CO_2 and thus Earth would probably be substantially warmer without clouds.

5.3.3 Regional cloud forcing: cloud systems with large forcing

Five-year averages of the C_s (Figure 5.5a) and regional C_1 (Figure 5.5b.) reveal the major climatic regimes and organized cloud systems of Earth. Maxima in C_1 are

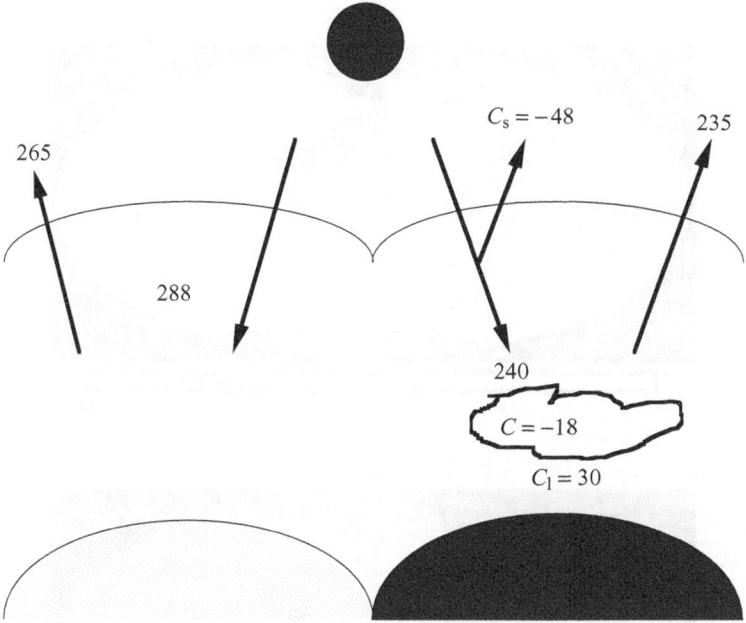

Figure 5.4. Global average clear-sky radiation budget (left panel) and cloud radiative forcing (right panel) from ERBE. The values (W m^{-2}) are for five-year averages between 1985 and 1989. From Ramanathan *et al.*, 1989; Harrison *et al.*, 1990; Collins *et al.*, 1994.

found in regions of moderately thick ($\geqslant$ 1 km) upper-troposphere cirrus clouds, such as deep convective regions in the tropics and jet-stream cirrus in mid and high latitudes. These clouds absorb longwave energy from the warmer regions below and re-radiate to space at the colder upper-troposphere temperature, thus leading to a large reduction in OLR. Maxima in C_s are found in regions of optically thick low-level clouds (stratus and marine strato-cumulus), precipitating cloud systems (extra-tropical cyclones) and deep convective-cirrus systems of the tropics (e.g., Ramanathan *et al.*, 1989; Harrison *et al.*, 1991).

- *Deep convective-cirrus systems.* Maxima in C_1 of 60 to 100 W m^{-2} found over the convectively disturbed regions of the tropics, including the tropical western Pacific, the eastern equatorial Indian Ocean, and the equatorial rain forest regions of South America and Africa. Seasonally (not shown here) the monsoon cloud systems and the ITCZ cloud systems exhibit comparably large values of C_1. The strong greenhouse heating by these upper-tropospheric cloud systems is accompanied by a correspondingly large cooling due to reflection of solar radiation.
- *Extra-tropical storm track cloud systems.* In addition to the convective and ITCZ tropical clouds, persistent bright clouds that reflect more that 75 W m^{-2} of solar radiation ($C_s <$ -75 W m^{-2}) are found polewards of about 35° in the Pacific, Atlantic, and the Indian

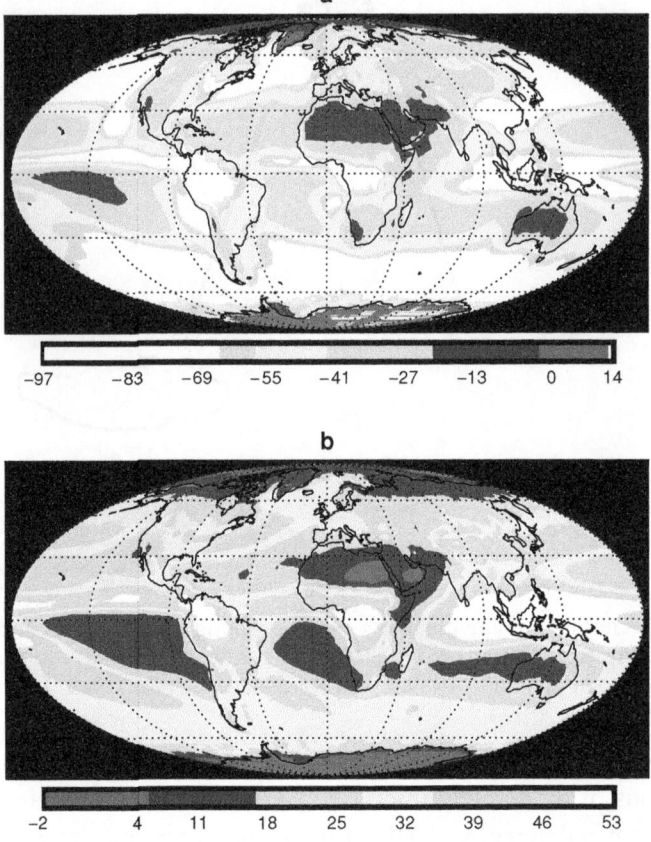

Figure 5.5. Shortwave (a) and longwave (b) cloud radiative forcing (in W m^{-2}) from ERBE (1985–1989).

oceans. Cloud systems associated with extra-tropical cyclones and the stratus that follows the passage of these cyclones are responsible for this cooling.

- *Marine strato-cumulus coastal systems.* These sub-tropical cloud systems also contribute more than 75 W m^{-2} to C_s and extend as far as a thousand kilometers from the western coasts of North America, South America, and South Africa.

An intriguing feature of Figures 5.5 and 5.6 is that in tropical regions where the clouds significantly affected the longwave and shortwave fluxes (Figure 5.5), the longwave and shortwave cloud-forcing terms nearly cancel each other (Figure 5.6). This feature has defied an explanation so far (see Kiehl, 1994 for a partial resolution of this issue). *Does the near cancelation at the TOA imply a negligible role in tropical climate?*

Not necessarily, according to Stephens and Webster (1979) and Ramanathan (1987), who showed that these clouds alter significantly the vertical (between the

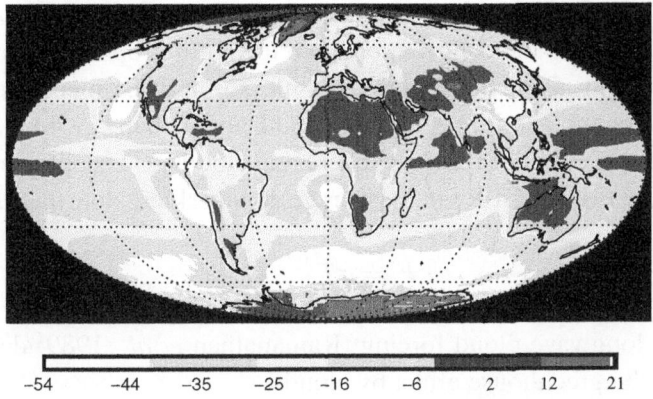

-54   -44   -35   -25   -16   -6   2   12   21

Figure 5.6. Net cloud radiative forcing ($= C_s + C_1$), in W m^{-2} (ERBE, 1985–1989).

ocean and atmosphere) and horizontal (longitudinal and meridional) heating gradients in the ocean and the atmosphere. A significant fraction of the longwave cloud forcing converges into the troposphere, because the clouds reduce the OLR escaping to space, but, the back radiation from the base of these clouds are absorbed by the intervening moist tropical atmosphere. The negative shortwave forcing is mostly due to reduction of solar radiation to the sea surface; in other words, clouds, like a translucent mirror, shield the surface from solar radiation by reflecting it back to space.

5.4 Atmospheric greenhouse effect: global and regional averages

Consider first a one-dimensional system with the surface emitting like a black body. The clear-sky outgoing longwave radiation (F_c) and G_a are related by Equation (5.3), where T_s is the surface temperature, and G_a, by definition is given by

$$F_c = \sigma T_s^4 - G_a \qquad (5.3)$$

Equation (5.4), where τ_v represents the optical depth between TOA and z' and the

$$G_a = \int_{\infty}^{o} dz' \int_{4 \, \mu m}^{500 \, \mu m} [1 - e^{-\tau_v(z',\infty)}] \frac{dB_v(Z')}{dz'} dv \qquad (5.4)$$

expression in square parentheses thus represents the absorptance. From Equation (5.4), we note that G_a is the reduction in the outgoing longwave radiation due to the presence of the atmosphere. From Equation (5.4), it is seen that the reduction in OLR depends on two factors: τ_v, the optical depth; and the vertical temperature gradient (note: $dB_v/dZ \equiv [dB_v/dT]^*[dT/dz]$). Without a radiatively active atmosphere, i.e., $\tau_v = 0$, $G_a = 0.0$ and OLR would be identically equal to

σT_s^4. Likewise, without a vertical temperature gradient, $dB_v/dz = 0$, and $G_a = 0$. It is obvious G_a includes the contribution from the entire troposphere and the stratosphere with weight given to all regions. Unless otherwise mentioned the results summarized here are taken from Inamdar and Ramanathan (1998).

The greenhouse effect of the atmosphere and clouds (G) is obtained from Equation (5.5), where F is the OLR for the average cloudy skies. Note that $G = G_a + C_1$,

$$F = \sigma T_s^4 - G \tag{5.5}$$

where C_1, the longwave cloud forcing (Ramanathan *et al.*, 1989) denotes the enhancement of the greenhouse effect by clouds.

Global-annual mean values. The global-annual means for the surface temperature and the different radiative flux parameters extracted from ERBE are shown in Figure 5.7. The ERBE clear-sky classification scheme is known to be inefficient over the ice-covered surfaces. Hence, while Figure 5.7a shows the results for the entire planet, Figure 5.7b shows results for the ice-free regions. The ice-covered portion of Earth constitutes approximately 6% of the globe. It is apparent from Figure 5.7b that the global surface temperatures have a warmer bias of about 4 K due to the exclusion of ice surfaces. Other uncertainties in the data are discussed in Inamdar and Ramanathan (1998).

The global average G_a is 131 W m^{-2} or the normalized g_a is 0.33, i.e., the atmosphere reduces the energy escaping to space by 131 W m^{-2} (or by a factor of 1/3). The ocean regions have a slightly larger greenhouse effect (0.35 for ocean vs. 0.33 for land) compared with the land (Figure 5.7b). In order to get another perspective on the results shown in Figure 5.7, we note that a doubling of CO_2 (holding the surface and atmospheric temperature fixed) will enhance G_a by about 4 W m^{-2}. We should note that the g_a shown in Figure 5.7 includes the greenhouse effect of water vapor and all other greenhouse gases including CO_2, O_3, and several trace gases.

Regional values. The regional distribution of the annual mean greenhouse effect and column water vapor is shown in Figure 5.8. Figure 5.8a shows the normalized greenhouse effect, $g_a = G_a/\sigma T_s^4$. A large fraction of the spatial variation in G_a is due to the variations in T_s. The contributions from variations in T_s are essentially removed in the normalized greenhouse effect (g_a), such that regional variations of g_a (shown in Figure 5.8a) reveal the effects of variations in atmospheric humidity and lapse rates. As shown in Figure 5.8, g_a increases from pole to equator largely due to the corresponding increase in humidity (shown in Figure 5.8b; also see Raval and Ramanathan 1989 and Stephens and Greenwald, 1991). Furthermore, for the same latitude zone, the continental g_a values are significantly lower than the ocean values indicating that the land regions are drier than the oceanic regions.

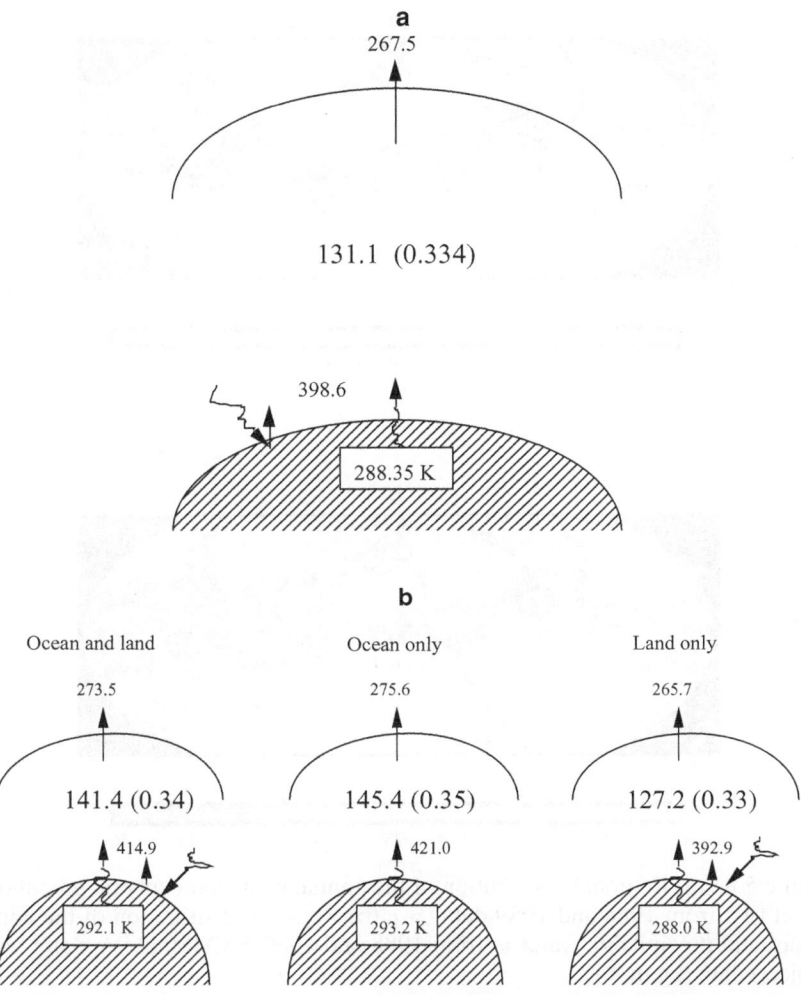

Figure 5.7. (a) Global average of surface temperature, surface emission, outgoing longwave radiation (OLR), and atmospheric greenhouse effect derived from ERBE, GEOS data (1985–89) and Salisbury and D'Aria (1992) surface emissivity tables. The results include all regions of the globe including sea ice and permanent ice. Uncertainties: surface emission, ± 3 W m^{-2}; TOA flux, ± 5 W m^{-2}; GA, ± 6 W m^{-2}.(b) Global average of surface temperature, surface emission, outgoing longwave radiation, and atmospheric greenhouse effect derived from ERBE, NMC-blended sea-surface temperature (SST) (for the oceans), NMC station surface temperatures (for the land), and surface emissivities based on Salisbury and D'Aria (1992) vegetation-index tables, for the period 1988–89. Results exclude ice-covered regions (*c.* 6% of Earth's surface). Uncertainties: surface emission, ± 3 W m^{-2}; TOA flux, ± 5 W m^{-2}; GA, ± 5 W m^{-2}.

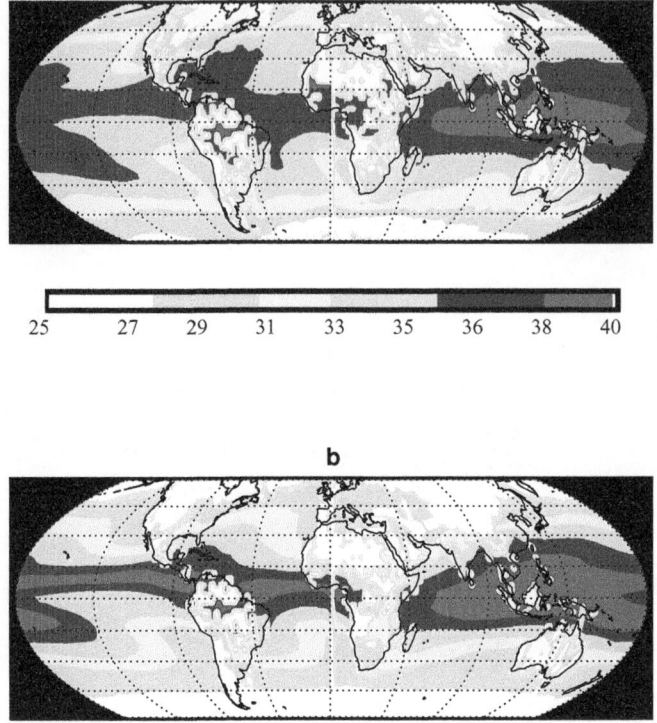

Figure 5.8. (a) Regional distribution of the annual mean atmospheric greenhouse effect (g_a) from 1988 and 1989 data (%). (b) Regional distribution of the annual mean total precipitable water w from 1988 and 1989 NVAP (NASA water vapor project) data (kg m^{-2}).

As shown by Raval and Ramanathan (1989) and Stephens (1990), geographical variations in g_a are dominated by the dependence of g_a on water vapor (w) and T_s, i.e., g_a increases with w which in turn increases with T_s. The equator-to-pole decrease in g_a is largely due to the corresponding decrease in T_s and w.

The atmospheric dynamics also has a strong influence on g_a. We can distinctly see enhanced values of g_a (top panel) and w (bottom panel) in the deep convective regions of the western Pacific and the ITCZ, surrounded by lower values of g_a in the sub-tropical high-pressure belt with strong subsidence. The subsidence over the desert regimes of North Africa (Sahel desert), South Africa (Kalahari desert), Asia (Gobi desert), South America, and Australia are also quite marked with lower values of g_a. These regions are characterized by very low surface emissivities (significantly below 1) and low precipitable water amounts (Figure 5.8b) causing an anomalously low greenhouse effect. In summary Figure 5.8 reveals a remarkable consistency between regional variations in g_a and w, which suggests that variations in water vapor rather than lapse rates contribute to regional variations in g_a.

5.5 Understanding water-vapor feedback using radiation budget data: an example

The response of water vapor and its greenhouse effect (i.e., g_a) to surface and atmospheric temperature changes is a fundamental issue in climate dynamics. Since water-vapor distribution is determined by both thermodynamics and dynamics, this response can depend on the spatial scales and the time scales of the temperature variations. Ideally we need data that can be integrated over spatial scales large enough to average over several dynamical processes such as the ascending and descending branches of the tropical circulation (Soden, 1997). With respect to time scales we need a spectrum of scales ranging from seasonal to decadal scales to examine the dependency on time scales. Such long-term datasets are not yet available. However with the five-year ERBE dataset we can examine the annual time scales over the entire tropics and the planet. In particular, as shown by Inamdar and Ramanathan (1998), water-vapor distribution, g_a, and T_s undergo large seasonal variations even when averaged over the entire tropics or the globe. Before summarizing the results, we will clarify an oft-held misconception that only the lower layers of the troposphere contribute to G_a.

Figure 5.9 shows the sensitivity parameters $dF_c/d(\ln w)$ and $dG_a/d(\ln w)$ as a function of altitude. In each layer of the atmosphere, we change the water-vapor concentration by 1% and estimate the change in OLR and G_a. The figure clearly demonstrates that both G_a and OLR (i.e., F_c) receive comparable contribution from all layers of the troposphere. We next give the following background to aid the interpretation of the results presented later.

The fundamental longwave climate-feedback parameter is dF/dT_s, and from Equation (5.3), Equation (5.6) is obtained.

$$dF/dT_s = 4\sigma T_s^3 - dG/dT_s = 4\sigma T_s^3 - (dG_a/dT_s + dC_1/dT_s) \qquad (5.6)$$

The water-vapor feedback effect is contained in dG_a/dT_s; the cloud feedback effect is contained in dC_1/dT_s and lapse-rate changes will influence both dG_a/dT_s and dC_1/dT_s. Considering first dG_a/dT_s, we obtain Equation (5.7) from Equation (5.3). For a uniform change in surface

$$dG_a/dT_s = 4\sigma T_s^3 - dF_c/dT_s \qquad (5.7)$$

and atmospheric temperature (i.e., without lapse-rate feedback) and without any change in water-vapor amount (i.e., no water-vapor feedback), $dF_c/dT_s =$

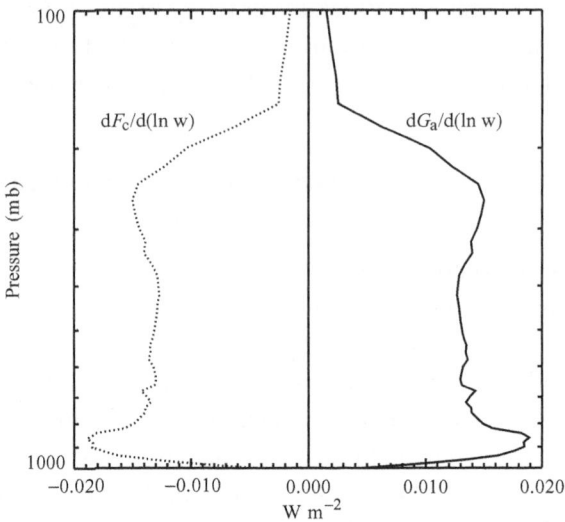

Figure 5.9. Sensitivity of OLR and G_a to height-dependent perturbations in water-vapor mixing ratio. Sensitivity (per unit % change in w) has been derived by averaging the sensitivity over a range of RH perturbations between -20 to 20% from the standard tropical profile.

3.3 W m^{-2} K^{-1} (Ramanathan *et al.*, 1981). Thus for lapse rate or water vapor to exert a positive feedback, requires $dF_c/dT_s < 3.3$ for positive feedback; ≈ 3.3 for no feedback; $\geqslant 3.3$ W m^{-2} K^{-1} for negative feedback.

Radiative-convective models with fixed relative humidity assumption yield (e.g., see Ramanathan *et al.*, 1981) $dF_c/dT_s \approx 2$ W m^{-2} K^{-1}.

To rephrase the above criteria in terms of Equation (5.7), we note that for the global average $T_s = 289$ K, $4\sigma T_s^3 = 5.47$ W m^{-2} K^{-1} and we obtain from Equation (5.7) $dG_a/dT_s > 2.2$ for positive feedback; ≈ 2.2 for no feedback; < 2.2 W m^{-2} K^{-1} for negative feedback.

The observed annual cycles of T_s, g_a, and precipitable water (w) are shown respectively in Figure 5.10a,b, and c for the entire tropics (30° N–30° S); in Figures 5.11a–c for the globe (90° N–90° S). The precipitable water, w, is resolved into three tropospheric layers: w_1 for lower (surface–700 mb), w_2 for middle (700–500 mb) and w_3 for upper (500–300 mb) troposphere.

For the tropics, T_s peaks in March/April, while for 90° N–90° S, T_s peaks in July. We can qualitatively interpret the phase of the annual cycle as follows. The tropical annual cycle is dominated by the coupled ocean–atmosphere system and as a result, the temperature response lags behind the forcing by a maximum of about three months ($\pi/2$); thus, with the solar insolation peaking in December 21, the temperature peaks in late March as shown in Figure 10. In addition to the annual

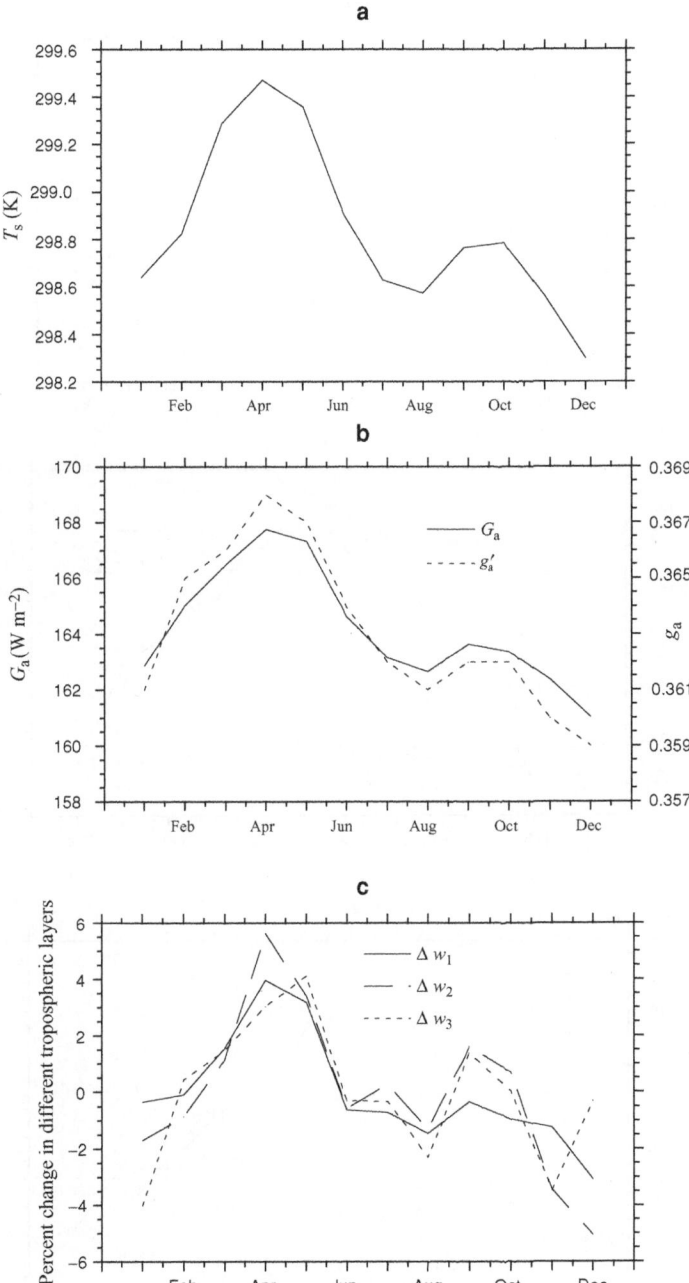

Figure 5.10. Annual cycles for the tropics (30° N–30° S). (a) surface temperature (T_s) obtained from the same sources as in Figure 5.7; (b) atmospheric greenhouse effect (G_a) and the normalized atmospheric greenhouse effect g_a shown by dashed line; and (c) percent change in precipitable water in three tropospheric layers, namely lower (w_1 between surface and 700 mbar), middle (w_2 between 700 and 500 mbar), and upper (w_3 between 500 and 300 mbar).

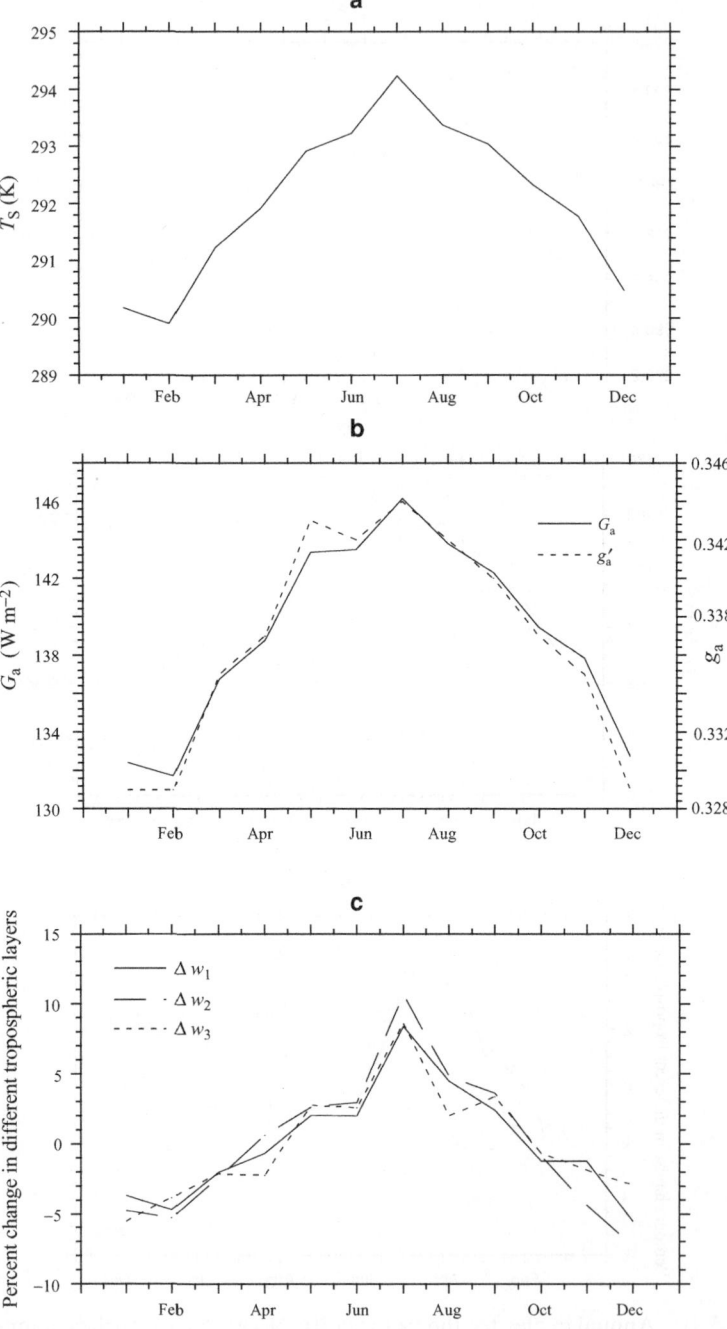

Figure 5.11. Same as Figure 5.10, but for 90° N–90° S.

cycle, the figure also reveals a semi-annual cycle, whose amplitude is comparable to the annual-cycle amplitude.

The extra-tropical and global annual cycle is most likely dominated by the hemispherical asymmetry in the land fraction. During the northern-hemisphere summer (June, July, and August), the large land masses warm rapidly (with about a one-month lag) which dominates the hemispherical and global mean response; however, during the southern-hemisphere summer, the relatively smaller fraction of land prevents a corresponding response. Thus, the globe is warmest during June/July and is coldest during December/January.

Both Figures 5.10 and 5.11 reveal that g_a, w_1, w_2, w_3, and T_s are positively correlated. The figures again reveal the consistency between the radiation-budget data and the water-vapor data. When w and g_a are correlated with T_s, the best correlation coefficient is obtained for a phase lag of less than a month (as shown later). This near-zero phase lag rules out the possibility of variations in g_a or w driving variations in T_s; were this to be the case, T_s should lag behind the forcing by at least more than a month. On the other hand, since convective time scales are less than a month, it is reasonable to expect that variations in g_a and w are driven by variations in T_s without much phase lag. The deduction is that the correlation coefficient between g_a and T_s or that between w and T_s is the feedback parameter, valid at least for annual time scales.

Scatter plots of G_a versus T_s for the tropics (30° N–30° S) and globe (90° N–90° S) are shown in Figure 5.12, and similar plots of F_c versus T_s are shown in Figure 5.13. The feedback term dG_a/dT_s derived from the annual cycle of monthly mean values are summarized in Figure 5.14 (ocean and land). The $G_a - T_s$ correlation was performed by successively including larger domains extending in increments of five degrees on either side of the equator (e.g., 5° N–5° S, 10° N–10° S, and so on). The period 1985–87 was marked by ENSO, which peaked with the El-Niño event in 1987. Since the annual-cycle signals were weak during this ENSO, we employ only the years 1988–89 for the correlation analysis here. The thick solid line depicted in Figure 5.14 has been derived using the annual cycle from station data, while the dashed line is derived employing the GEOS surface-temperature data.

Focusing first on the clear-sky sensitivity values (as opposed to the all-sky values shown by solid circles), we infer the following features from Figure 5.14.

- Between 10° N and 10° S (not shown), dG_a/dT_s (7–10 W m^{-2} K^{-1}) exceeds the blackbody-emission value of about 6 W m^{-2} K^{-1}, thus reproducing the so-called super greenhouse effect inferred from latitudinal variations in G_a and T_s (Raval and Ramanathan, 1989) and from El-Niño-induced variations (Ramanathan and Collins, 1991). The large value of dG_a/dT_s is due to the increase in frequency of convection with T_s (Waliser et al.,

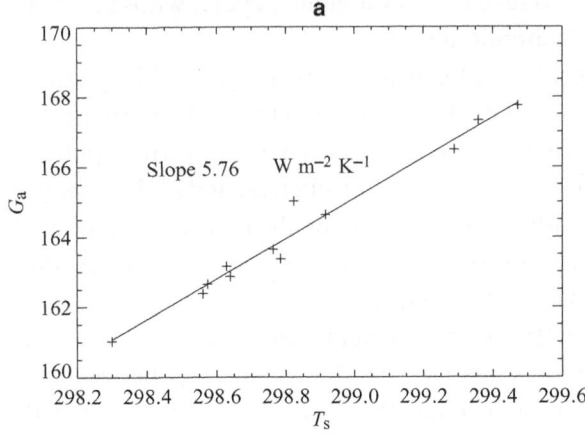

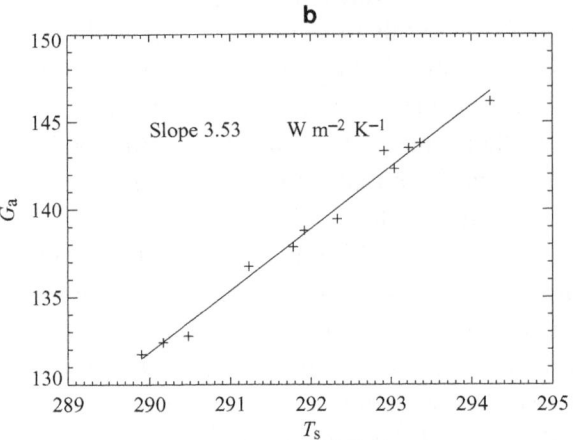

Figure 5.12. Scatterplots of G_a versus T_s using the annual cycles shown in Figures 5.10 and 5.11 for the two domains (i.e., 30° N–30° S and globe). The slope dG_a/dT_s, representing the water-vapor feedback sensitivity parameter, obtained from the least square fit of the data, is inset in the figures.

1993) and the subsequent increase in mid and upper tropospheric water vapor (Hallberg and Inamdar, 1993). We have to bear in mind, however, that both temperature effect (as represented by the Planck and lapse-rate feedbacks) and moisture feedbacks contribute to the observed dG_a/dT_s, as can be inferred from Equation (5.4); see also Inamdar and Ramanathan, 1994.

• Away from the equatorial regions, dG_a/dT_s decreases rapidly and asymptotes to the global mean value of about 3.5 W m^{-2} K^{-1}. The enhanced trapping in the equatorial regions is compensated by enhanced emission to space from the sub-tropics and the extra-tropics. The enhanced emission in the sub-tropics is most likely due to the drying effect of deep convection (as evidenced from the all-sky flux changes discussed later); in the

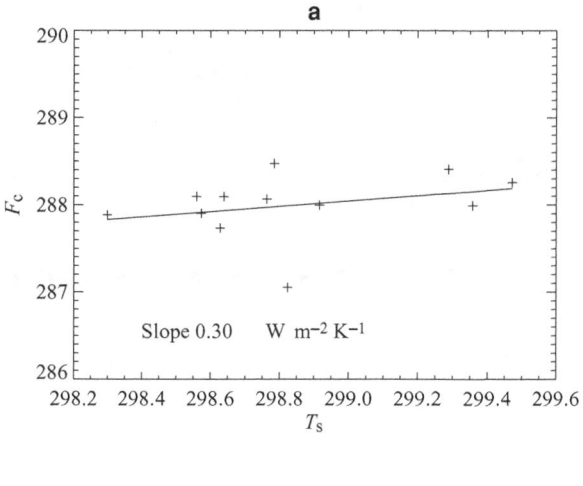

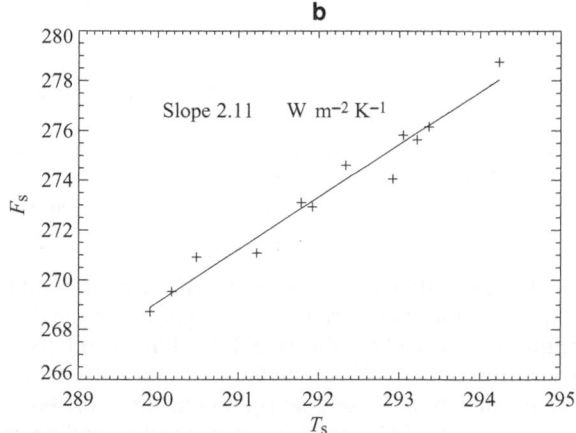

Figure 5.13. Same as in Figure 5.12 but for F_c versus T_s.

extra-tropics, on the other hand, temperature changes are confined largely to the northern-hemisphere land regions, which are not as effective as oceanic regions in enhancing water vapor in the atmosphere (this point is discussed in detail later). The magnitude of the feedback, as well as the corresponding variation in water vapor is consistent with values we expect for the fixed relative-humidity models, as is indeed seen by comparing the model values shown in Figure 5.14.

- Upon comparing with the all-sky values, we see that the cloud longwave forcing feedback term does not contribute to the global sensitivity. This does not, however, imply that clouds do not change. In fact, in the equatorial regions, the all-sky sensitivity is much larger thus indicating a large increase in convective clouds. In the sub-tropical regions, however, the drying effect of increased equatorial convection decreases the sensitivity to very close to clear-sky values.

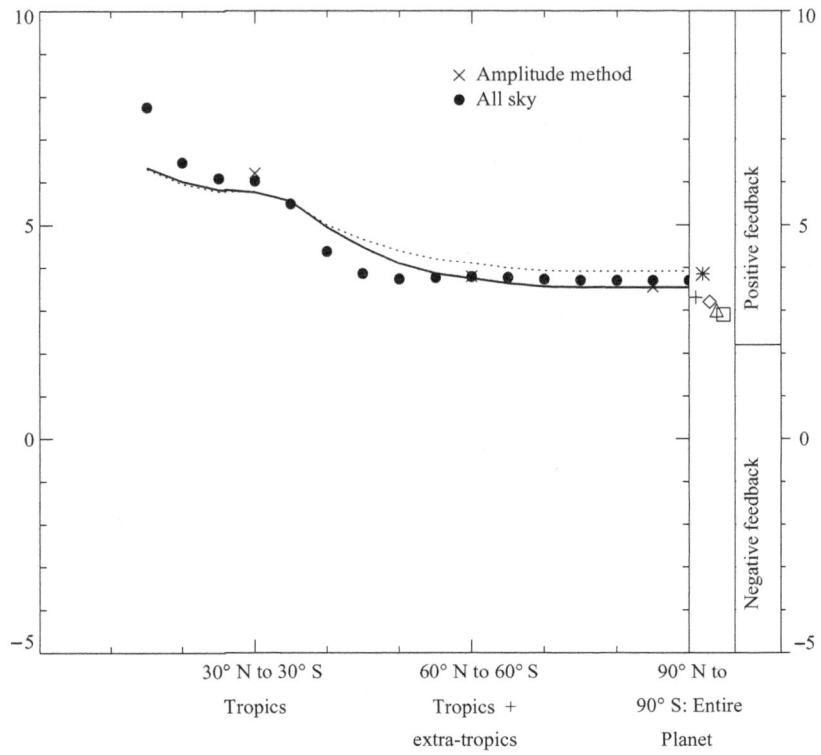

Figure 5.14. Feedback sensitivity parameter dG_a/dT_s for different latitude ranges (ocean plus land). The solid line depicts results, which use the data for the land-surface temperatures (1988–1989), while the dashed line employs the GEOS data. The results derived from the amplitude method ($\times$, see text), and also the ERBE all-sky OLRs ($\bullet$) are shown in the second right box, while the box at the extreme right marks the range of dG_a/dT_s into regions of positive and negative feedback. Other data points are as follows: $\square$, Arrhenius (1896); $\ast$, Manabe and Wetherald (1967); $\lozenge$, Mitchell, results reported in Cess *et al.*, 1989; $+$ Raval and Ramanathan (1989); $\triangle$ Cess *et al.* GCM intercomparison study (1989).

The data, however, reveal the sub-tropical drying effect of deep convection, qualitatively similar to that suggested by Lindzen (1990); but unlike Lindzen's mechanism, the drying effect is not sufficiently large and the positive water-vapor feedback dominates in the tropics. One possible reason is the failure of Lindzen's mechanism to account for the large super greenhouse effect in the low latitudes and the advection of moisture from the moist-convective regions to the dry subsidence regions.

However, our results do not necessarily confirm the positive feedback resulting from the fixed relative humidity models for global warming, for the present results are based on annual cycle. We need additional tests with decadal time-scale data for a rigorous test. Nevertheless, the analysis confirms that water vapor has a positive feedback effect for global-scale changes on seasonal to inter-annual time scales. In addition, the results presented here provide a starting point for validating general circulation models.

5.6 Discussion: where do we go from here?

The results presented here clearly demonstrate that radiation-budget data, when integrated with correlative climate data, can provide valuable insights into the physics and dynamics of climate feedback and forcing. We restrict our comments that are germane to the topic of this paper, i.e., radiative forcing and feedback. First, we need improvements in data on several fronts.

5.6.1 Needed improvements in observations

Radiation budget data. We need to reduce the uncertainty in the retrieval of fluxes (needed for climate studies) from the radiances measured by satellites. These uncertainties can be of the order of 50 W m^{-2} for instantaneous values and about 5 W m^{-2} for time and spatial mean values. In addition, we need to improve the identification of clear-sky values from the cloudy pixels. Both these issues require high-resolution (few kilometers or less) cloud data. The recently launched CERES radiation-budget instrument onboard the TRMM (Tropical Rainfall Mapping Mission) and Terra (a NASA–EOS satellite) has such capabilities, and higher quality data are expected soon. Furthermore, all of the currently available radiation-budget data have severe diurnal sampling bias (sampling the planet at one or two local times each day). We need to launch the instruments on geostationary satellites to cover all times of day for the entire planet.

Cloud cover and properties. Recall that the cloud forcing approach avoids entirely the need for cloud-fraction information. While this is a strength of the approach and enabled us to make progress, we need the cloud data to make further progress towards a physical understanding of how clouds influence the radiative forcing of the planet. The newly launched MODIS instrument on Terra will yield higher quality data on cloud optical depths and cloud fraction. However, much progress in instrumental capability is needed in this area.

Vertical structure of water vapor and its greenhouse effect. A central issue in the water-vapor feedback problem has been the role of vertical structure of variability in the water-vapor distribution and the associated sensitivity of climate. One of the greatest uncertainties in the feedback issue, according to the IPCC (1995) document, is that related to the redistribution of water vapor. Divergent views prevail on the variability of tropospheric humidity, thus masking the actual sensitivity of the water-vapor greenhouse effect to height-dependent changes in the concentration of water vapor. The latter may also represent uncertainties in the measurement of tropospheric humidity, especially in the upper troposphere. Lindzen (1990) argued in favor of enhanced climate sensitivity to the upper-tropospheric humidity to advocate his drying hypothesis. He suggested that it is the upper-tropospheric humidity that is the most important and stressed the need to evaluate the sensitivity of the radiative budget of the atmosphere to height-dependent changes of moisture. The study by Shine and Sinha (1991), however, suggested that the atmospheric response to increased concentrations of greenhouse gases is closer to a constant relative change at all altitudes thus translating into a larger absolute change in the lower troposphere. Hence the peak response is obtained in the lower troposphere ($\sim$ 850 mb) due to the continuum absorption. Spencer and Braswell (1997) concluded that fluctuations in the clear-sky outgoing longwave flux are most sensitive to the upper tropical free troposphere where the sensitivity is greatest at the lowest humidities. A recent study (Schneider *et al.*, 1999) suggested that the largest contribution to the sensitivity is from layers between 450 and 750 mb, while the smallest contribution is from layers above 230 mb.

These apparently glaring differences can be traced to a consideration of different structures for the variability of the tropospheric water vapor. Part of the reason for the uncertainty has been the non-availability of reliable measurements in the upper troposphere to characterize this variability. Seasonal humidity changes (Inamdar and Ramanathan, 1998) derived from the NASA Water Vapor Project data (Randel *et al.*, 1996) reveal a nearly unchanged relative-humidity level in the lower troposphere for the tropics whereas a significant moistening occurs in the mid to upper troposphere dominated by the deep convective regions of tropics. These results are also supported by the decadal-scale tropical humidity changes reported by Hense *et al.* (1988) and Flohn and Kapala (1989). Other studies, especially Schroeder and Mcguire (1998) and Spencer and Braswell (1997) along with Lindzen *et al.* (1995) suggest that the drying in the arid regions is so overwhelmingly large as to more than offset the moistening elsewhere. Since these uncertainties are likely to prevail at least near term, it becomes more essential to turn to other diagnostic products like OLR and the atmospheric greenhouse effect measured from satellites to resolve the issue. The CERES instrument with its window channel (8–12 μm) is expected to aid significantly in this diagnostic study. This would enable us to partition the atmospheric greenhouse effect into contributions from the continuum, vibration–rotation, and pure rotation bands of water vapor.

5.6.2 *Unresolved and emerging issues*

Aerosol–cloud interactions. In a fundamental sense, cloud drops are simply aerosols (an aerosol growing explosively in the presence of near-saturation relative humidity becomes a cloud droplet). The fine-particle aerosol concentration is increasing significantly due to human activities and it is clear that cloud properties in the northern hemisphere are being altered. This increase is hypothesized to increase the number of cloud drops, suppress rainfall, and also decrease the daytime cloud fraction. We need a careful study of the cloud and aerosol forcing from satellite radiation-budget data to sort out these complex issues. The first step is to quantify the aerosol forcing at TOA from satellite data. New approaches have been proposed (Haywood *et al.*, 1999; Satheesh and Ramanathan, 2000) towards this goal. We also need to develop a better framework for sorting the data. For example, our ideas of clear skies and overcast skies in sorting the data may not hold under rigorous scrutiny. It is likely that a better approach is to think of the problem as varying from very low aerosol content (clear skies) to optically thick (overcast) aerosol content.

Surface and atmospheric forcing. We urgently need to extend the TOA forcing approach to consider the surface forcing and the atmospheric forcing individually. The so-called excess or anomalous absorption issue (resurrected recently by Cess *et al.*, 1995; Ramanathan *et al.*, 1995, and Pelewski and Valero, 1995, see Chapter 7) has highlighted the need for the surface forcing. What is the physics behind this excess absorption? None of the published studies to-date has identified the source for this excess absorption. Until we understand the physics for its source, we must remain skeptical. The main message to infer from the recent studies is that we lack accurate data to answer some fundamental questions. How much solar energy reaches the surface of the planet? How do clouds regulate the surface solar insulation? Considerable work is needed to develop radiation-budget instruments for surface-based measurements.

Cloud feedback. This is still an unresolved issue (see Chapter 8). The few results we have on the role of cloud feedback in climate change is mostly from GCMs. Their treatment of clouds is so rudimentary that we need an observational basis to check the model conclusions. We do not know how the net forcing of -18 W m^{-2} will change in response to global warming. Thus, the magnitude as well as the sign of the cloud feedback is uncertain. Cloud radiative forcing effects are concentrated regionally (Figure 5.5). The data reveal three regions of major interest for future study.

- *Storm-track cloud systems over mid-latitude oceans.* These cloud systems associated with extra-tropical cyclones along with the persistent oceanic stratus, reflect more than 75% of solar radiation ($C_s < -75$ W m^{-2}), and contribute about 60% of the -18 W m^{-2} net global cloud radiative forcing (Weaver and Ramanathan, 1997). Anthropogenic activities can directly perturb the radiative forcing of these clouds. The emission of anthropogenic SO_2 and its subsequent conversion to sulfate aerosols can increase the number of cloud drops, which can enhance the shortwave cloud forcing by as much as -5 to -10 W m^{-2} (Kiehl and Briegleb, 1993).

- *Deep convective cloud systems in the tropics.* Maxima in C_1 and in C_s of 60 to 100 W m^{-2} (Figure 5.5) are found over the convectively disturbed regions of the tropics, including the tropical western Pacific, eastern equatorial Indian Ocean, equatorial rainforest regions of South America and Africa. These cloud systems have been linked with a number of important climate feedback effects including the western Pacific warm pool cirrus thermostat (Ramanathan and Collins, 1991), cirrus albedo effect on global warming (Meehl and Washington, 1993) and increased convective-cirrus cloudiness with a tropical warming (Wetherald and Manabe, 1988; Mitchell and Ingram, 1992).
- *Marine stratocumulus coastal and sub-tropical cloud systems.* These sub-tropical cloud systems also contribute more than 75 W m^{-2} to C_s and extend as far as 1000 km from the western coasts of North America, South America, and South Africa. Cloudiness is shown to increase with increase in low-level static stability (Klein *et al.*, 1993). Furthermore, the effect of sulfate aerosols in enhancing the cloud albedo is expected to be large for these low-level cloud systems.

References

Arrhenius, S. (1896). On the influence of carbonic acid in the air upon the temperature of the ground. *Phil. Mag.* **41**, 237–76.

Augustsson, T. and V. Ramanathan (1977). A radiative-convective model study of the CO_2 climate problem. *J. Atmos. Sci.* **34**, 448–51.

Barkstrom, B. R. (1984). The Earth Radiation Budget Experiment (ERBE). *Bull. Amer. Meteor. Soc.* **65**, 1170–85.

Cess, R. D. (1971). A radiative transfer model for planetary atmospheres. *J. Quant. Spectrosc. Radiat. Transfer* **11**, 1699–710.

(1974). Radiative transfer due to atmospheric water vapor: global considerations of the Earth's energy balance. *J. Quant. Spectrosc. Radiat. Transfer* **14**, 861–71.

(1975). Global climate change: an investigation of atmospheric feedback mechanisms. *Tellus* **27**, 193–8.

(1976). Climate change: an appraisal of atmospheric feedback mechanisms employing zonal climatology, *J. Atmos. Sci.* **33**, 1831–43.

(1982). The thermal structure within the stratospheres of Venus and Mars. *Icarus* **17**, 561–9.

Cess, R. D. and S. Khetan (1973). Radiative transfer within the atmospheres of the major planets. *J. Quant. Spectrosc. Radiat. Transfer* **13**, 995–1009.

Cess, R. D. and V. Ramanathan (1972). Radiative transfer in the atmosphere of Mars and that of Venus above the cloud deck. *J. Quant. Spectrosc. Radiat. Transfer* **12**, 933–45.

Cess, R. D. and S. N. Tiwari (1972). Infrared radiative energy transfer in gases. In *Advances in Heat Transfer*, New York, NY, Vol. 8, pp. 229–82.

Cess, R. D., G. L. Potter, J. P. Blanchet *et al.* (1989). Interpretation of cloud-climate feedback as produced by 14 atmospheric general circulation models. *Science* **245**, 513–16.

(1990). Intercomparison and interpretation of climate feedback processes in nineteen atmospheric general circulation models. *J. Geophys. Res.* **95**, 16,601–16,615.

Chamberlin, T. C. (1899). An attempt to frame a working hypothesis of the cause of glacial periods on an atmospheric basis. *J. Geol.* **7**, 609–21.

Clough, S. A., F. X. Kneizys, and R. W. Davies (1989). Line shape and the water vapor continuum. In *IRS '88: Current Problems in Atmospheric Radiation*.
(1992). Line shape and the water vapor continuum. *Atmos. Res.* **23**, 229–41.

Collins, W. D., W. C. Conant, and V. Ramanathan (1994). Earth Radiation Budget, Clouds and Climate Sensitivity. In *The Chemistry of the Atmosphere: its Impact on Global Change*, ed. Jack, G. Calvert, Oxford, Blackwell Scientific Publishers, pp. 207–15.

Drayson, S. R. (1967). Atmospheric transmission in the CO_2 bands between 12 m and 18 m. *Appl. Opt.* **5**, 385–91.

Elsaesser, W. M. (1942). *Heat Transfer by Infrared Radiation in the Atmosphere*. Harvard Meteorological Studies 6, Cambridge, MA, Harvard University Press.

Flohn, H. and A. Kapala (1989). Changes in tropical sea–air interaction processes over a 30-year period. *Nature* **338**, 244–5.

Fourier, J.-B. J. (1827). Mémoire sur les Températures du Globe Terrestre et des Espaces Planétaires. *Mem. l'Inst. Fr.* **7**, 570–604.

Goody, R. M. (1964). *Atmospheric Radiation I: Theoretical Basis*. London, Clarendon Press.

Hallberg, R. and A. K. Inamdar (1993). Observation of seasonal variations of atmospheric greenhouse trapping and its enhancement at high sea surface temperature. *J. Clim.* **6**, 920–31.

Harrison, E. F., P. Minnis, B. R. Barkstrom *et al.* (1990). Seasonal variation of cloud radiative forcing derived from the Earth Radiation Budget Experiment. *J. Geophys. Res.* **95**, 18687–703.

Hartmann, D. L., V. Ramanathan, A. Berroir, and G. E. Hunt (1986). Earth radiation budget data and climate research. *Rev Geophys.* **24**, 439–68.

Haywood, J. M., V. Ramaswamy, and B. J. Soden (1999). Tropospheric aerosol climate forcing in clear-sky satellite observations over the oceans. *Science* **283**, 1299–303.

Hense, A., P. Krahe, and H. Flohn (1988). Recent fluctuations of tropospheric temperature and water vapor content in the tropics. *Meteorol. Atmos. Phys.* **38**, 215–27.

Hitschfeld, W. and J. T. Houghton (1961). Radiative transfer in the lower stratosphere due to the 9.6 micron band of ozone. *Quart. J. Roy. Meteor. Soc.* **87**, 562–77.

Inamdar, A. K. and V. Ramanathan (1994). Physics of greenhouse effect and convection in warm oceans. *J. Climate* **5**, 715–31.
(1998). Tropical and global scale interactions among water vapor, atmospheric greenhouse effect, and surface temperature. *J. Geophys. Res.* **103**, 177–94.

IPCC (1995). *The Science of Climate Change*. Cambridge, Cambridge University Press.

Kaplan, L. D. (1959). A method for calculation of infrared flux for use in numerical models of atmospheric motion. *The Atmosphere and the Sea in Motion*. New York, NY, Rockfeller Institute Press, pp. 170–77.
(1960). The influence of CO_2 variations on the atmosphere heat balance. *Tellus* **12**, 204–8.

Kiehl, J. T. (1994). On the observed near cancellation between longwave and shortwave cloud forcing in tropical regions. *J. Clim.* **7**, 559–65.

Kiehl, J. T. and B. P. Briegleb (1993). The relative roles of sulphate aerosols and greenhouse gases in climate forcing. *Science* **260**, 311–14.

Kiehl, J. T. and V. Ramanathan (1983). CO_2 radiative parameterization used in climate models: comparison with narrow band models and with laboratory data. *J. Geophys. Res.* **88**, 5191–202.

Langley, S. P. (1884). Researches on solar heat and its absorption by the Earth's atmosphere. *A Report of the Mount Whitney Expedition*. Professional Papers of the Signal Service No. 15, Washington, Government Printing Office.

(1889). The temperature of the moon. *Mem. Nat'l. Acad. Sci.* **4**, Part II, 107–212.

Lindzen, R. S. (1990). Some coolness concerning global warming. *Bull. Amer. Meteor. Soc.* **71**, 288–99.

Lindzen, R. S., B. Kirtman, D. Kirk-Davidoff, and E. K. Schneider (1995). Seasonal surrogate for climate. *J. Clim.* **8**, 1681–4.

Luther, F. M., R. G. Ellingson, Y. Fourquart *et al.* (1988). Intercomparison of radiation codes in climate models (ICRCCM): longwave clear-sky results – A workshop summary. *Bull. Amer. Meteor. Soc.* **69**, 40–8.

Manabe, S. and R. F. Strickler (1964). On the thermal equilibrium of the atmosphere with a convective adjustment. *J. Atmos. Sci.* **21**, 361–85.

Manabe, S. and R. T. Wetherald (1967). Thermal equilibrium of the atmosphere with a given distribution of relative humidity. *J. Atmos. Sci.* **24**, 241–59.

McClatchey, R. A., W. S. Benedict, S. A. Clough *et al.* (1973). AFCRL atmospheric absorption line parameters compilation. Bedford, MA, Air Force Cambridge Research Laboratory, Report AFCRL-TR-73-0096.

Meehl, G. A. and W. M. Washington (1995). Cloud albedo feedback and the super greenhouse effect in a global coupled GCM. *Clim. Dyn.* **11**, 399–411.

Michelson, A. A. (1895). On the broadening of spectral lines. *Astrophys. J.* **2**, 251–61.

Mitchell, J. F. B. and W. J. Ingram (1992). Carbon dioxide and climate – mechanisms of changes in cloud. *J. Clim.* **5**, 5–21.

Möller, F. (1963). On the influence of changes in CO_2 concentration in air on the radiation balance of Earth's surface and on climate. *J. Geophys. Res.* **68**, 3877–86.

Ohmura, A. and H. Gilgen (1993). Re-evaluation of the global energy balance. *Geophys. Monogr.* **75**, IUGG15: 93–110.

Ramanathan, V. (1981). The role of ocean–atmosphere interactions in the CO_2 climate problem. *J. Atmos. Sci.* **38**, 918–30.

(1987). The role of Earth radiation budget studies in climate and general circulation research. *J. Geophys. Res.* **92**, 4075–95.

Ramanthan, V. and W. Collins (1991). Thermodynamic regulation of ocean warming by cirrus clouds deduced from observations of the 1987 El Niño. *Nature* **351**, 27–32.

Ramanathan, V., R. D. Cess, E. F. Harrison *et al.* (1989). Cloud-radiative forcing and climate: results from the earth radiation budget experiment. *Science* **243**, 57–63.

Randel, D. L., T. H. Vonder Haar, M. A. Ringerud *et al.* (1996). A new global water vapor dataset. *Bull. Amer. Meteor. Soc.* **77**, 1233–46.

Raval, A. and V. Ramanathan (1989). Observational determination of the greenhouse effect. *Nature* **342**, 758–61.

Revercomb, H. E., H. Buijs, H. B. Howell *et al.* (1988). Radiometric calibration of IR Fourier transform spectrometers: solution to a problem with the high-spectral resolution Interferometer Sounder. *Appl. Opt.* **27**, 3210–18.

Roberts, R. E., L. M. Biberman, and J. E. A. Selby (1976). Infrared continuum absorption by atmospheric water vapor in the 8–10 μ window. *Appl. Opt.* **15**, 2085–90.

Rothman L. S., R. R. Gamache, A. Golman *et al.* (1987). The HITRAN database. *Appl. Opt.* **26**, 4058–97.

Salisbury, J. W. and D. M. D' Aria (1992). Emissivity of terrestrial materials in the 8–14 micron atmospheric window. *Remote Sens. Environ.* **42**, 83–106.

Satheesh, S. K. and V. Ramanathan (2000). Large differences in tropical aerosol forcing at the top of the atmosphere and Earth's surface. *Nature* **405**, 60–3.

Schneider, E. K., B. P. Kirtman, and R. S. Lindzen (1999). Tropospheric water vapor and climate sensitivity. *J. Atmos. Sci.* **56**, 1649–58.

Schneider, S. H. (1972). Cloudiness as a global climate feedback mechanism: the effects of a radiation balance and surface temperature of variations in cloudiness. *J. Atmos. Sci.* **29**, 1413–22.

Schroeder, S. R. and J. P. McGuire (1998). Widespread tropical atmospheric drying from 1979 to 1995. *Geophys. Res. Lett.* **25**, 1301–4.

Shine, K. P. and A. Sinha (1991). Sensitivity of the Earth's climate to height-dependent changes in the water vapour mixing ratio. *Nature* **354**, 382–4.

Soden, B. J. (1997). Variations in the tropical greenhouse effect during El Niño. *J. Climate* **10**, 1050–5.

Soden, B. J. and S. R. Schroeder (2000). Decadal variations in tropical water vapor: a comparison of observations and a model simulation. *J. Clim.* **13**, 3337–41.

Spencer, R. W. and W. D. Braswell (1997). How dry is the tropical free troposphere? Implications for global warming theory. *Bull. Amer. Meteor. Soc.* **78**, 1097–1106.

Stephens, G. L. (1990). On the relationship between water vapor over the oceans and sea surface temperature. *J. Clim.* **3**, 634–45.

Stephens, G. L. and T. J. Greenwald (1991). The Earth's radiation budget and its relation to atmospheric hydrology. 1. Observations of the clear sky greenhouse effect. *J. Geophys. Res.* **96**, 15311–24.

Stephens, G. L. and P. J. Webster (1979). Sensitivity of radiative forcing to variable cloud and moisture. *J. Atmos. Sci.* **36**, 1542–56.

Tyndall, J. (1861). On the absorption and radiation of heat by gases and vapours, and on the physical connexion of radiation, absorption, and conduction. *Philosophical Magazine* **22**, 169–94, 273–85.

Waliser, D. E., N. E. Graham, and C. Gautier (1993). Comparison of the highly reflective cloud and outgoing longwave radiation datasets for use in estimating tropical deep convection. *J. Clim.* **6**, 331–53.

Weaver, C. P. and V. Ramanathan (1997). Relationships between large-scale vertical velocity, static stability, and cloud radiative forcing over northern hemisphere extratropical oceans. *J. Clim.* **10**, 2871–87.

Wetherald, R. T. and S. Manabe (1988). Cloud feedback processes in a general-circulation model. *J. Atmos. Sci.* **45**, 1397–1415.

6

A model study of the effect of Pinatubo volcanic aerosols on stratospheric temperatures

V. RAMASWAMY

Geophysical Fluid Dynamics Laboratory, Princeton Forrestal Campus 201, Princeton, NJ

S. RAMACHANDRAN

Physical Research Laboratory, Ahmedabad, India

GEORGIY STENCHIKOV AND ALAN ROBOCK

Department of Environmental Sciences, Rutgers University, New Brunswick, NJ

6.1 Introduction

As discussed in Chapters 1 and 4, radiation forcing from stratospheric aerosols contributes to the variability of the climatic system. This chapter provides a detailed analysis of how these aerosols affect Earth's climate system. Stratospheric aerosols, in the aftermath of intense volcanic eruptions, can perturb substantially the climate of the stratosphere and surface–troposphere system (IPCC, 1995). Typically, the particulates appearing initially are comprised of silicates (e.g., ash). These are large (diameter greater than a few microns) and tend to fall off rapidly (within a few months) (Robock *et al.*, 1995). The conversion of the sulfur-containing gases injected into the stratosphere to sulfate aerosols occurs in a few weeks to months. This results in a loading of the stratosphere with submicron sulfate aerosols having a residence time of one to two years. The radiative effects depend on the type, size, and shape of the particulates. While in the first several weeks, the radiative effects may be dominated by the ash particles, over the longer term (after a few months and up to ~ two years) and of considerable relevance to climate, the sulfate aerosols dominate the effects. We focus on the effects due to sulfate aerosols in this study.

Sulfate aerosols possess absorption bands in both the solar and longwave spectra, and have significant scattering ability in the solar spectrum. It is now well understood from a number of investigations over the past three decades that the radiative effects due to these aerosols consist of an increase of the planetary albedo, i.e., less radiative energy available to the surface-troposphere system, and an enhancement of the radiative heating of the stratosphere owing to an increase in the near-infrared (IR) absorption of the incoming solar radiation and an increase in the longwave

Frontiers of Climate Modeling, eds. J. T. Kiehl and V. Ramanathan.
Published by Cambridge University Press. © Cambridge University Press 2006.

convergence (e.g., Harshvardhan and Cess, 1976; Hansen *et al.*, 1978; Cess *et al.*, 1981; Pollack and Ackerman, 1983; WMO, 1990; Lacis *et al.*, 1992; Hansen *et al.*, 1996; Stenchikov *et al.*, 1998). There have been three major volcanic events of climatic consequence since 1960, all erupting in the low latitudes: Mt. Agung in 1963 (Hansen *et al.*, 1978), El Chichon in 1982 (Pollack and Ackerman, 1983) and Mt. Pinatubo in 1991 (Hansen *et al.*, 1996). In this chapter, we focus on the consequences due to the radiative heating of the lower stratosphere.

Observations provide ample evidence of a lower-stratospheric warming in the tropics following intense volcanic eruptions in the low latitudes (see summary in Ramaswamy *et al.*, 2001). While only radiosonde data was available in the case of the Agung eruption, the availability of satellite, lidar, and radiosonde data has enabled an improved space–time coverage of the temperature change following the El Chichon eruption and even more so following the Pinatubo eruption.

Theoretical and modeling studies based on fundamental principles have provided results that are broadly confirmed by observations, at least in a qualitative sense if not a precise quantitative manner, regarding the warming of the low-latitude lower stratosphere in the months after an intense eruption. Harshvardhan and Cess (1976) provided a simple mathematical formulation for the warming of the lower strato-sphere, pointing out the importance of the non-grey nature of gaseous absorption and temperature contrast between surface–troposphere and lower stratosphere in determining the sign and magnitude of the effect. Later studies have constructed more physically elaborate models, with the interest in the main centered on the peak warming resulting from the volcanic aerosol loading, and its impacts, if any, on long-term temperature trends. Thus, Hansen *et al.* (1978) showed a comparison of a model simulation with observation for the Agung case, while Pollack and Ackerman (1983) and the WMO (1990) show results for the El Chichon eruption case. Hansen *et al.* (1996) demonstrate the good correspondence over a broad low latitude region for the Pinatubo eruption. In the case of Pinatubo in particular, it has been possible for the first time to use observed aerosol optical properties' evolution, compute the temperature response using a model, and match it against observations (Hansen *et al.*, 1996; Kirchner *et al.*, 1999; Ramachandran *et al.*, 2000).

The stratospheric temperature change caused by volcanic aerosols lasts only for one to two years but it can affect the dynamical circulation in the stratosphere, in particular affecting the polar winter vortex (Graf *et al.*, 1994; Kodera, 1994; Kirchner *et al.*, 1999), the middle- and high-latitude northern-hemisphere conti-nental wintertime surface temperatures (Robock and Mao, 1992), as well as exert an influence on the decadal temperature trends (IPCC, 1995).

While the broad nature of the perturbation to the stratosphere is well understood as a result of the various observations and modeling studies, considerably more attention is required to evaluate (a) the degree to which the stratospheric temperature response is unambiguously attributable to the aerosols, through a comparison of model simulations with observations; and (b) the role of the "internal" dynamical variability of the climate system in masking the quantification of the signal due to the volcanic perturbations.

In this study, we employ a general circulation model (GCM) that has a higher vertical resolution than used in several previous studies, and that resolves the solar and longwave spectra with well-calibrated radiative algorithms (Ramachandran *et al.*, 2000). We use this model to simulate the temperature changes occurring following the Pinatubo eruption, in particular the response in the lower stratosphere. In contrast to most previous studies that have examined only the mean warming from model integrations, we perform an ensemble of GCM integrations designed to address the issue of statistical significance of the stratospheric temperature change due to the aerosols. By doing so, we demonstrate that, while there exists a definitive signature of the aerosol perturbation of the stratosphere, there also exists considerable ambiguity in interpreting quantitatively several aspects of the space–time changes in the global lower stratosphere, and in attributing the variations in the observed record solely to the Pinatubo aerosols.

6.2 Model

The model used is the GFDL SKYHI 40-layer GCM with a latitude–longitude resolution of $3 \times 3.6°$. It is a finite-difference grid model (Hamilton *et al.*, 1995; Schwarzkopf and Ramaswamy, 1999), with the top level at 80 km. The vertical resolution is ~ 1 km in the troposphere, ~ 2 km in the stratosphere, and ~ 3 km in the mesosphere. The GCM employs the cloud prediction scheme of Wetherald and Manabe (1988) wherein each layer is assumed to be fully cloud covered when relative humidity predicted by the model exceeds 100%. Clouds are grouped into three categories in terms of the radiative properties: 'high' clouds occur between 100 and 440 hPa, 'middle' clouds between 440 and 680 hPa, and 'low' clouds between 680 and 1000 hPa. The optical properties of the clouds are based on Slingo (1989), with the effective radius of all clouds set to 10 μm. The optical depths are fixed in the model integrations and are 1, 3, and 12 for the high, middle, and low clouds, respectively. The model has fixed sea-surface temperatures, prescribed as a function of month, latitude, and longitude. Versions of this model have been employed to study several troposphere–stratosphere–mesosphere problems (e.g., Fels *et al.*, 1980; Mahlman and Umscheid, 1984; Mahlman *et al.*, 1994; Ramaswamy *et al.*, 1996). See Hamilton *et al.* (1995) for a more complete description of the basic SKYHI model's climatology. One particularly novel feature of this model compared to several other atmospheric GCMs is its vertical resolution in the stratosphere, and its ability to resolve gravity waves which lead to improvements in the simulation of stratospheric temperatures in the northern hemisphere (Mahlman and Umscheid, 1984; Hamilton *et al.*, 1995).

6.3 Radiative transfer algorithm

The shortwave radiation scheme is based on Freidenreich and Ramaswamy (1999) and is calibrated against "benchmark" results. The shortwave (SW) spectrum, divided into 25 bands, ranges from 0.17 to 4 μm, with 11 in the ultraviolet (UV), 4 in the visible and 10 in the near-IR. It accounts for absorption by CO_2, H_2O, O_3, and O_2, and also accounts for Rayleigh scattering. The algorithm employs the delta-Eddington approximation together with an exponential-sum fit treatment of gaseous absorption and an appropriate 'adding' method to combine the inhomogeneous layers (Ramaswamy and Bowen, 1994).

The longwave-radiation algorithm follows Schwarzkopf and Ramaswamy (1999). The recent updates include accounting for the effects of CH_4, N_2O, halocarbons and the foreign-broadened H_2O continuum. The spectrum covers the region from 4.55 μm to infinity. The longwave gaseous and aerosol radiative properties are approximated over eight separate spectral bands. Aerosols and clouds are treated as absorbers in the longwave spectrum.

6.4 Aerosol optical properties

A computation of the radiative forcing and stratospheric response due to the Pinatubo aerosols requires spectral-, space-, and time-dependent single-scattering properties (extinction optical depth, single-scattering albedo and asymmetry factor). Based on observations from a wide variety of satellite instruments, as well as lidars and balloons, Stenchikov *et al.* (1998) (see also Ramachandran *et al.*, 2000) have derived zonal-and-monthly-mean aerosol optical properties. These are employed in the SKYHI GCM to compute the perturbation due to the Pinatubo aerosols. The calculated optical depth at 0.55 μm wavelength from June 1991 to May 1993 is shown in Figure 6.1 for selected latitude zones and global mean. There was a rapid build-up of the aerosols in the low latitudes over the first $\sim$ six months, followed by a slow decay, with the optical depth estimated to be > 0.1 even two years after the eruption. The peak values occurred in winter and spring 1992, and then remained high through summer 1992 before diminishing towards the end of 1992 and beginning of 1993. At 60° N, the build-up occurs later than at the low latitudes, partly as a result of transport and partly due to *in situ* gas-to-particle conversion. During spring 1992, the optical depth at 60° N is comparable to that in the tropics. The pattern of evolution is quite different at the high latitudes compared to the tropics. The global mean values essentially track the low-latitude evolution.

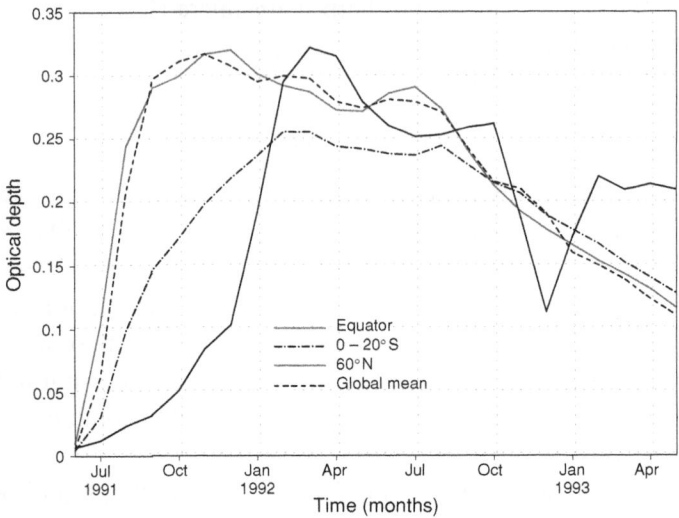

Figure 6.1. Evolution of the monthly-averaged stratospheric aerosol optical depth ($\lambda = 0.55$ μm) over selected domains following the eruption of Mt. Pinatubo volcano in June 1991.

Following the observations, the Pinatubo aerosols are introduced in the GCM between 10 and 200 hPa. This corresponds to 11 vertical layers in the model.

6.5 Radiative heating of stratosphere

As an example of the instantaneous radiative perturbation caused by the Pinatubo aerosols, Figure 6.2 illustrates the vertical profile of the heating rates in the visible (including UV), near-IR, longwave, and total spectra, as computed for October 1991. The meteorological condition used for the computation is that corresponding to the middle of the month. For the aerosol radiative-heating computation, the state of the atmosphere is held fixed, with insertion of the aerosols being the only perturbation.

In October 1991, the aerosols had extended to cover $\sim$ 10–100 hPa over $\sim$ 60° N–S (Ramachandran *et al.*, 2000). In the visible, the scattering by the aerosols causes an enhanced absorption above the aerosol layer by O_3 and O_2. Some enhancement due to increased pathlengths caused by aerosols and subsequent absorption by O_3 is responsible for the increase in heating in the topmost aerosol layer. Below the top of aerosol layer, there is a slight decrease in the absorption owing to reduction in UV transmission and thus the gaseous absorption. In the high latitudes, despite the small aerosol optical depth (Figure 6.1), there is a reduction in transmission leading to lesser heating in the bulk of the aerosol layer.

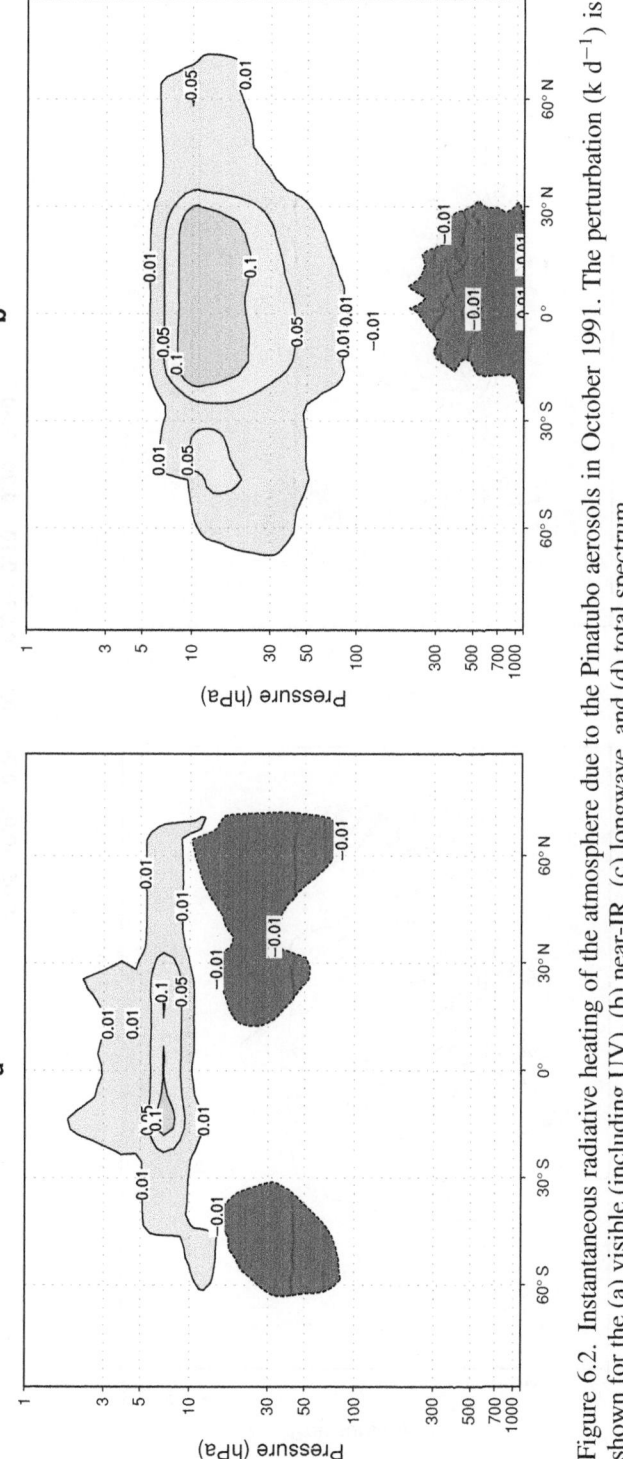

Figure 6.2. Instantaneous radiative heating of the atmosphere due to the Pinatubo aerosols in October 1991. The perturbation (k d⁻¹) is shown for the (a) visible (including UV), (b) near-IR, (c) longwave, and (d) total spectrum.

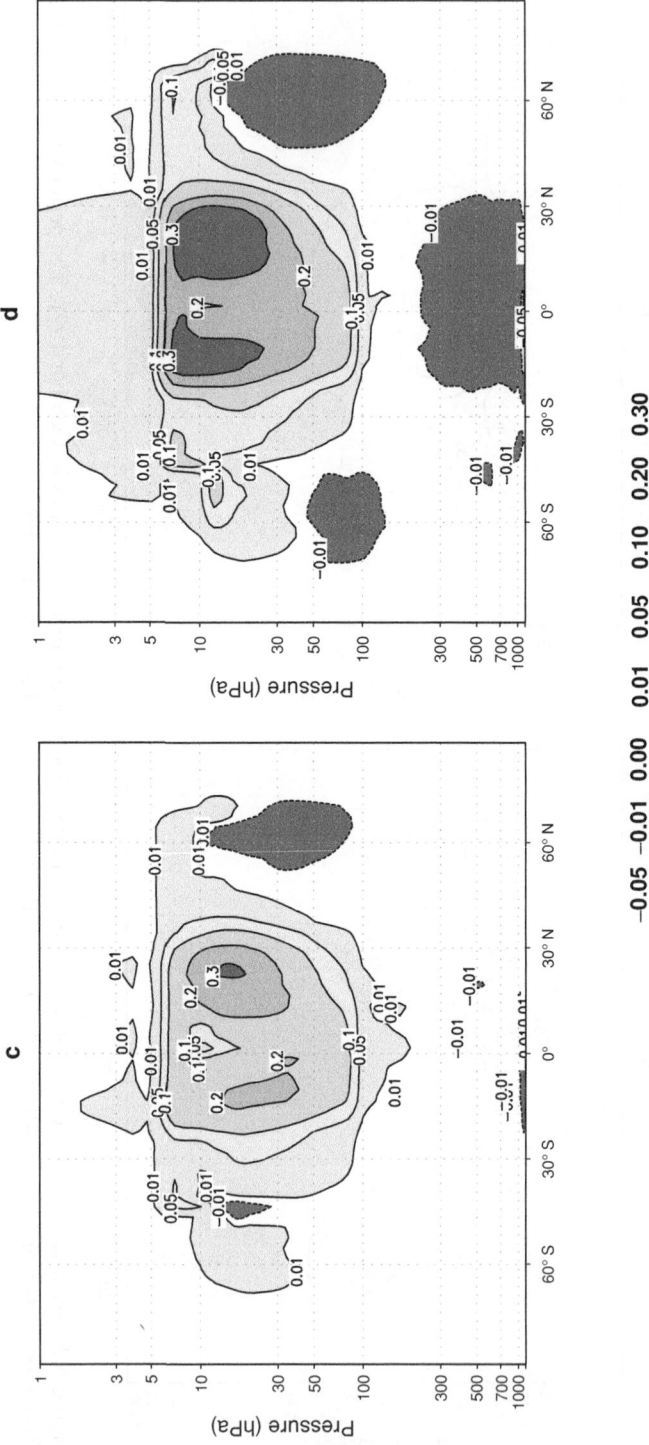

Figure 6.2. *(cont.)*

158

The near-IR absorption by the aerosol leads to an increase in heating throughout the top and middle portions of the aerosol layer; the maximum occurs in the tropics with non-negligible values extending into the mid latitudes. In the tropics, the increased near-IR aerosol absorption causes a reduction in transmission in that spectral region resulting in a reduction in the heating due to water vapor in the lower troposphere.

The longwave-radiative effect is one of a heating due to absorption of the up-welling longwave radiation from the troposphere by the aerosols and re-emission at the lower stratospheric temperatures (Pollack and Ackerman, 1983; WMO, 1990; Stenchikov *et al.*, 1998). This extends throughout the depth of the aerosol location. It is particularly pronounced in the middle layers of the aerosols in the tropics, where there occurs a substantial contrast in the temperature of the upwelling radiation from the warm surface–troposphere system and that of the lower stratosphere; this leads to the aerosol absorption dominating over the emission at the local lower-stratospheric temperatures. The peak heating can be two to three times the near-IR effect and is centered slightly lower in altitude than that in the near-IR.

The total heating effect is one of a warming of the tropical stratosphere from ~ 5 to 100 hPa. The longwave effects are comparable to or outweigh the solar in the low latitudes (30° N–S). The warming extends to the 30–60° degree region, too; the warming here spans a lesser altitude range, and near-IR and longwave make almost comparable contributions. There are small domains of a slight radiative cooling owing mainly to the solar contribution. The total heating perturbation is in striking contrast to that at the top-of-the-atmosphere, which is dominated by a negative solar (especially visible) forcing, with the magnitude of the positive longwave forcing being less than half of the solar forcing. This is a reversal of the roles played out with regards to the radiative heating of the tropical lower stratosphere. The quantitative aspects of both these forcings depend substantially on the distributions of clouds (Ramachandran *et al.*, 2000).

6.6 Stratospheric temperature response

Integrations with the SKYHI GCM were performed for the period June 1991 to May 1993, with fixed but seasonally- and spatially-varying sea-surface tempera-ture climatology. An ensemble of six control (i.e., unforced) and six corresponding Pinatubo aerosol perturbation experiments starting from the same initial conditions were performed. The number of ensemble members is greater than that examined in Ramachandran *et al.* (2000). Each of the six sets of initial conditions are dif-ferent and are arbitrarily taken from different years of a long unforced run of the model. These yield six independent statistical realizations. The difference between each pair constitutes the model's response for that realization. The average of the meteorological fields over the six responses yields the simulated-ensemble mean response. Both the mean values and the standard deviation are compared against observations. Note that, in contrast to the ensemble runs, the real atmosphere has gone through only one potential realization.

We compare the model responses with the NCEP re-analyses (Kalnay *et al.*, 1996). The latter, which we shall refer to as observations in this paper, cover the 30-year period, January 1968 to December 1997. The NCEP temperature anomalies for the June 1991 to May 1993 period were calculated with respect to a 24-year (1968–1997, excluding 1982–1984 and 1991–1993, i.e., the El Chichon and Pinatubo years, respectively) mean (Ramachandran *et al.*, 2000).

6.6.1 Vertical structure of zonal-mean change

The evolution of the ensemble mean model-simulated temperature for selected seasons is compared against observations (Figure 6.3). During the first winter, at ∼ 30–50 hPa, in the 10–30° N–S degree latitude band, the model compares well with the observations albeit with a positive bias, both in terms of the mean value and statistical significance of the estimate. There is a ∼ 1–2 K warming although the model values are more spatially uniform over the 30° N–40° S region than the observations. Poleward of these regions, the significance drops off in both model and observations owing to high inter-annual variability in the middle and high latitudes. Above ∼ 20 hPa, the model-simulated warming and significance is not evident in the observations.

There is a tendency for the model's simulation of statistically significant warming to persist through spring 1992 in the low-latitude lower stratosphere whereas, in the observations, it is less distinct and has shrunk in area relative to the first winter. In the second winter, the warming has diminished in both model and observation throughout the 30–100 hPa region, but there is still a substantial area between ∼ 10 and 50 hPa that continues to be significant in the model. There is a highly significant area appearing in the observation centered at the equator; this is likely not due to the aerosols and is most probably due to the manifestation of the quasi-biennial oscillation (QBO) (Kirchner *et al.*, 1999). By spring 1993, neither the model nor the observation exhibits a statistically significant warming practically anywhere.

The equator-to-pole gradient of the temperature response in the northern hemisphere is quite different between the model and observations. There is a strong polar-cooling anomaly during the first winter in the ensemble-mean model result whereas there is a warming in the observation. During the second winter, there is a cooling anomaly in both model and observations, although stronger in the former.

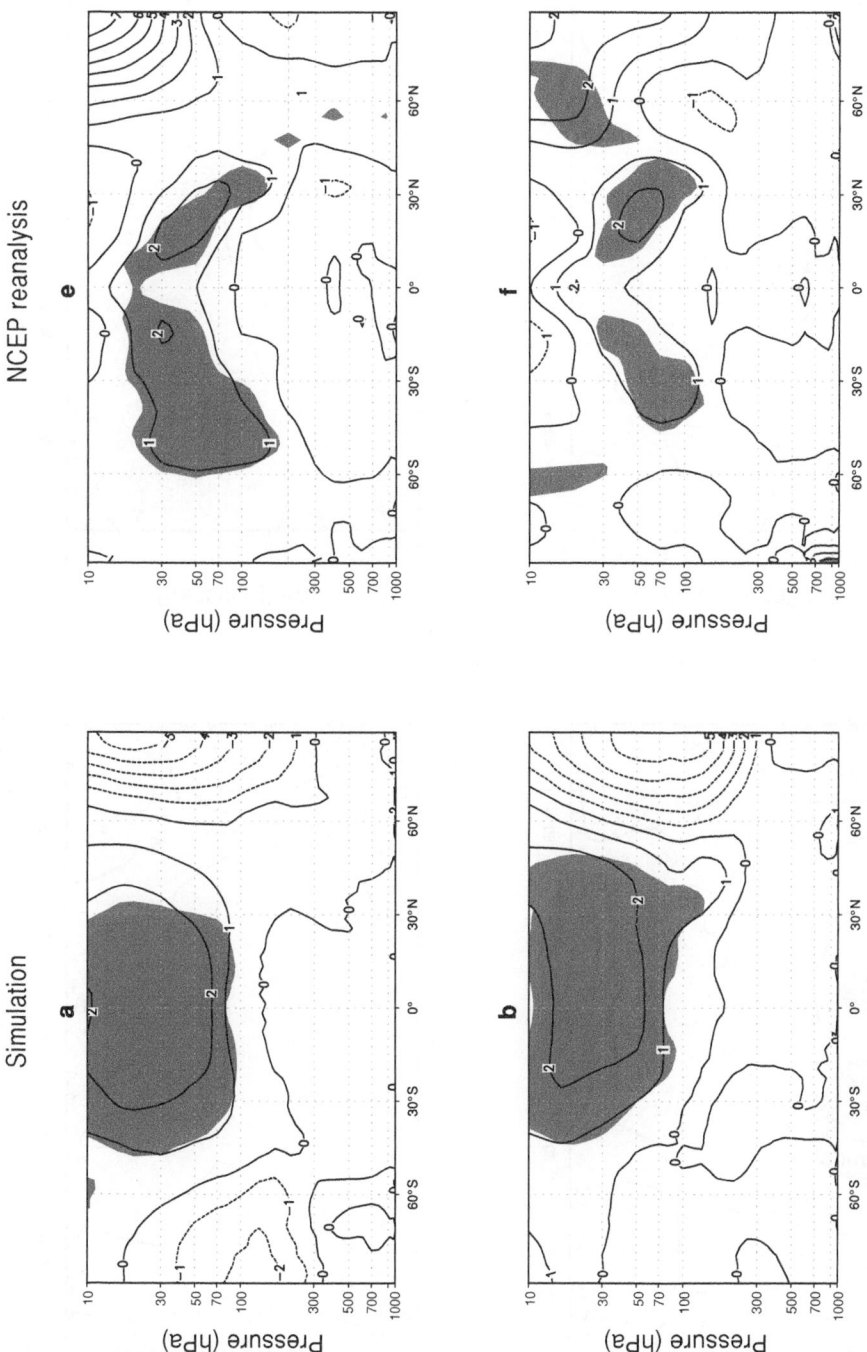

Figure 6.3. Comparison of the simulated and NCEP reanalyses temperature anomalies (K) for the three-month average during northern winter 1991/1992, (a) and (e), spring 1992, (b) and (f), winter 1992/1993, (c) and (g), and spring 1993, (d) and (h), respectively. The model anomalies are means of differences occurring in six ensemble runs (see Section 6.6). Shaded areas denote domains where the anomalies are significant at the 90% (dark) confidence levels.

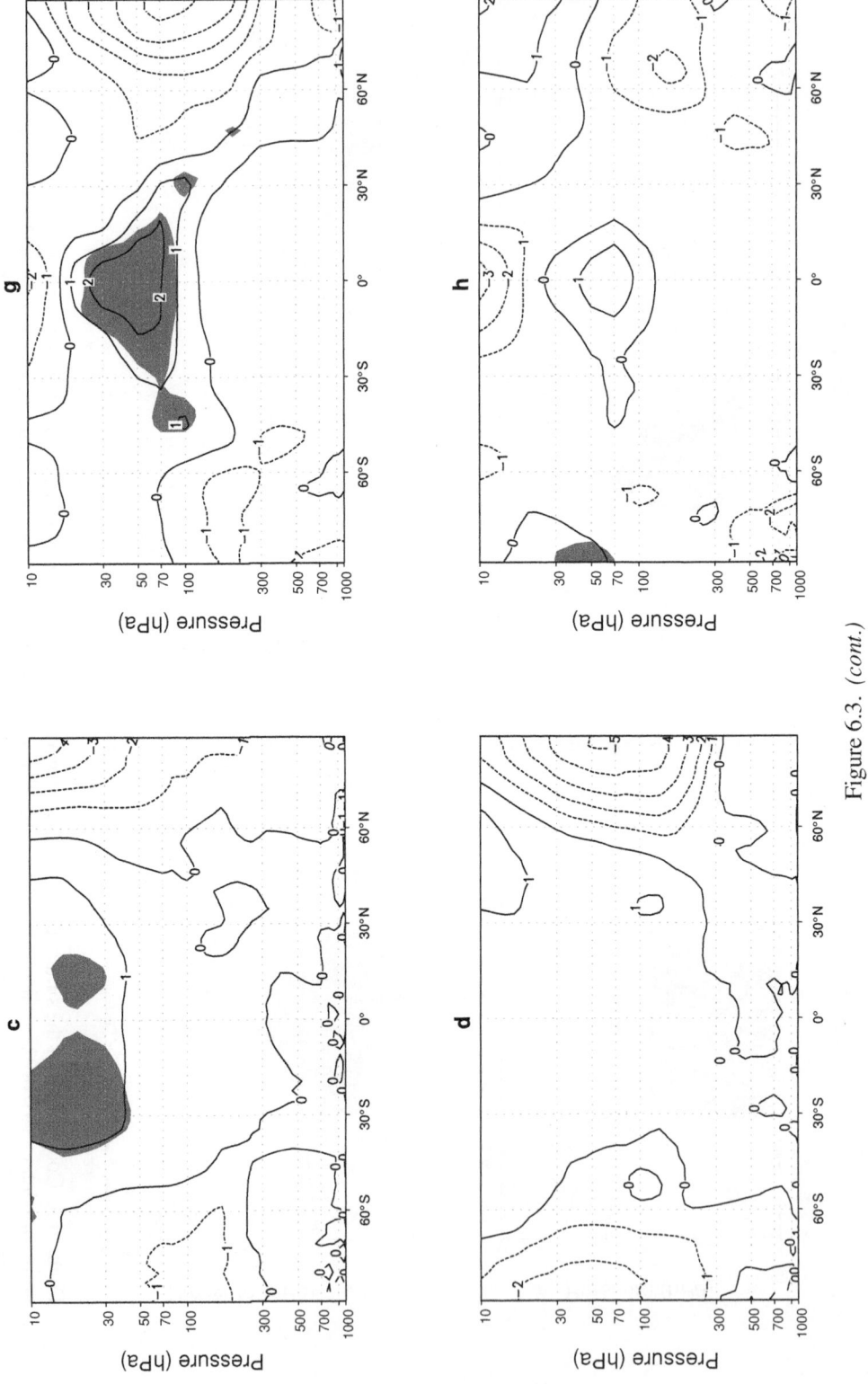

Figure 6.3. (cont.)

162

The features at the polar latitude indicate that it is an extremely difficult proposition to compare the effects between model and observations, especially during winters.

In the observations, the initial warming seen at low latitude has been attributed to Pinatubo aerosols (e.g., IPCC, 1995). The agreement between model and observations in the period of around three to six months after the eruption has been shown in several model studies (e.g., Hansen *et al.*, 1996). However, in a deeper probing of the quantitative aspects, the spatial and temporal structure of the tropical warming differ in some respects from the observation, leading to the notion that other forcings may be occurring in the tropics besides the volcanic aerosols, or else the aerosol input parameters are not quite correct. If there is indeed another mechanism unaccounted for in the model, this could in principle be an external forcing or an internal dynamical variation.

Some evidence that there could be an internal climate variation in the actual atmosphere comes from the fact that a QBO effect is manifest in the observation around the time of the eruption (Kirchner *et al.*, 1999). In addition, there was a dynamical readjustment of ozone concentrations in the tropics as a result of the strong thermal perturbation following the Pinatubo eruption (Ramachandran *et al.*, 2000). Further, as aerosols were formed in the middle- and high-latitude stratosphere, there could have been an ozone depletion in these regions which would yield a radiative forcing. In fact, Kirchner *et al.* (1999) and Stenchikov *et al.* (2002; 2004) show that the inclusion of considerations of ozone changes and QBO-like effects can improve the model simulations relative to observations.

6.6.2 Tropical versus global mean change

We next consider the difference between the ensemble mean model and observation anomalies in the tropics (which are large and can be statistically significant; Figure 6.3), and compare them with those occurring in the global mean in Figure 6.4.

The global mean analysis offers a test of whether the change seen in the observation is purely due to an external radiative forcing. A global mean change in stratospheric temperature at any pressure level has to be principally due to an external radiative forcing. This is because the global mean stratosphere's thermal structure is based on a balance between the solar-radiative heating and longwave cooling. Even if there were a lack of a pure thermal equilibrium at any pressure level, the dynamical motions in the stratosphere caused by some internal rearrangement have to approximately cancel when the global means are evaluated (J. Mahlman, personal communication, 2000). There could be, however, transient dynamical effects that may yield a residual in the global mean, although analyses of long-term GCM integrations indicate that the standard deviation of inter-annual variability in the global mean is no more than ~ 0.2 K (Orris, 1997). While the year-to-year

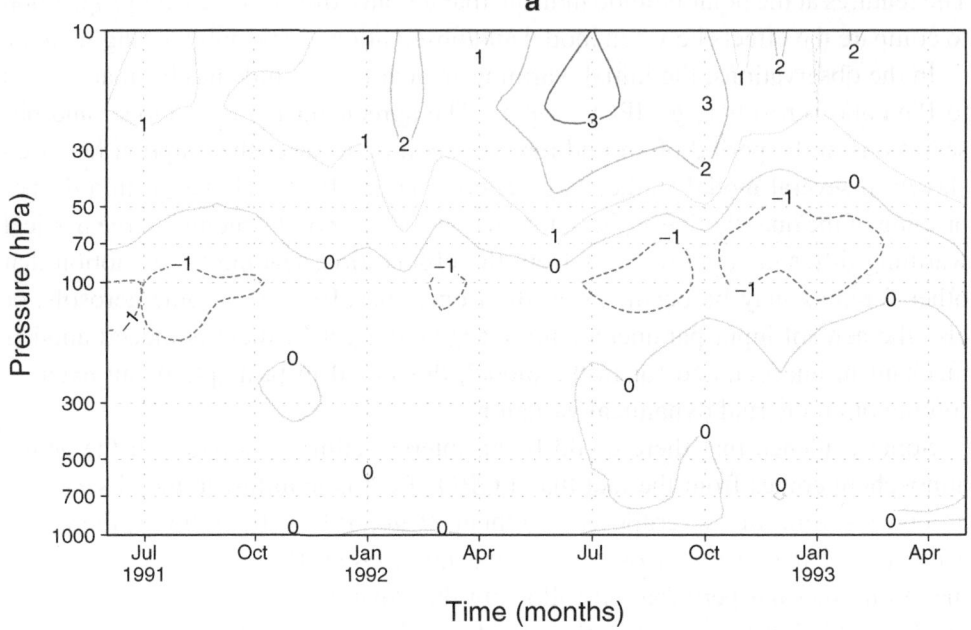

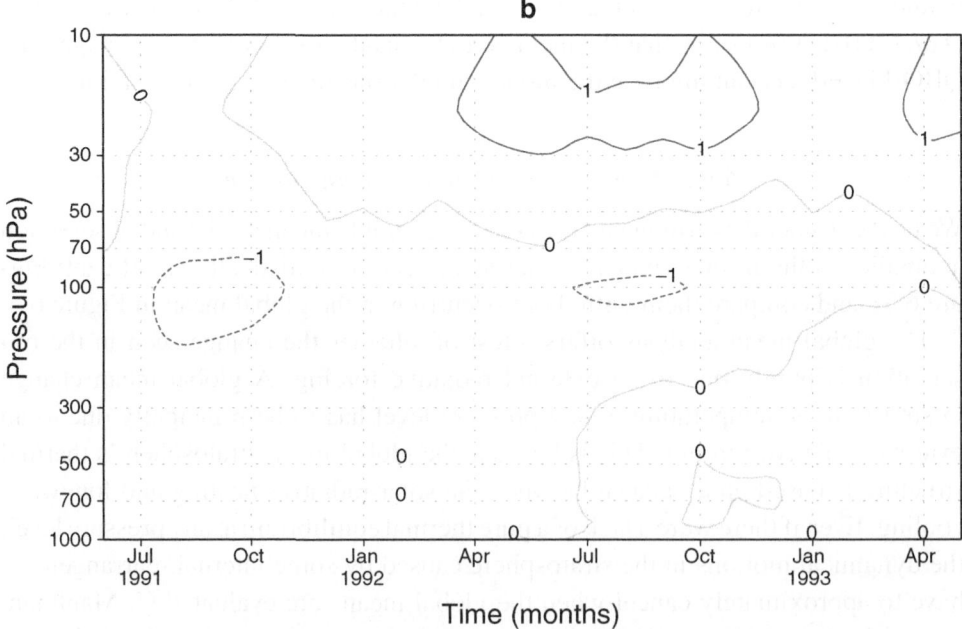

Figure 6.4. Zonal mean difference between simulated temperature anomalies (mean of the differences in the six ensemble runs) and NCEP re-analyses as a function of altitude for (a) the low-latitude 30° N–30° S region and (b) global mean.

variations in the global mean deduced from available satellite observations suggest somewhat larger values, these are still smaller than those obtained for various latitude belts.

Although the evolution of the aerosols, and the large ($\sim$ 30–75 d) radiative time constant of the lower stratosphere, imply that the response would lag the forcing applicable at any instant, a fundamental aspect remains: the global mean response would depend principally on the applied external forcing considering an appropriately long time mean. The same arguments cannot be made about spatial extents less than the global mean and thus any particular domain of the globe (Orris, 1997). More importantly, the statement will certainly not be true for the low latitudes which, as already noted, were influenced substantially by the QBO during the time period considered.

Figure 6.4a shows the 30° N–30° S spatially-averaged ensemble-mean temperature response from the model and compares it with observations. The differences are fairly substantial. The zero value at 70 hPa could be fortuitous or could be due to a true cancelation that is reproduced well by the model. The differences seen in the 30–70 hPa region, especially after winter 1992, are likely due to the QBO which is not simulated by or accounted for in the model.

Figure 6.4b shows the corresponding comparison for the global mean. At 50 hPa, throughout the simulated period, there is almost no difference, confirming that the temperature response here is most likely due to the aerosol external forcing, and which is mimicked well by the model. The fair correspondence (better than $\sim$ 0.5 K) extends over almost all altitudes ($\sim$ 10–50 hPa) until about early spring 1992. Thereafter, above and below 50 hPa, biases up to $\sim$ 1 K are manifest, especially above $\sim$ 30 hPa and at $\sim$ 100 hPa. The temporal gradients and vertical inhomogeneity manifest in the evolution of the tropical lower-stratosphere (30–70 hPa) temperature change are much weaker when the global mean is considered. The 70 hPa zero value and the negative bias below seen for the tropics (Figure 6.4a) is reproduced in the global mean. The differences higher up ($P < 50$ hPa) could be due to too much aerosol-related forcing causing more warming in the simulation, while below ($\sim$ 70–100 hPa) there could be a lesser aerosol-related forcing prescribed than present in the actual global mean atmosphere. The dipole in the bias suggests that the vertical profile of the aerosol profile evolution could be a causal factor; if so, this may require a re-examination of the inferred profile so that a higher accuracy can be attained in the simulations.

6.6.3 Characteristics of the individual member integrations

While the above discussions have been with regards to the ensemble-mean temperature response, it is of interest to inquire into the individual ensemble simulations and examine the degree of similarity between each of them and the actual atmosphere response. Figure 6.5 shows a scatterplot comparing the tropical temperature change at 50 hPa emerging out of each one of the six ensemble runs with the observations. The model has 12 zones encompassing the 30° N–30° S region and the zonal mean for each of these is plotted. Also shown is the comparison for the ensemble-mean temperature change. A time mean is performed for each case, over the first year (Figure 6.5a) and over the second year (Figure 6.5b). For each of the ensemble runs, there tends to be a better correlation with the observed change during the first year than during the second year; in fact, during the second year, there is an almost complete absence of correlation for all the ensemble members as well as the mean. However, even during the first year, when the aerosol forcing is larger, each member has a different degree of correspondence with the observations. This reflects the variability as a result of starting from different initial meteorological conditions. This is somewhat of a surprise for the low latitudes which traditionally are assumed to have substantially less variability relative to, say, the high latitudes. That the tropical stratosphere is not exactly quiescent verifies the inferences of Orris (1997). The variability in the low latitudes is a factor affecting the statistical significance of the temperature anomalies there. In turn, this affects the robustness of the attribution of the low-latitude temperature changes manifest in observations solely to the volcanic aerosols. When considering the ensemble mean, the different latitude belts in the 30° N–30° S exhibit varying degrees of correspondence with the observations.

The variability of the anomalies amongst the ensemble members over the entire two-year period is amplified in Figure 6.6; this figure shows that, during specific months, individual ensemble-member responses can differ by as much as 4 K in the tropics (0–20° S). Even with respect to the ensemble mean, individual ensemble runs can differ by 2 K at specific times. At high latitude (60° N), this feature tends to be even more pronounced, with variations of as much as 20 K during winter and spring, thus guaranteeing the absence of a statistically significant signal and the failure of the aerosol-induced perturbation to register a detectable and attributable signal there (Figure 6.3). Even the deviations from the ensemble mean can be substantial. It is of interest to note that some of the individual peaks or troughs appearing for the low-latitude anomaly evolution tend to be matched approximately by an incursion in the opposite direction at 60° N. Thus, in the global mean, there tends to be an offset such that the pattern of evolution of temperature change for the ensembles resemble one another fairly well. The inter-ensemble differences for any month in the global mean are less than ~ 0.7 K, and the difference of an individual member from the ensemble mean is less than that by almost one-half. The inter-ensemble difference is much less than in the case of the high latitudes, and also less than when the relatively quiescent tropics are considered. Nevertheless, the ensemble

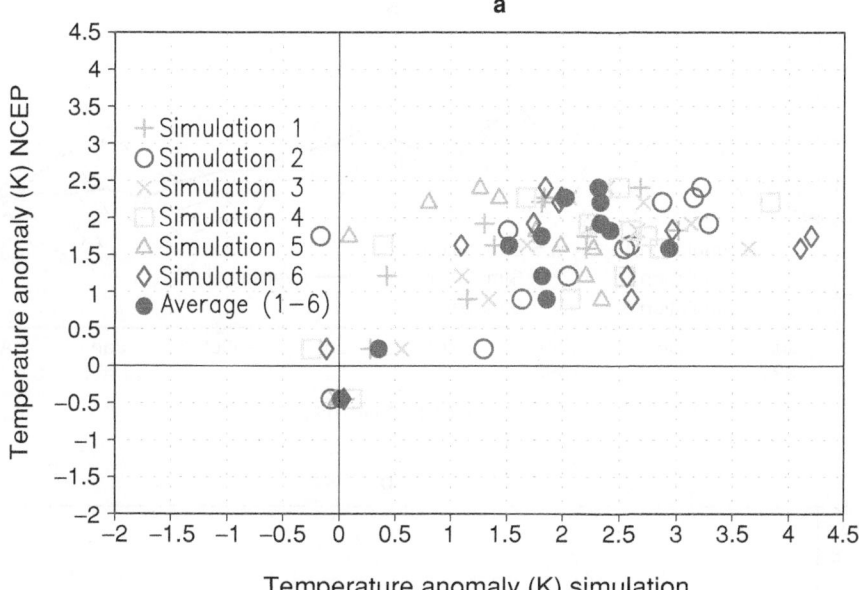

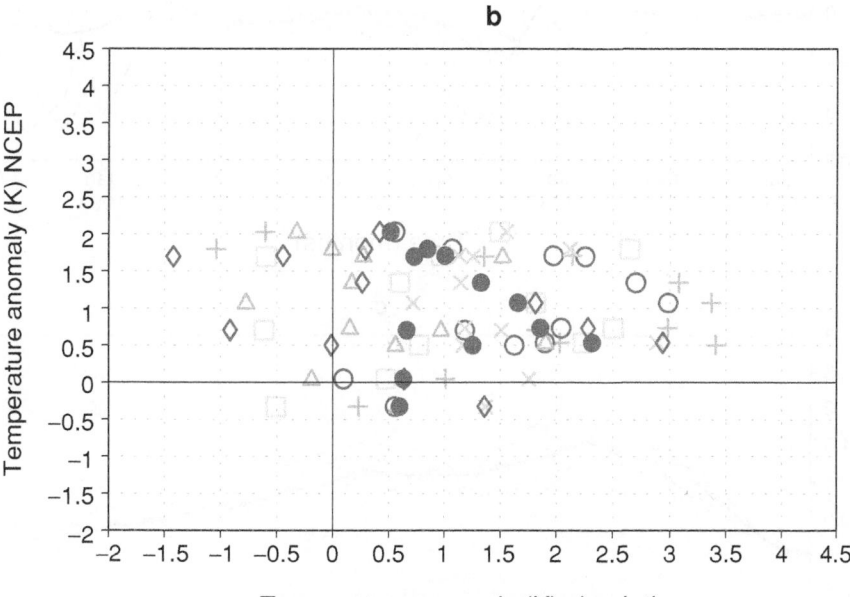

Figure 6.5. Scatterplot of simulated temperature change at 50 hPa over the belt from 30° N to 30° S versus NCEP estimates averaged for (a) the first (June 1991–May 1992) and (b) second (June 1992–May 1993) year after the eruption. Each of the six ensemble members and the ensemble mean is considered for each of the 12 latitude zones comprising the 30° N–30° S zone in the model.
This figure is available for download in colour from
www.cambridge.org/9780521791328

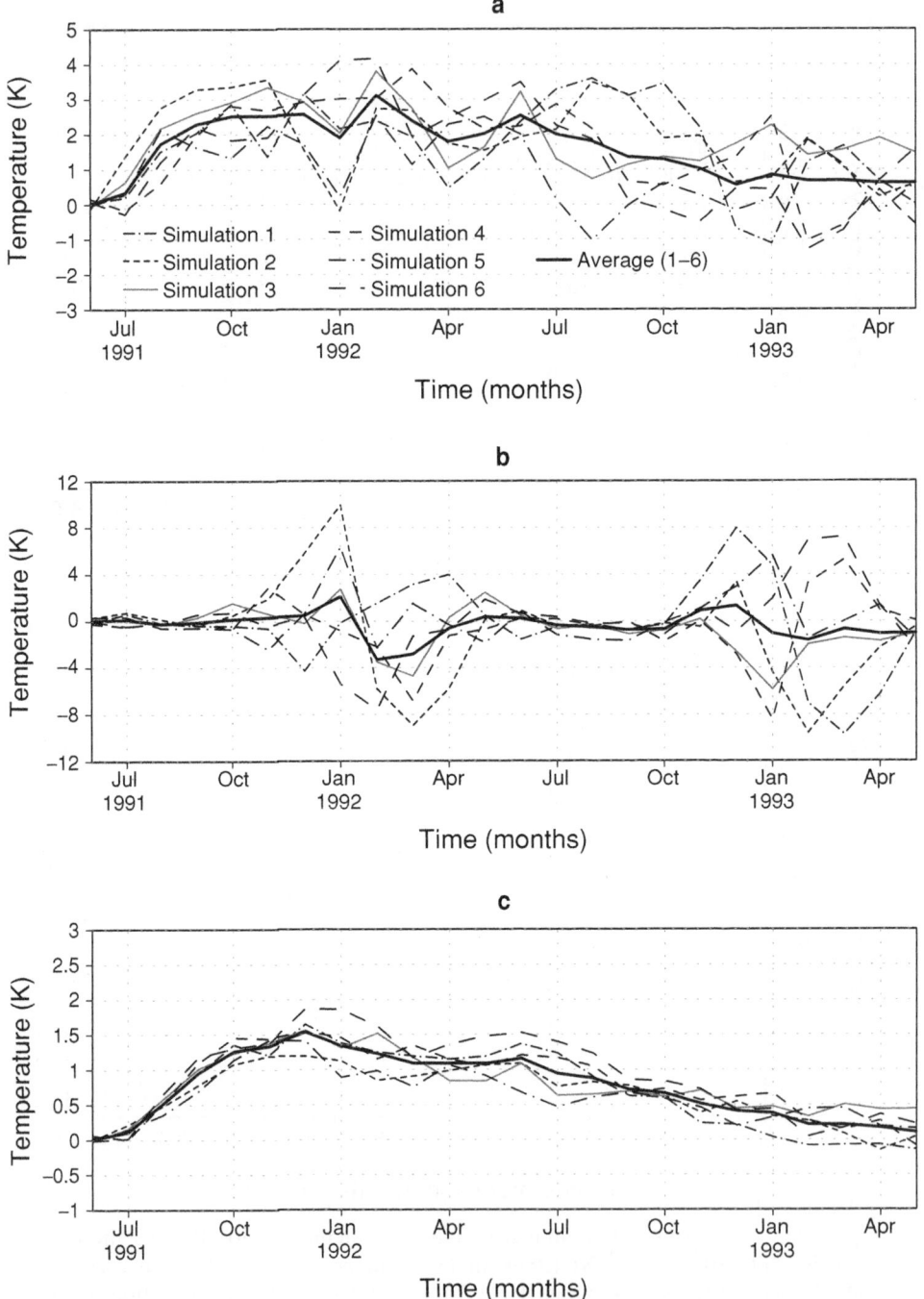

Figure 6.6. Evolution of the temperature anomalies in each of the six GCM ensemble integrations for the 50 hPa layer at (a) 0–20° S, (b) 60° N, and (c) global mean.

members and their mean tend to track the observed global mean pattern fairly well (Ramachandran *et al.*, 2000). This feature of the global mean is consistent with the findings of Orris (1997) whose investigation focused on the inter-annual variability of long-term integrations of a version of the unforced SKYHI GCM, in contrast to the present consideration of a large but transitory forcing.

Finally, we consider another important feature arising due to the warming of the tropical lower stratosphere. This concerns the change in the equator-to-pole temperature gradient in the wake of the Pinatubo eruption. The meridional gradient has a particular significance in climate-change considerations. It has been noted that the warming in the low latitudes due to volcanic aerosols affects the planetary wave propagation which, in turn, perturbs its interactions with the zonal-mean flow and thereby determines how cold the northern-polar stratosphere is during winter and how far it departs from radiative equilibrium conditions during the winter/spring time (Kodera, 1994; Kodera and Yamazaki, 1994). As seen above, the aerosols warm the tropical lower stratosphere and this, by itself, would strengthen the meridional temperature gradient. This would cause planetary waves to refract more equatorward than they would otherwise. This, in turn, would make the flow more zonal and stronger in the polar winter stratosphere, causing colder conditions there and an intensification of the polar vortex. A stronger winter/springtime northern-polar vortex has considerable relevance for the prospects of heterogeneous chemical depletion of ozone. Further, these features can lead to the propagation of the zonal wind anomalies into the troposphere thereby causing changes at lower altitudes as well (Graf *et al.*, 1994; Kodera, 1994; Kirchner *et al.*, 1999).

Figures 6.7a and b show, respectively, the first and second winter season-averaged temperature changes for the northern hemisphere occuring in each of the six model ensemble runs. Each ensemble's perturbed run is differenced from the corresponding control run to obtain the anomalies. It is clear that the stratosphere can evolve quite differently depending on the initial meteorological conditions, with differing equator-to-pole temperature gradients and significant year-to-year variability in the polar temperatures.

Figure 6.7 shows a warming of the lower stratosphere from ~ 0–$40°$ N, with an enhancement of the low-to-mid-latitude temperature gradient in the lower stratosphere in all the runs, more pronounced during the first winter (compare with corresponding panels in Figures 6.7a and b). When considering the equator-to-pole gradient during the first winter, a warming of the tropical- and a perturbation cooling of the polar-stratosphere happens for only four of the six ensemble runs; in runs 2 and 5, the equator-to-pole gradient during the first winter is actually reversed and exhibits a more complicated pattern than in the other runs. The precise structure, in particular the latitude zone where the transition occurs from a warming at low

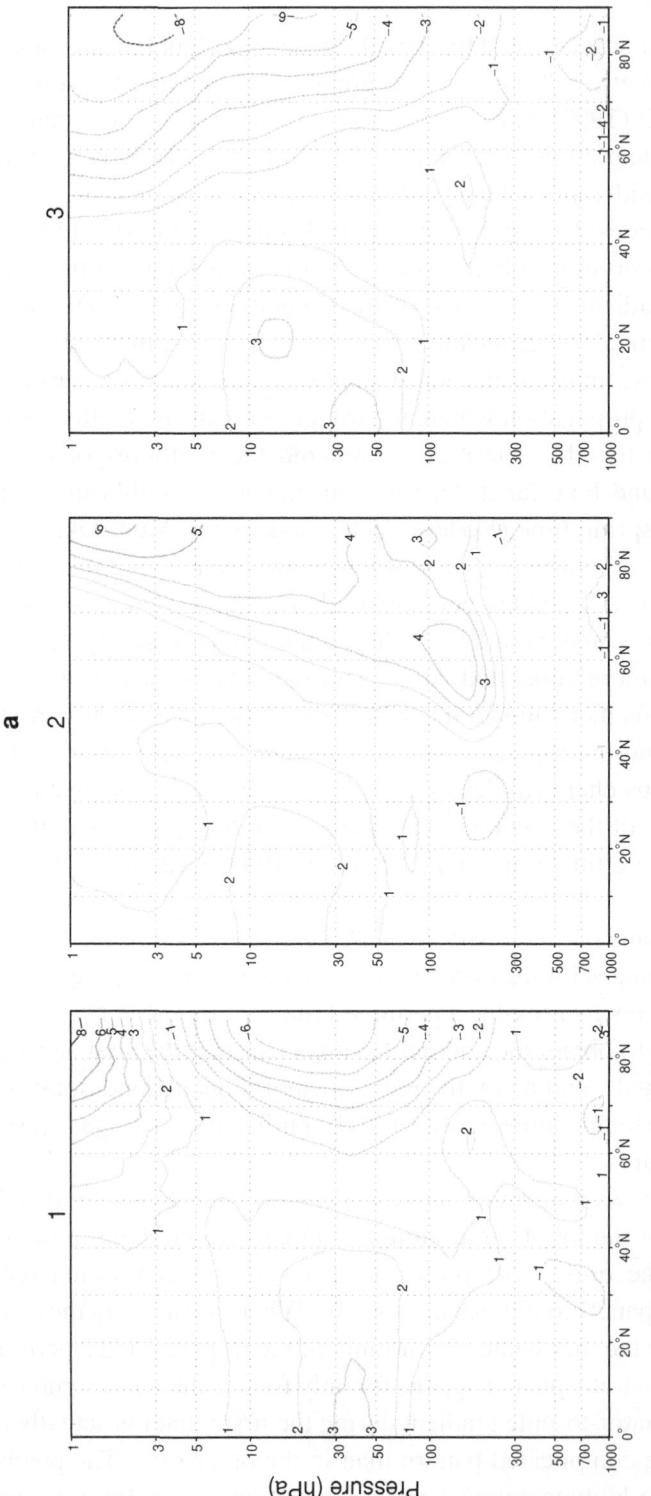

Figure 6.7. Height-latitude temperature anomalies (K) for winter 1991/1992 (a) and 1992/1993 (b) in the northern hemisphere, as arising for each of the six ensemble integrations (1–6).

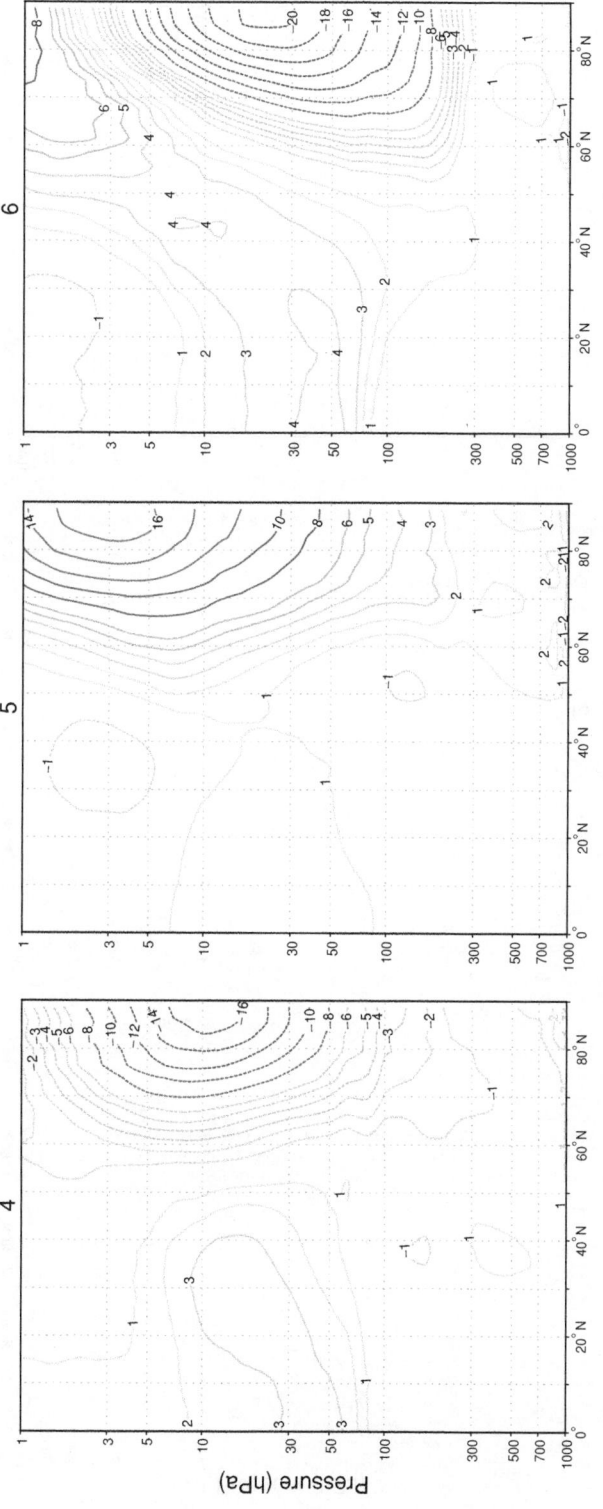

Figure 6.7. *(cont.)*

171

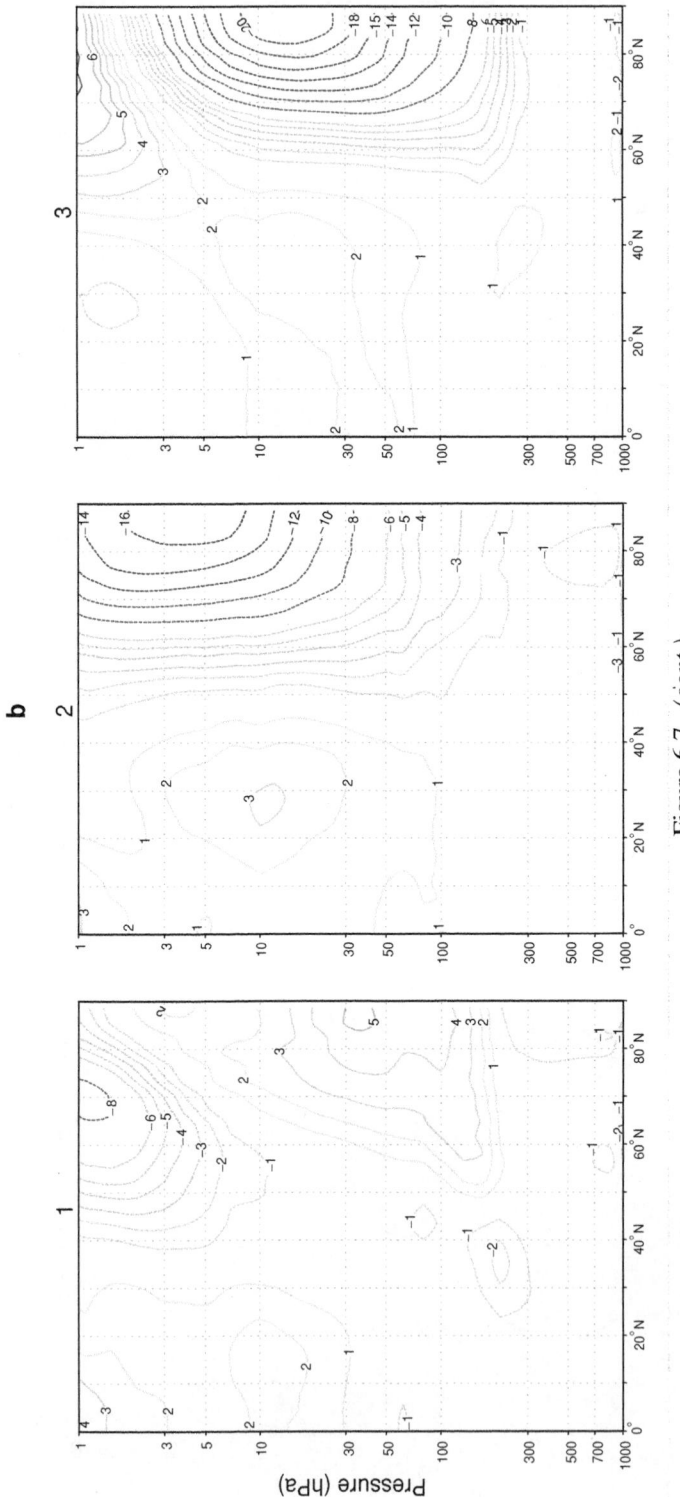

Figure 6.7. (*cont.*)

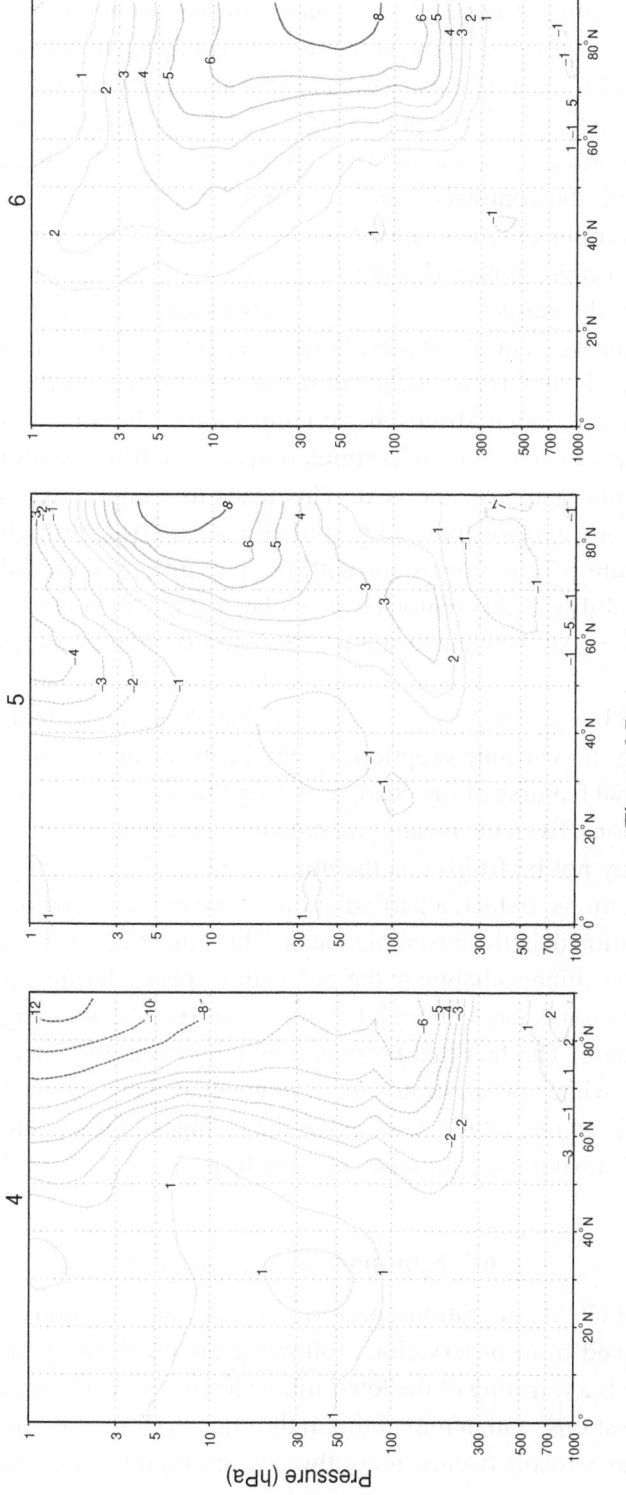

Figure 6.7. (cont.)

latitudes to a cooling at the higher latitudes, in the altitude regime from 100 to 30 hPa, varies considerably amongst the runs. This implies differing characteristics of the dynamical response in the various runs, with accompanying differences in the location and intensity of the polar vortex. Run 6 in particular exhibits very cold temperatures in the polar regions. The intricacy in the polar winter stratosphere is consistent with the contrasts seen in the temperature evolution at 60° N during winter for the various ensemble runs (Figure 6.6).

During the second winter (Figure 6.7b), the low-latitude lower-stratospheric warming due to the aerosols is weaker (see also Figure 6.3). There is still a semblance of an enhancement of the meridional gradient in most of the runs. However, now runs 1, 5, and 6 exhibit a reversal of the gradient while some other runs (2 and 3) exhibit quite cold polar stratospheric temperatures. In general, each ensemble member can vary considerably in its simulation of wintertime conditions in the high latitudes from one winter to the next. This leads to a large inter-annual variability that dwarfs any aerosol-induced perturbations occurring at high latitudes. The features in Figure 6.7 are consistent with the present-day knowledge of the high dynamical variability of this region of the globe, and the attendant difficulties in robust interpretations of the changes taking place at the polar latitudes (Ramaswamy *et al.*, 2001).

Unless there is a detailed knowledge available about the meteorological conditions preceding the volcanic eruption, it may be exceedingly difficult to simulate well the principal features of the changes during the polar winter, as per the present model integrations. Even the means gathered from a very large number of ensemble integrations may not be fruitful as the atmosphere undergoes only one of several potential realizations. In fact, a particular ensemble member could offer a far more realistic simulation than the ensemble mean. Thus, there is considerable ambiguity in discerning the climate change in the polar stratosphere during winter, even if the aerosol perturbation is large enough to cause a substantial warming in the tropical lower stratosphere. The fact that there are ambiguities in the interpretation of the features in the winter polar stratosphere imply that there would be corresponding uncertainties in the details concerning the changes in the polar vortex and the propagation of zonal wind anomalies into the troposphere.

6.7 Summary and discussion

An ensemble of GCM integrations, performed using the evolution of aerosol optical properties derived from observations following the eruption of Mt. Pinatubo, reveals that there is a warming of the low-latitude lower stratosphere that is consistent with the observations, with a maximum in the equatorial region; the peak warming there due to the aerosols occurs about three to six months after the eruption. The warming confirms that our understanding of the radiative mechanisms leading to it are sound, confirming the many earlier studies on major volcanic eruptions. However, in considering the space–time evolution of temperature change, there remain quantitative gaps. At low latitudes, the tracking of the evolution of temperature change in the lower stratosphere is roughly similar to observations only up to about nine months and, considering the region 10–100 hPa, the time span of good agreement (to within about 1 K) is limited to about six months (see also Ramachandran *et al.*, 2000). After that time, the presence of internal dynamical variations in the actual atmosphere and that are not accounted for in the model, cause both a quantitative as well as a qualitative departure from observation. The most likely factor is the quasi-biennial oscillation, followed by dynamically-induced ozone changes in the wake of the volcanic-aerosol perturbation.

In contrast, the global mean evolution is tracked quite well by the model, with the difference from observations being less than 1 K over most of the height (10–100 hPa)–time domain. Since the global mean stratospheric-temperature response is essentially governed by the externally applied radiative forcing, the good agreement signifies that the external forcing and stratospheric sensitivity of the model (viz., radiative response time) are being captured well. However, the slight difference appearing above (positive bias) and below (negative bias) 50 hPa implies that the vertical profile of the aerosol-property representation may need to be improved further to obtain a greater accuracy. An additional external forcing due to changes in global ozone may also need to be taken into account. Some effects due to differences in the transient dynamical effects between the simulations and the real atmosphere could also be a factor. Overall, the multiplicity of observations from diverse platforms has enabled a proper test of the ability of a GCM to capture the evolution of the stratospheric temperature response to a volcanic-aerosol injection. In this test, the global mean version serves to confirm the rudimentary theoretical knowledge that the increase in the solar near-IR and longwave-radiative convergences caused by the aerosols leads to a significant transitory warming, consistent with Pollack and Ackerman (1983), the WMO (1990) and Hansen *et al.* (1996). The global mean simulation also captures well the decay of the aerosol loading as manifest by the evolution of the warming seen in the lower stratosphere.

The ensemble-model runs reveal that it is a particularly challenging task unambiguously to identify the spatial signals due to the volcanic perturbation in the observed record. The initial conditions driving the model are very important in determining the change in the equator-to-pole temperature gradient despite the fact that the tropical lower-stratospheric warming is captured by all of the ensemble members in a broad sense. The vertical profile of change between 30 and 90 hPa also differs quantitatively from one ensemble run to another. The most pronounced differences are seen to occur in the high latitudes during the winter. In some runs, the polar winter temperatures become quite cold whereas in some others there can be a warming, i.e., not only can there be a difference in the magnitude but even the sign could differ. These features are driven by the internal variability of the model and are not directly associated with the volcanic perturbation. Thus, the meteorological conditions prevailing during the aftermath of a volcanic eruption are just as important as, and are perhaps more important than, the severity of the event (and hence aerosol loading) in determining the perturbation at the high latitudes and the equator-to-pole gradient in the lower-stratospheric temperature change. Further, for any ensemble run, one winter can be quite different from the next one, as evidenced for the second winter versus the first in each of the six runs. This winter variability prevailing at the high latitudes presents a difficulty in generalizing the effects at these latitudes. As the changes in meridional gradient have been hypothesized to affect the polar stratospheric temperature and winds, which in turn are believed to lead to a stronger vortex and downward propagation of anomalies into the troposphere (Kodera, 1994), the present GCM integrations indicate that the prevailing meteorological conditions, particularly in the troposphere, are probably as crucial as the aerosol characteristics.

The variations seen amongst the ensemble members, coupled with the fact that all of them have a tropical lower-stratospheric warming, serve a cautionary note in the attribution of the temperature changes observed in the aftermath of an eruption solely to the volcanic aerosols. Even in the global mean, where only the radiative perturbations would be considered to be the most important, there are small discrepancies that suggest the forcing picture, dominated by the aerosols, may not be a complete one. For domains less than the global mean, the interpretation and attribution problems in the case of the Pinatubo eruption become even more difficult, particularly for periods beyond about the first six months following an eruption. This is true even for the tropics, which turn out not to be quiescent places, at least in the present context. Moreover, if internal dynamical variations are occurring there, the probability of an aerosol signal being unambiguously identified beyond the first year when the optical depths diminish in magnitude, may well be impossible.

References

Cess, R. D., J. A. Coakley, and P. M. Kolesnikov (1981). Stratospheric volcanic aerosols: a model study of interactive influences upon solar radiation. *Tellus* **33**, 444–52.

Fels, S. B., J. D. Mahlman, M. D. Schwarzkopf, and R. W. Sinclair (1980). Stratospheric sensitivity to perturbations in ozone and carbon dioxide: radiative and dynamical response. *J. Atmos. Sci.* **37**, 2265–97.

Freidenreich, S. M. and V. Ramaswamy (1999). A new multiple band solar radiative parameterization for GCMs. *J. Geophys. Res.* **104**, 31389–409.

Graf, H.-F., J. Perlwitz, and I. Kirchner (1994). Northern hemisphere tropospheric mid-latitude circulation after violent volcanic eruptions. *Contrib. Atmos. Phys.* **67**, 3–13.

Hamilton, K., R. J. Wilson, J. D. Mahlman, and L. J. Umscheid (1995). Climatology of the GFDL SKYHI troposphere–stratosphere–mesosphere general circulation model. *J. Atmos. Sci.* **52**, 5–43.

Hansen, J., W.-C. Wang, and A. Lacis (1978). Mt. Agung provides test of a global climate perturbation. *Science* **199**, 1065–8.

Hansen, J., M. Sato, R. Ruedy *et al.* (1996). A Pinatubo climate modeling investigation. In *The Mount Pinatubo Eruption Effects on the Atmosphere and Climate*, eds. G. Fiocco, D. Fua, and G. Visconti, NATO ASI Series, Vol. 142, Berlin, Springer-Verlag, pp. 233–72.

Harshvardhan and R. D. Cess (1976). Stratospheric aerosols: effect upon atmospheric temperature and global climate. *Tellus* **28**, 1–10.

IPCC Climate Change (1994). *Radiative Forcing of Climate Change and an Evaluation of the IPCC IS92 Emission Scenarios*, eds. J. T. Houghton, L. G. M. Filho, J. Bruce *et al.*, Cambridge, Cambridge University Press.

Kalnay, E., M. Kanamitsu, R. Kistler *et al.* (1996). The NCEP/ NCAR 40-year reanalysis project. *Bull. Amer. Meteor. Soc.* **77**, 437–71.

Kirchner, I., G. Stenchikov, H.-F. Graf, A. Robock, and J. Antuna (1999). Climate model simulation of winter warming and summer cooling following the 1991 Mount Pinatubo volcanic eruption. *J. Geophys. Res.* **104**, 19039–55.

Kodera, K. (1994). Influence of volcanic eruptions on the troposphere through stratospheric dynamical processes in the northern hemisphere winter. *J. Geophys. Res.* **99**, 1273–82.

Kodera, K. and K. Yamazaki (1994). A possible influence of recent polar stratospheric coolings on the troposphere in the recent northern hemisphere winter. *Geophys. Res. Lett.* **21**, 809–12.

Lacis, A., J. Hansen, and M. Sato (1992). Climate forcing by stratospheric aerosol. *Geophys. Res. Lett.* **19**, 1607–10.

Mahlman, J. D. and L. J. Umscheid (1984). Dynamics of the middle atmosphere: successes and problems of the GFDL SKYHI general circulation model. In *Dynamics of the Middle Atmosphere*, eds., J. R. Holton and T. Matsuno, Tokyo, Terra Scientific, pp. 501–25.

Mahlman, J. D., J. P. Pinto, and L. J. Umscheid (1994). Transport, radiative, and dynamical effects of the antarctic ozone hole: a GFDL "SKYHI" model experiment. *J. Atmos. Sci.* **51**, 489–508.

Orris, R. (1997). Ozone and temperature: a test of the consistency of models and observations in the middle atmosphere. Ph.D. thesis, Princeton University, NJ.

Pollack, J. and T. Ackerman (1983). Possible effects of the El Chichon cloud on the radiation budget of the northern tropics. *Geophys. Res. Lett.* **10**, 1057–60.

Ramachandran, S., V. Ramaswamy, G. Stenchikov, and A. Robock (2000). Radiative impact of the Mt. Pinatubo volcanic eruption: lower stratospheric response. *J. Geophys. Res.* **105**, 24409–29.

Ramaswamy, V. and M. M. Bowen (1994). Effect of changes in radiatively active species upon the lower stratospheric temperatures. *J. Geophys. Res.* **99**, 18909–21.

Ramaswamy, V., M. D. Schwarzkopf, and W. J. Randel (1996). Fingerprint of ozone depletion in the spatial and temporal pattern of recent lower-stratospheric cooling. *Nature* **382**, 616–18.

Ramaswamy, V., M.-L. Chanin, J. Angell *et al.* (2001). Stratospheric temperature trends: observations and model simulations. *Rev. of Geophys.* **39**, 71–122.

Robock, A. and J. Mao (1992). Winter warming from large volcanic eruptions. *Geophys. Res. Lett.* **12**, 2405–8.

Robock, A., K. Taylor, G. Stenchikov, and Y. Liu (1995). GCM evaluation of a mechanism for El Niño triggering by the El Chichon ash cloud. *Geophys. Res. Lett.* **22**, 2369–72.

Schwarzkopf, M. D. and V. Ramaswamy (1999). Radiative effects of CH_4, N_2O, halocarbons and the foreign-broadened H_2O continuum: a GCM experiment. *J. Geophys. Res.* **104**, 9467–88.

Slingo, A. (1989). A GCM parameterization of the shortwave radiative properties of water clouds. *J. Atmos. Sci.* **46**, 1419–27.

Stenchikov, G., K. Hamilton, A. Robock, V. Ramaswamy, and M. D. Schwartzkopf (2004). Arctic oscillation response to the 1991 Pinatubo eruption in the SKYHI general circulation model with a realistic quasi-biennial oscillation. *J. Geophys. Res.* **109**, D03112.

Stenchikov, G., K. Hamilton, A. Robock *et al.* (2002). Arctic oscillation response to the 1991 Mount Pinatubo eruption: effects of volcanic aerosols and ozone depletion. *J. Geophys. Res.* **107**, 4803.

Stenchikov, G., I. Kirchner, A. Robock *et al.* (1998). Radiative forcing from the 1991 Mount Pinatubo volcanic eruption. *J. Geophys. Res.* **103**, 13837–57.

Wetherald, R. T. and S. Manabe (1988). Cloud feedback processes in a general circulation model. *J. Atmos. Sci.* **45**, 1397–1415.

WMO (1990). *Report of the International Ozone Trends Panel: 1988, Global Ozone Research and Monitoring Project.* Geneva, International Ozone Trends Panel, WMO, Report No. 18, ch. 6.

7

Unresolved issues in atmospheric solar absorption

WILLIAM D. COLLINS

National Center for Atmospheric Research, Boulder, CO

7.1 Introduction

7.1.1 The importance of accurate estimates of solar absorption

The absorption of solar radiation by Earth's atmosphere and surface is the fundamental forcing of the climate system (Chapter 1). The net solar energy absorbed by the Earth–atmosphere system can be accurately quantified with satellite measurements of top-of-the-atmosphere (TOA) insolation and planetary albedo. These observations have determined that the net global annually-average solar radiation absorbed by the Earth is 239 W m^{-2} (Harrison *et al.*, 1990; Collins *et al.*, 1994) with an uncertainty of approximately 2 W m^{-2}. This absorbed solar radiation can be written as the sum of the solar radiation absorbed in the atmosphere and by the planetary surface. Despite numerous investigations over the last 40 years, the partitioning of the absorption between the surface and atmosphere is still very uncertain (for reviews, see Stephens and Tsay, 1991 and Ramanathan and Vogelmann, 1997). The discrepancy between estimates from models and observational studies is known as anomalous or enhanced shortwave absorption. The enhanced absorption is usually associated with clouds, although some investigators find evidence for "anomalous" absorption in clear-sky regions (Li *et al.*, 1995; Li and Moreau, 1996; Arking, 1996; Kato *et al.*, 1997).

The issues of the existence and magnitude of enhanced solar absorption has been examined intensively since Ramanathan *et al.* (1995), Cess *et al.* (1995), and Pilewskie and Valero (1995) found evidence for the anomaly in the tropical Pacific. Cess *et al.* (1995) also found evidence for enhanced absorption at Barrow, Alaska; Boulder, Colorado; Cape Grim, Tasmania; and Wisconsin. Including the original papers (Cess *et al.*, 1995; Ramanathan *et al.*, 1995), Cess has authored or

Frontiers of Climate Modeling, eds. J. T. Kiehl and V. Ramanathan.
Published by Cambridge University Press. © Cambridge University Press 2006.

co-authored 10 studies on this subject from 1995 through 1999. These include the paper which motivated this study (Kiehl *et al.*, 1995) and a number of observational analyses (Cess and Zhang, 1996; Cess *et al.*, 1996; Zhang *et al.*, 1997; Jing and Cess, 1998; Yu *et al.*, 1999). Dr. Cess was one of the principal investigators in the field experiment conducted by the Atmospheric Radiation Measurement (ARM) program called the ARM Enhanced Shortwave Experiment (ARESE). His analysis of measurements from stacked aircraft above and below cloud fields observed during ARESE supported the conclusions of these earlier studies (Valero *et al.*, 1997; Cess *et al.*, 1999). The basic findings from Dr. Cess's work are that absorption of shortwave radiation in cloudy atmospheres is much larger than models predict, while models of clear-sky radiative transfer are in excellent agreement with observations.

There are several reasons why this field has generated so much research activity and controversy. A number of independent analyses suggest that the difference between the modeled and observed global annually averaged shortwave absorption is roughly 20–25 W m^{-2} (Cess *et al.*, 1995; Cess *et al.*, 1996). If these estimates are correct, then the bias in absorbed radiation is the largest error by far in global energy budgets calculated from general circulation models (GCMs). It is still not clear whether the anomalies are due to errors in the basic physics underlying the models, errors introduced by approximations in the models, underestimation of absorption by known atmospheric constituents (e.g., water vapor), or the omission of radiatively important constituents. There is also no known physical mechanism that can explain the magnitude and apparent ubiquity of the enhancement, although absorption by nitrogen compounds produced in convective storms may help explain some of the signal (Solomon *et al.*, 1999). The optical properties of clouds computed from Mie theory cannot explain the size of the anomaly unless cloud particles are much larger than observed (Wiscombe *et al.*, 1984). Estimates of cloud properties from remote sensing seem to support the conventional theory for cloud optics (King *et al.*, 1990). The estimation of absorption in cloudy regions from surface or aircraft data is notoriously difficult because of complex spatial and temporal variations in cloud geometry (Fouquart *et al.*, 1990). In some cases, multiple investigators have analyzed the same datasets and yet have reached diametrically opposing conclusions (e.g., Li *et al.*, 1999; Valero *et al.*, 2000).

The principle unresolved issues regarding enhanced shortwave absorption are the magnitude of the effect, the physical mechanisms responsible for the enhancement beyond conventional radiative-transfer theory, and the effects of the enhancement on our understanding of the climate system. In this chapter we will discuss the impact of enhanced absorption on simulations of the tropical Pacific. There are several reasons for focusing on the tropical Pacific. First, this region acts as a "flywheel" for climate since the energy exported from the Pacific basin to higher latitudes is a significant fraction of global meridional energy transport (Webster, 1994). Second, the

Table 7.1. *Surface and TOA cloud forcing for the tropical Western Pacific*

Level	Time period	Forcing (W m^{-2})
TOA	Climatological[a]	−66[b]
Surface	Climatological[a]	−100[b]
	COARE[c]	−102[d]
	COARE[c]	−99[e]

[a] Climatological estimates are annual averages over ocean points within 10° N to 10° S, 140° E to 170° E. [b] Ramanathan *et al.* (1995). [c] The COARE estimates are averages for November, 1992–February, 1993 at selected sites within the intensive flux array. [d] Waliser *et al.* (1996). [e] Chou and Zhao (1997).

inter-annual oscillations of the coupled ocean–atmosphere system known as El Niño and La Niña alter weather systems and precipitation patterns worldwide (Philander, 1990). Therefore significant changes in the energy budget of this region will affect global climate. Third, the tropical western Pacific (TWP) is a site of frequent deep convection, and this convection generates extensive cirrus-cloud decks (Webster, 1994). In addition, the TOA insolation is maximized over equatorial regions. The combination of peak insolation and persistent cloud cover makes the TWP an ideal area for studying shortwave absorption in clouds. Finally, recent satellite programs and field experiments have provided the datasets necessary to characterize the TOA and surface energy budgets of the region. The most extensive data are from the Tropical Ocean Global Atmosphere (TOGA) Coupled Ocean-Atmosphere Response Experiment (COARE) (Webster and Lukas, 1992). These observations can be used to evaluate the fidelity of simulations of the TWP from climate models.

The magnitude of the solar absorption in the TWP atmosphere can be readily determined from these observations. The difference between all-sky and clear-sky fluxes is known as shortwave cloud forcing (e.g., Ramanathan *et al.*, 1989). Two sets of estimates for the TOA and surface shortwave cloud forcing are shown in Table 7.1. The cloud forcing has been estimated for climatological conditions and for the TOGA COARE observing period. The climatological and TOGA values are consistent with each other and show that the surface shortwave forcing is approximately 50% larger than the TOA forcing. Conventional models of radiative transfer yield a 10% difference between the TOA and surface with peak values no larger than 20% (Lubin *et al.*, 1996). The difference between the surface and TOA forcing gives the amount of additional shortwave absorption in the atmosphere introduced by clouds. From the observations summarized in Table 7.1, the additional absorption is between 30 and 35 W m^{-2}. The models for the TWP indicate that the additional absorption in clouds should be less than 13 W m^{-2} and probably closer to 6 W m^{-2}. To put this discrepancy in context, one of the goals of TOGA COARE was to estimate each term in the time-averaged surface energy budget to an accuracy of 10 W m^{-2}. It is clear from Table 7.1 that the various observational estimates are consistent to within this experimental tolerance. However, the discrepancy between the modeled and observed all-sky surface fluxes and atmospheric absorption are roughly an order of magnitude larger than the differences between the observational values.

7.1.2 Implications for modeling of the climate

Estimates of the global net surface insolation from GCMs are larger than observational estimates by 20–42 W m^{-2} (Garratt, 1994; Wild *et al.*, 1995; Zhang *et al.*, 1998). These biases occur despite the fact that GCMs are generally tuned to reproduce the zonal-mean annually-averaged TOA shortwave fluxes measured by satellite. If these differences could be reduced or eliminated while maintaining agreement with satellite measurements of planetary albedo, the effects on the atmospheric and oceanic circulation would be substantial (Kiehl, 1994). Perhaps the most important consequence is that the additional absorption would alter meridional heat transport in the atmosphere and oceans. Enhancing shortwave absorption in the tropical atmosphere by 25 W m^{-2} would increase the meridional transport of moist static energy by roughly 50% (Kiehl, 1994). Because the sum of meridional transport of heat in the ocean and atmosphere is constrained by the net radiative flux at TOA, the meridional ocean transports would decrease by an equal amount.

In addition, the surface-heat balance of the tropical oceans is dominated by the terms for insolation and latent heat flux (Monin, 1986). If the insolation is actually 25 W m^{-2} lower than GCM estimates, the latent heat fluxes would have to decrease by a similar amount to maintain surface-energy balance. The reduction of evaporation by this amount would have profound effects on the mean state of the tropical troposphere (Kiehl *et al.*, 1995). The decrease would presumably reduce the convectively available potential energy, and would therefore tend to decelerate the Walker circulation and hydrological cycle.

The bias in GCM estimates of insolation could conceivably be caused by errors in the representation of cloud properties, the omission of radiatively important atmospheric constituents, or the approximations used to parameterize radiative transfer. A portion of the bias may be related to neglecting absorptive aerosols in most GCMs (Cusack *et al.*, 1999; Wild and Ohmura, 1999). Introduction of absorptive aerosols into GCMs increases the atmospheric absorption by approximately 5 W m^{-2} (Cusack *et al.*, 1999). The corresponding reduction in global, annual mean absorbed radiation at Earth's surface is 7 W m^{-2}. At least in these simulations, the effects of aerosols are not sufficient to explain the large differences in surface insolation between GCMs and observations. The results are not definitive since the simulated absorption is sensitive to the vertical profile and parameterization of hygroscopic growth for the aerosols.

Part of the bias may also be related to the uncertainties in the optical properties of ice clouds. These uncertainties are introduced by limitations on *in situ* measurements of cirrus optical properties and by the variability of cirrus-particle geometries. However, model experiments to test the sensitivity to optical properties of ice clouds increased absorption by only 2.9 W m^{-2} (Ho *et al.*, 1998). The sensitivity of the anomaly to the simulation of cloud physical and microphysical properties and to the parameterization of radiative transfer have been examined in Collins (1998). In this study, the underestimation of solar absorption is related to differences in the spectral partitioning of reflected solar radiation derived from models and satellite observations. The differences are unchanged when the GCM cloud distribution is replaced with cloud distributions retrieved by the International Satellite Cloud Climatology Project (ISCCP). The differences are also insensitive to more exact treatments of cloud–radiative interactions (Collins, 1998). In summary, the differences between insolation values derived from measurements and GCMs are still unexplained and are dominated by large biases under cloudy conditions.

The effects of enhancing shortwave absorption in cloudy atmospheres have been tested in uncoupled atmospheric GCMs (AGCMs). Kiehl *et al.* (1995) increase cloud absorption in the NCAR Community Climate Model 2 (CCM2) by introducing ad hoc changes to the single-scattering albedo of cloud particles. They emphasize that their experiment is a sensitivity study and that their modifications to the cloud optics are not a physically based, unique solution for increasing absorption. The enhanced absorption tends to stabilize the tropical convective atmosphere by heating the column in cloudy regions and causes a 3–4 K warming of the upper tropical troposphere. The increased stability reduces convective activity in the TWP and results in a weaker Walker circulation.

The primary objective of this study is to examine the effects of enhanced absorption on a coupled simulation of the climate. The integrations are performed using the NCAR Climate System Model (CSM), a suite of models for the atmosphere, ocean, sea ice, and land surfaces which interact through a flux coupler (Boville and Gent, 1998). The simulation of the TWP with the standard CSM has several features that differ significantly from the observed climate state. These differences may be related to the omission of enhanced shortwave absorption from the model physics (Kiehl, 1998). The other objective of this study is to test whether the biases are reduced or eliminated when solar absorption is increased in cloudy regions. It should be noted that the increased absorption used here is not consistent with current radiative-transfer theory applied to standard models of atmospheric state and composition.

The dominant errors in the CSM simulation of the TWP are the overestimation of the net surface-heat budget, latent-heat flux, and insolation; underestimation of SST in the mean state; and a rapid transient reduction in sea-surface temperature (SST) at the beginning of the coupled integration. The SSTs in the central Pacific are 2 K colder than standard SST climatologies (Shea *et al.*, 1992) (Figure 7.1). Kiehl (1998) suggests that the biases in surface insolation together with dynamical constraints on the TWP surface-heat budget require an overestimation of latent-heat fluxes to satisfy these constraints. The TOA and surface-energy budgets in the uncoupled CCM and fully coupled CSM simulations are compared against observations in Table 7.2. In the uncoupled simulation, the term with the largest differences relative to observations is the surface insolation, and it is overestimated by 52 W m^{-2}. The net surface-heat budget is 26 W m^{-2} larger than observed. In the coupled simulation, all the fluxes are quite close to the values from the uncoupled model except for the latent-heat flux, which has increased by 19 W m^{-2}. The increase in latent heat flux is achieved by accelerating surface easterlies (Figure 7.2), which leads to greater Ekman pumping and a reduction of modeled SSTs in the central Pacific to values 2 K below observed. The introduction of enhanced shortwave absorption would lower the surface insolation to near the observed values and thus presumably reduce the dynamical response of the model to solar heating of the ocean boundary layer.

In the CSM simulations discussed here, the solar absorption is enhanced through an ad hoc adjustment of shortwave diabatic heating called "generic enhanced absorption" (GEA). The CSM experiments should be interpreted as sensitivity studies rather than as definitive simulations of the climate with anomalous absorption. It is certainly plausible that other methods for introducing enhanced absorption may give qualitatively different results. Although the results suggest that the mean climate state is significantly improved by the increased solar absorption, this is not sufficient justification for introducing GEA into standard community models. Current and future versions of the NCAR CSM will include standard radiative parameterizations until specific physical mechanisms for enhanced absorption (if any) can be demonstrated and justified on theoretical and experiment grounds.

The GEA method, model configuration, and configuration of the model integration are described in Section 7.2. The response of the uncoupled atmospheric model to the introduction of GEA is examined in Section 7.3. The surface fields from the uncoupled atmospheric model are used as boundary conditions for an uncoupled integration of the ocean model. The ocean model is run to equilibrium with the surface forcing before coupling the ocean and atmosphere (Boville and Gent,

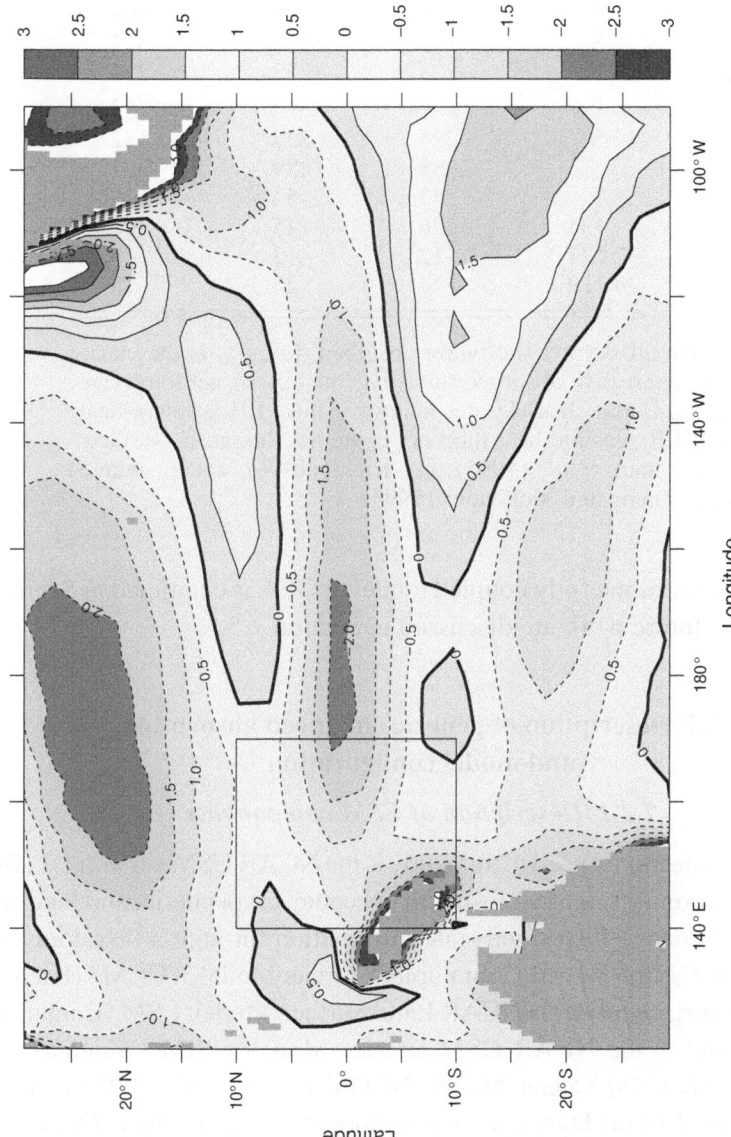

Figure 7.1. Difference in annual mean SST between the CSM (years 11–60) and the observational estimates (Shea *et al.*, 1992). The box indicates the tropical Pacific warm pool (10° N to 10° S, 140° E to 170° E). From Kiehl (1998) with permission.

This figure is available for download in colour from
www.cambridge.org/9780521791328

185

Table 7.2. *Warm-pool energy budgets (10° N to 10° S,*
140° E to 170° E)

Level	Flux[a]	CCM3	CSM	Observed[b]
TOA	S	318	319	309
	S_{clr}	373	374	373
	F	−223	−233	−225
	F_{clr}	−288	−289	−285
Surface	S	234	237	182
	S_{clr}	288	289	282
	F	−52	−54	−49
	LH	−126	−145	−107[c]
	SH	−12	−13	−8[c]
	Net	44	25	18

[a] S is the all-sky net shortwave absorbed flux; S_{clr} is the clear-sky net shortwave absorbed flux; F is the all-sky net longwave flux; F_{clr} is the clear-sky net longwave flux; LH is latent-heat flux; SH is sensible-heat flux; net is the net flux at the surface.
[b] Ramanathan *et al.* (1995) and Kiehl (1998), except where noted. [c] Zhang and McPhaden (1995).

1998). The behavior of the fully coupled model to GEA is examined in Section 7.4. Conclusions and future work are discussed in Section 7.5.

7.2 Description of generic enhanced absorption and model configuration

7.2.1 Description of CSM components

The coupled climate model used in this study is the NCAR CSM version 1.1 (Boville and Gent, 1998), comprised of atmospheric, oceanic, cryospheric, and land surface models which exchange fluxes and state information through a flux coupler. The atmospheric model is the NCAR Community Climate Model 3, CCM3 (Kiehl *et al.*, 1998), the land-surface model is NCAR Land Surface Model, LSM (Bonan, 1998), and the ice model is the NCAR CSM Sea-Ice Model, CSIM (Weatherly *et al.*, 1998). The NCAR CSM Ocean Model, NCOM (Gent *et al.*, 1998) is a variant of the Geophysical Fluid Dynamics Laboratory (GFDL) Modular Ocean Model (MOM) code with modifications for parameterizing mesoscale eddies (Gent *et al.*, 1995). The flux coupler is described in Bryan *et al.* (1996). The CCM and LSM are run at T42 spectral truncation with 18 levels in the vertical for the atmosphere. The NCOM and CSIM are run at 2° × 2° horizontal resolution.

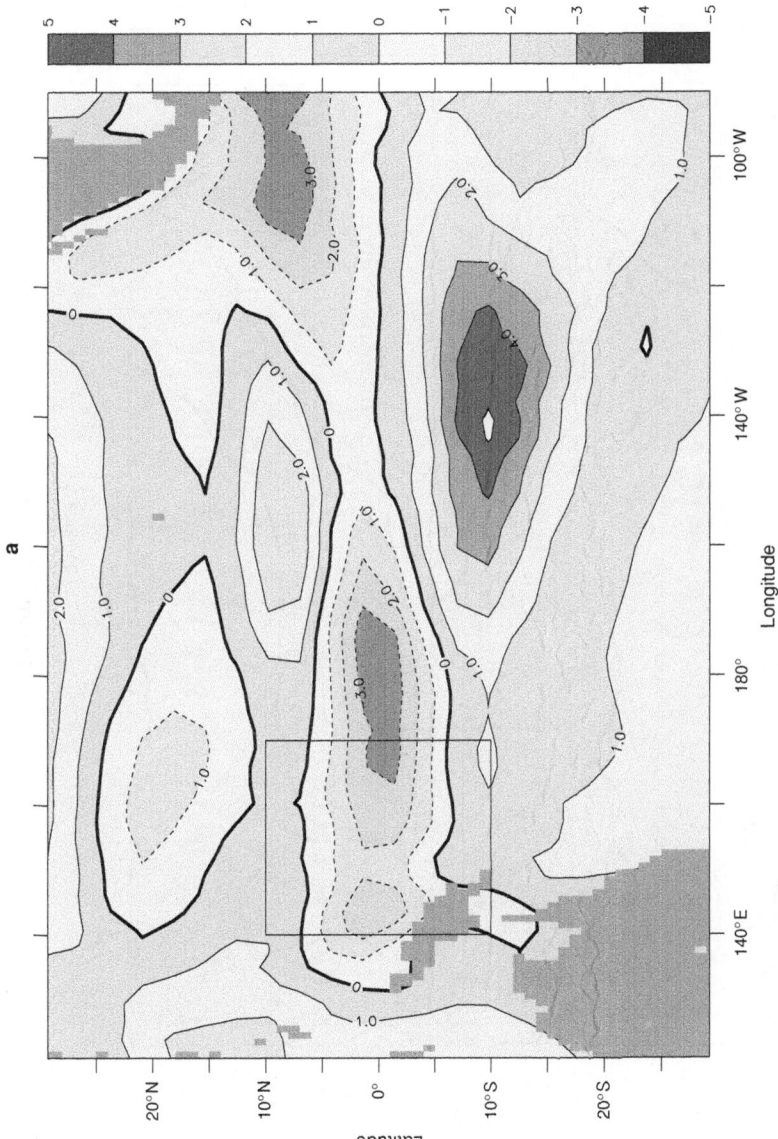

Figure 7.2. (a) Annual mean difference in zonal surface wind between the CSM and CCM3. Units are m s^{-1}. (b) Difference in annual mean latent-heat flux between CSM and CCM. Units are W m^{-2}. From Kiehl (1998) with permission.

This figure is available for download in colour from
www.cambridge.org/9780521791328

187

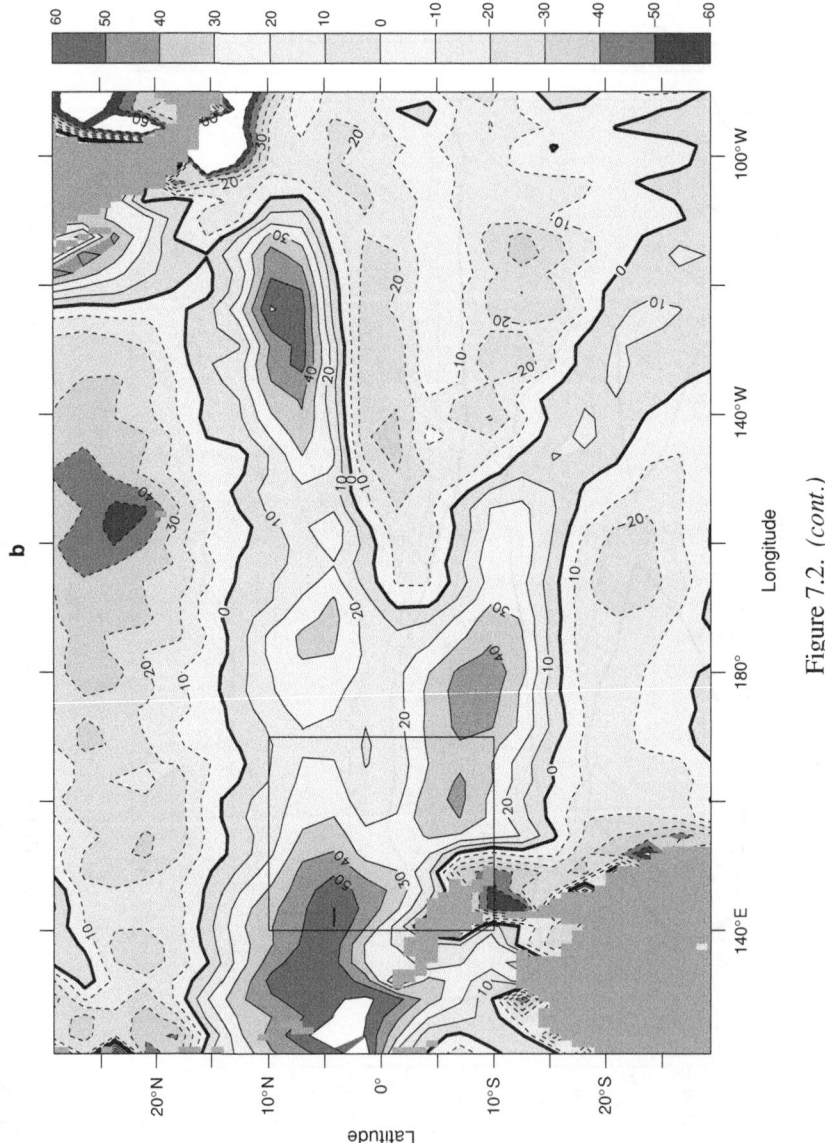

Figure 7.2. (*cont.*)

7.2.2 Description of generic enhanced absorption

In this study, the absorption of shortwave radiation is enhanced by explicitly increasing the differences between the all-sky and clear-sky shortwave-flux convergences. Any change in cloud radiative properties to increase absorption would also increase the shortwave-flux convergence in cloudy regions, and therefore we call our approach "generic enhanced absorption." Clearly the formulation of GEA does not address the issue of specific and plausible physical mechanisms for enhanced shortwave absorption. In addition, other methods for increasing absorption such as changes to single-scattering albedos or spectral resolution (Section 7.1) could lead to different radiative diabatic heating profiles than those generated by GEA. The characterization of the method as generic does not imply that the resulting heating profiles are universal.

In the current CCM, the clear-sky fluxes are computed by applying the delta-Eddington method to a two-layer approximation of the atmosphere (Kiehl *et al.*, 1996). The two layers represent the troposphere and stratosphere. Since GEA requires the capability to compute the differences between clear-sky and all-sky fluxes at the full 18-level vertical resolution of the model, the shortwave-radiative parameterization has been replaced with a generalized scheme developed by Bergman and Hendon (1998). This new scheme admits arbitrary assumptions regarding vertical cloud overlap in the calculation of the all-sky fluxes. It also computes corresponding clear-sky and all-sky flux profiles at full vertical resolution. The all-sky fluxes are computed using the random overlap assumption adopted in the standard CCM.

The radiation code computes upwelling and downwelling fluxes at the interfaces of each model layer. Let superscripts ↑ and ↓ represent the direction of propagation, subscripts i and $i + 1$ denote the upper and lower interfaces of a model layer at altitude z_i, and a and c indicate all-sky and clear-sky fluxes, respectively. For each vertical column, the clear-sky fluxes are computed under the same atmospheric conditions as the all-sky fluxes except that clouds are completely omitted. The all-sky and clear-sky shortwave flux convergences computed for a layer are given by Equations (7.1) and (7.2). Because of absorption by trace gases and cloud particles,

$$\delta F_a(z_i) = \left(F^{\downarrow}_{a,i} - F^{\uparrow}_{a,i} \right) - \left(F^{\downarrow}_{a,i+1} - F^{\uparrow}_{a,i+1} \right) \tag{7.1}$$

$$\delta F_c(z_i) = \left(F^{\downarrow}_{c,i} - F^{\uparrow}_{c,i} \right) - \left(F^{\downarrow}_{c,i+1} - F^{\uparrow}_{c,i+1} \right) \tag{7.2}$$

both $\delta F_c(z_i)$ and $\delta F_a(z_i)$ are positive. The all-sky net flux across the lower interface of layer i can be written in terms of the TOA net flux ($i = 0$) and flux convergences as in Equations (7.3) and (7.4).

$$F_{a,i+1}^{net} = \left(F_{a,i+1}^{\downarrow} - F_{a,i+1}^{\uparrow} \right) \tag{7.3}$$

$$= \left(F_{a,0}^{\downarrow} - F_{a,0}^{\uparrow} \right) - \sum_{j=0}^{i} \delta F_a(z_j) \tag{7.4}$$

Equation (7.4) can also be used to compute the net insolation at Earth's surface ($i = N$) in place of the usual delta-Eddington formulae.

In GEA, the clear-sky and all-sky fluxes are first computed at each grid box and time step using the standard code of Bergman and Hendon (1998). This radiative parameterization differs from the standard radiation codes in CCM in its treatment of geometrical cloud overlap. The solution for the TOA shortwave fluxes is left unchanged. This considerably simplifies the process of retuning the modified model to restore the radiative-energy balance at TOA required for coupled climate simulations. The profile of $\delta F_a(z_i)$ computed with the standard radiation code is replaced by $\Delta \tilde{F}_a(z_i)$. The tilde denotes the GEA value of the fluxes and flux convergences. The net fluxes $\tilde{F}_{a,i+1}^{net}$ at each layer interface are computed from Equation (7.3) with $\Delta \tilde{F}_a(z_i)$ substituted for $\delta F_a(z_i)$. Because $\Delta \tilde{F}_a(z_i)$ is generally larger than $\delta F_a(z_i)$ in cloudy model grid cells, the net surface insolation with GEA is generally smaller than in the standard calculation.

The expression for $\Delta \tilde{F}_a(z_i)$ is given by Equation (7.5). The first term on the

$$\Delta \tilde{F}_a(z_i) = \min \left\{ \delta F_c(z_i) + \lambda' \left[\delta F_a(z_i) - \delta F_c(z_i) \right], \left(F_{a,i}^{\downarrow} - F_{a,i}^{\uparrow} \right) \right\} \tag{7.5}$$

$$\lambda' = \begin{cases} \lambda \text{ if } \delta F_a(z_i) \geqslant \delta F_c(z_i) \\ \\ 1 \text{ if } \delta F_a(z_i) < \delta F_c(z_i) \end{cases} \tag{7.6}$$

right-hand side of Equation (7.5) represents the enhancement of the difference between all-sky and clear-sky flux convergence by a factor λ. When $\lambda = 1$, $\Delta \tilde{F}_a(z_i)$ reverts to the standard solution with no enhanced absorption. If $\delta F_c(z_i)$ is larger than $\delta F_a(z_i)$, Equation (7.6) insures that the standard solution is used. Under overcast conditions, $\delta F_c(z_i)$ can exceed $\delta F_a(z_i)$ in the lower troposphere because the clouds at higher altitudes greatly reduce the insolation that can be absorbed. The frequency of occurrence of $\delta F_c(z_i) > \delta F_a(z_i)$ reaches a maximum value of roughly 33% in the lowest model layers. The second term on the right-side side of Equation (7.4) insures that $\tilde{F}_{a,i+1}^{net} \geqslant 0$.

Solar radiation interacts with the atmosphere by altering the radiative diabatic heating rates and by warming the surface. The solar-heating rate in the prognostic thermodynamic equation is replaced by Equation (7.7), where g is the acceleration due to gravity, c_p is the specific heat of air, p_i and p_{i+1} are the atmospheric pressures at levels i and $i + 1$, respectively.

$$\tilde{Q}_{sol}(z_i) = \frac{g}{c_p} \frac{\Delta \tilde{F}_a(z_i)}{(p_{i+1} - p_i)} \qquad (7.7)$$

Equations (7.1)–(7.6) and a specification for λ (Section 7.2.3) completely determine the effects of GEA on the atmospheric model. The versions of CCM and CSM incorporating GEA will be indicated by CCM+GEA and CSM+GEA, respectively.

The coupled model requires several surface solar fluxes to specify the energy exchanged between the atmosphere, ocean, sea ice, and land. The required fluxes are the downwelling shortwave flux and the downwelling direct-beam fluxes and diffuse fluxes in visible and near-infrared wavelengths. Calculation of the surface fluxes depends on the surface albedo and the ratio of direct-beam to total downwelling spectral flux. For GEA, the simplest assumption is that these quantities are unchanged from their values in the standard calculation. The surface fluxes follow directly from this assumption and Equation (7.3) for the surface net fluxes after substitution of $\Delta \tilde{F}_a(z_i)$ for $\delta F_a(z_i)$.

7.2.3 Selection of values for GEA parameter λ

The amount of additional absorption in GEA is governed by the dimensionless parameter λ. For a fixed cloud distribution and atmospheric state, the additional absorption increases monotonically with λ. Although it would be possible to specify λ as a function of cloud type, geographic location, altitude, and/or season, the calculations are performed with a global, fixed value of λ. This is qualitatively consistent with previous studies that have detected anomalous absorption for a wide range of geographic locations and meteorological conditions (Cess *et al.*, 1995; Cess *et al.*, 1996). Other studies have found significant meridional and seasonal variability in the shortwave absorption by clouds (e.g., Li *et al.*, 1995; Li and Moreau, 1996). However, analysis of a global dataset of spectral albedos from the Nimbus-7 satellite shows anomalies relative to model calculations at all latitudes examined and during all seasons (Collins, 1998). The anomalies occur for clouds at all altitudes, although the magnitude of the differences between the models and observations has some dependence on cloud height. The results from the Nimbus-7 observations support using a single global constant for λ. For simplicity, a single value of λ is applied to all spectral intervals in the visible and near-infrared.

The value of λ used in these experiments is based upon a simple derivation. In the current CCM radiation code, the difference between the global annual mean atmospheric shortwave absorption under clear and all-sky conditions is less than 2 W m^{-2} (Kiehl, 1994). The magnitude of the additional absorption in clouds inferred from observational studies is roughly 25–30 W m^{-2} (Cess *et al.*, 1995; Cess *et al.*, 1996). In terms of the global annual mean flux convergences for clear and all-sky regions, the difference between the observations and CCM is given by Equation (7.8).

$$\sum_{z_i}\left[\delta F_a(z_i) - \delta F_c(z_i)\right]_{\text{obs}} \simeq 15 \sum_{z_i}\left[\delta F_a(z_i) - \delta F_c(z_i)\right]_{\text{CCM}} \qquad (7.8)$$

In GEA, $\delta F_a(z_i)$ is replaced with $\Delta \tilde{F}_a(z_i)$ to match the observational estimate on the left-hand side of Equation (7.8). From the expression for $\Delta \tilde{F}_a(z_i)$ given by Equation (7.5), it follows that λ should be set to approximately 15 in order to reproduce the observed difference in clear-sky and all-sky flux convergence. The experiments with CCM and CSM including GEA are run with $\lambda = 15$.

This derivation assumes that the global annual mean values for $\delta F_a(z_i)$ and $\delta F_c(z_i)$ from the standard radiation code in CCM+GEA are independent of λ. However, the significant changes in atmospheric and surface heating introduced by increasing λ can certainly affect the cloud distributions in the model. Therefore the global annual-mean value of $\delta F_a(z_i)$ from CCM+GEA will probably change as λ is varied, and the actual enhancement in solar absorption may differ from the simple derivation given here. As shown in Section 7.3, the atmospheric absorption in the uncoupled CCM3 AGCM is enhanced by 22 W m^{-2} when $\lambda = 15$.

7.3 Response of the uncoupled atmospheric model to GEA

7.3.1 Adjustment of atmospheric model for energy balance

In order to integrate the coupled climate model without significant secular drift, the global annually averaged radiative energy balance at TOA should be close to zero. This constraint follows from the first law of thermodynamics applied to the entire climate system. The standard CSM satisfies the energy-balance requirement to within ±0.5 W m^{-2}. By construction, GEA is designed so that the instantaneous TOA fluxes are identical to those from the standard radiation calculations. In the absence of changes to the climate due to GEA, the CCM+GEA should still be in radiative balance. However, the introduction of GEA alters the climate state of the model so that the net energy balance at TOA is -6.7 W m^{-2}. The negative sign indicates that the outgoing longwave radiation (OLR) exceeds the incoming net solar flux. Since the OLR is 236.8 W m^{-2} and this value is close to the observed OLR (Kiehl and Trenberth, 1997), the model is modified to increase the net solar flux. The balance is restored by two minor adjustments. First, the effective droplet radius for liquid clouds over land surfaces is increased from 5 to 10 μm so that it is identical to the radius for liquid clouds over oceans. This has the effect of reducing the cloud albedos over land because the larger drops are more absorptive and the cloud optical depths are smaller. The cloud optical depths are proportional to condensed water path and inversely proportional to the effective radius, so larger droplets result in smaller optical depths. Second, the optical depth for aerosols is reduced. Both of these changes increase the amount of solar radiation absorbed by the simulated climate system. After the introduction of the modifications to the effective radii and aerosol loading, the TOA energy balance is -0.65 W m^{-2}.

The components of the TOA energy budget from the standard CCM, the CCM+GEA, and from the Earth Radiation Budget Experiment (ERBE) are compared in Figure 7.3. The annually averaged zonal-mean TOA fluxes from CCM and CCM+GEA are very close at most latitudes, and the global annually-averaged all-sky fluxes agree to within 3 W m^{-2}. The good agreement between CCM and ERBE is not degraded by the introduction of GEA.

7.3.2 The climate simulation with GEA

The absorption in the atmosphere increases by 33% after the introduction of GEA. The annually averaged shortwave absorption is plotted as a function of latitude in Figure 7.4. The absorbed radiation reaches its maximum value in the tropics where the annual TOA insolation is greatest. In the original CCM, the clear-sky and all-sky absorbed solar fluxes are nearly identical for both zonally and globally averaged values. This does not imply that clouds do not absorb radiation in conventional atmospheric models. Instead, the additional absorption within clouds is balanced by reduced absorption by trace gases beneath the clouds. The all-sky absorption from CCM+GEA is significantly larger than the clear-sky absorption, with the largest differences occurring in the tropics. Since the net TOA shortwave flux in CCM+GEA is nearly identical to the original CCM (Figure 7.3), and since the fraction of this radiation absorbed by the atmosphere is 33% larger, the fraction transmitted to the surface is greatly reduced. The net surface shortwave flux decreases from a global annual mean value of 170 W m^{-2} in CCM to 149 W m^{-2} in CCM+GEA. This reduction eliminates most of the differences in surface shortwave flux between the atmospheric model and observational estimates of 142 W m^{-2} (Wild *et al.*, 1995).

The increased absorption can be expressed in terms of increase in the ratio R of the surface to TOA shortwave cloud forcing. In the CCM and in most conventional GCMs, this ratio is close to 1 (Figure 7.5); $R \simeq 1$ implies that the reduction in net shortwave radiation by clouds is nearly equal at the surface and TOA. Several observational studies have concluded that the ratio should be closer to 1.5 in the tropics and perhaps at higher latitudes as well (e.g., Ramanathan *et al.*, 1995; and Cess *et al.*, 1995). The value of R computed with CCM+GEA is 1.36, with

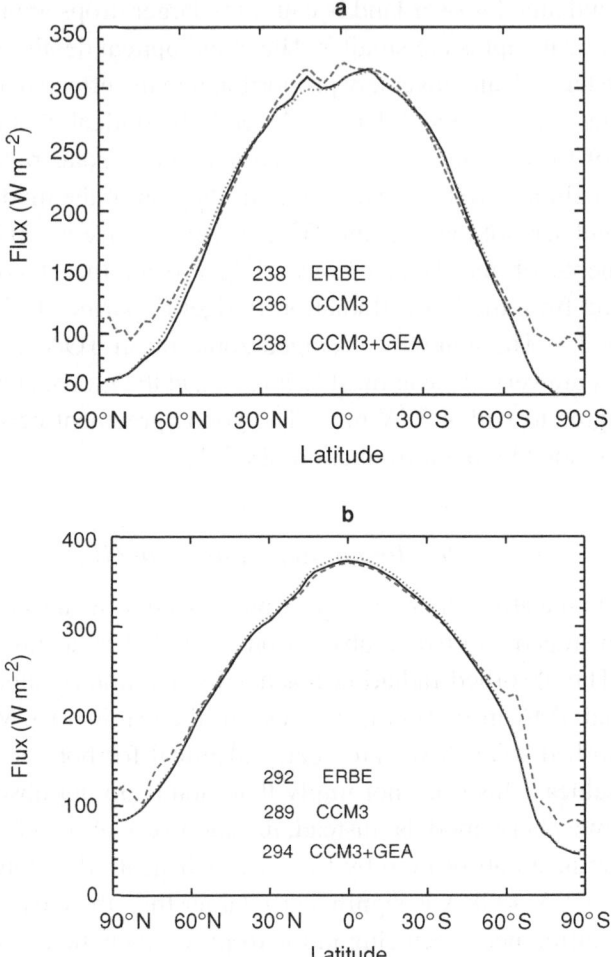

Figure 7.3. Zonal-mean annually averaged components of the TOA energy budget from ERBE (dashed), CCM3 (solid), and CCM3+GEA (dotted). (a) Net all-sky shortwave flux. (b) Net clear-sky shortwave flux. (c) Net all-sky longwave flux. (d) Net clear-sky longwave flux. The global annual averages for each quantity are given on each plot.

peak values in the tropics exceeding 1.4. The introduction of GEA has improved the agreement of the model with the observational studies which find enhanced shortwave absorption in tropical regions.

The shortwave radiative heating of the atmosphere increases significantly with the introduction of GEA. This is a direct consequence of the relationship between the heating rate and shortwave flux convergence given by Equation 7.7. The vertical profiles of the heating rates in CCM are shown in Figure 7.6a. Because the absorption of solar radiation adds internal energy to the atmosphere, the heating

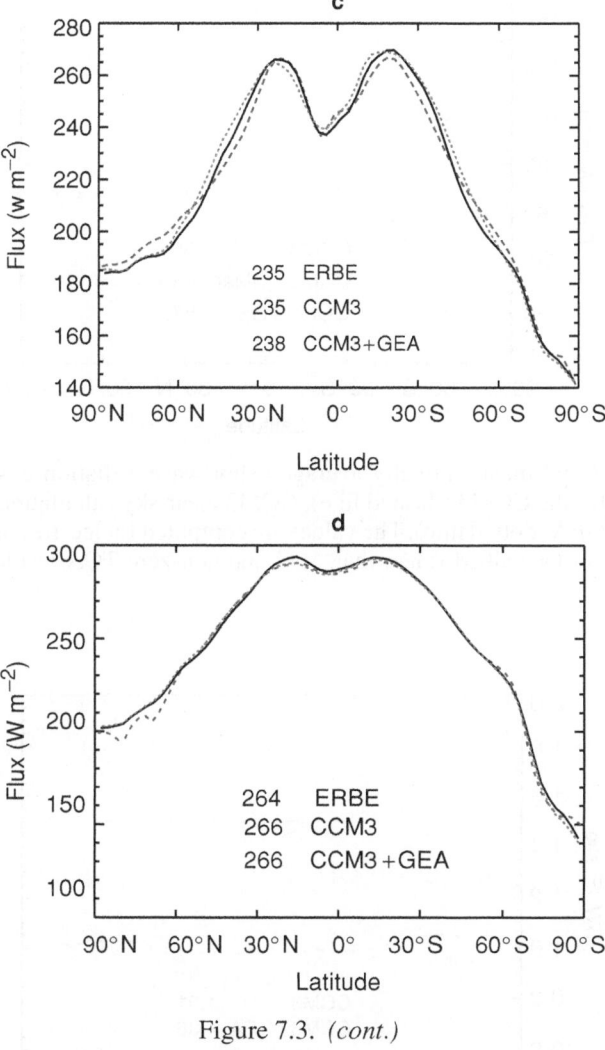

Figure 7.3. *(cont.)*

rates are positive throughout the troposphere and stratosphere. The corresponding heating rates for clear-sky conditions are very similar, and the annual-mean all-sky and clear-sky heating rates differ by less than 0.2 K d^{-1}. In CCM+GEA, the all-sky heating rates increase by as much as 1 K d^{-1} in the upper tropical troposphere (Figure 7.6b). Unfortunately there is very little observational data in this region of the troposphere for evaluating the realism of the larger heating rates. The largest increases are associated with the extensive upper-level cloud decks formed by tropical convection (Figure 7.8a). The changes in heating rates result in substantial changes in atmospheric structure.

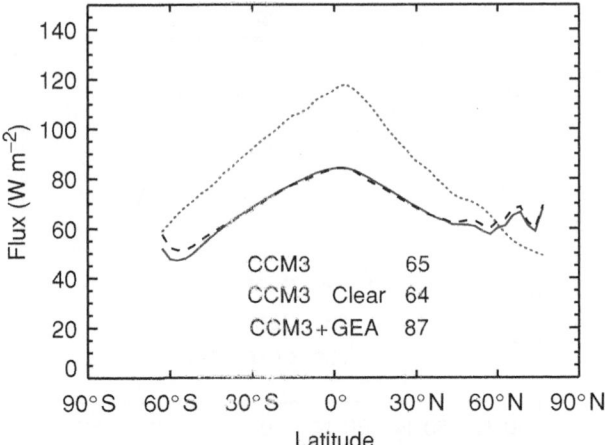

Figure 7.4. Zonal-mean annually averaged shortwave radiation absorbed in the atmosphere for the CCM3 (dashed line), CCM3 clear-sky calculations (solid line) and CCM3+GEA (dotted line). The values are computed for ice-free and snow-free regions with surface albedos less than 50% and non-zero TOA insolation.

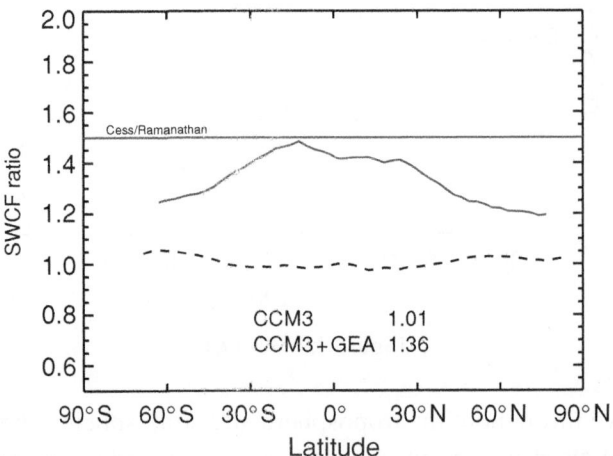

Figure 7.5. Ratio of zonal-mean annually averaged surface-to-TOA shortwave cloud forcing (SWCF) for CCM3 (dashed line) and CCM+GEA (solid line). The globally averaged cloud forcing ratios for each model are given. The ratio of 1.5 derived by Ramanathan *et al.* (1995) for the warm pool and by Cess *et al.* (1995) for a number of sites worldwide is also shown. The model values are computed for ice-free and snow-free regions with surface albedos less than 50% and non-zero TOA insolation.

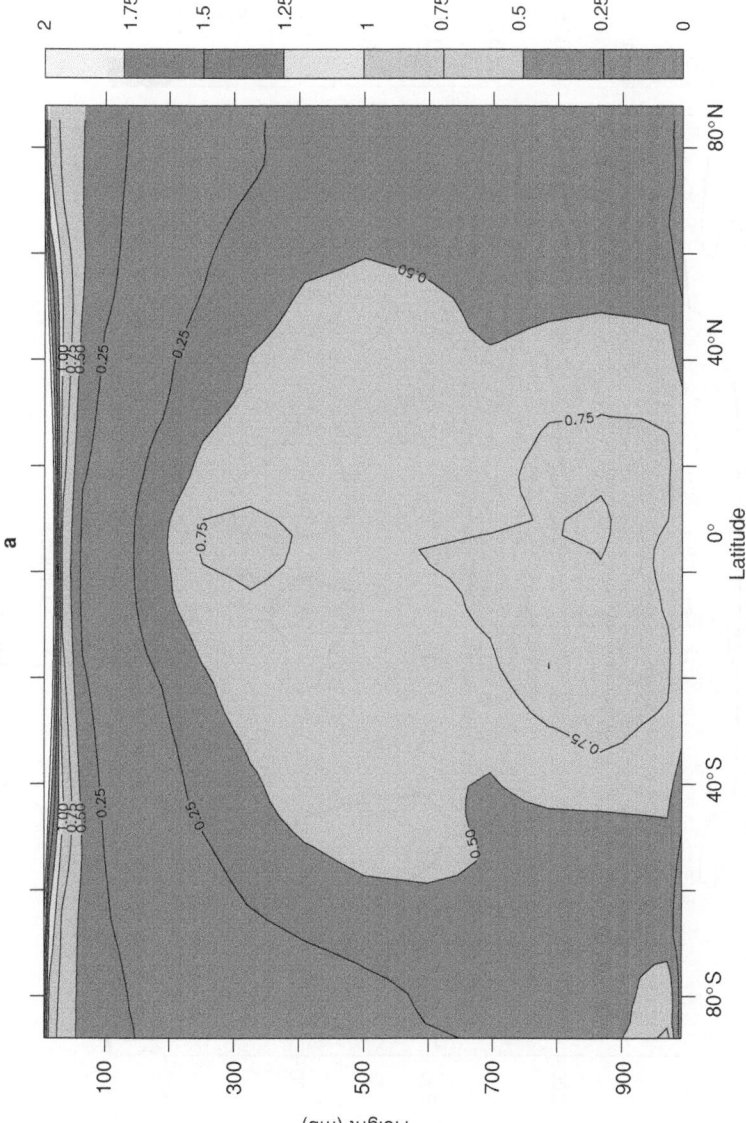

Figure 7.6. Zonal-mean annually averaged shortwave heating in K d^{-1}. (a) All-sky rate from CCM3. (b) Difference in all-sky rates between CCM+GEA and CCM3. Vertical coordinate is σ.

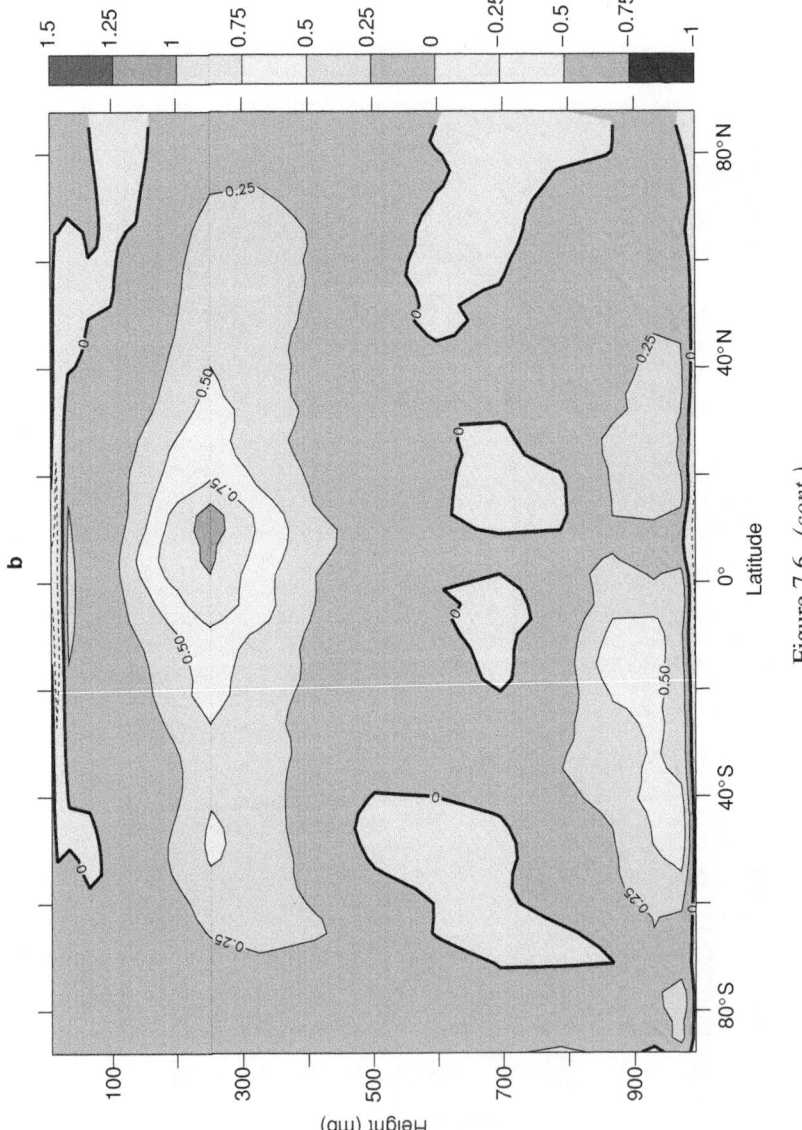

Figure 7.6. *(cont.)*

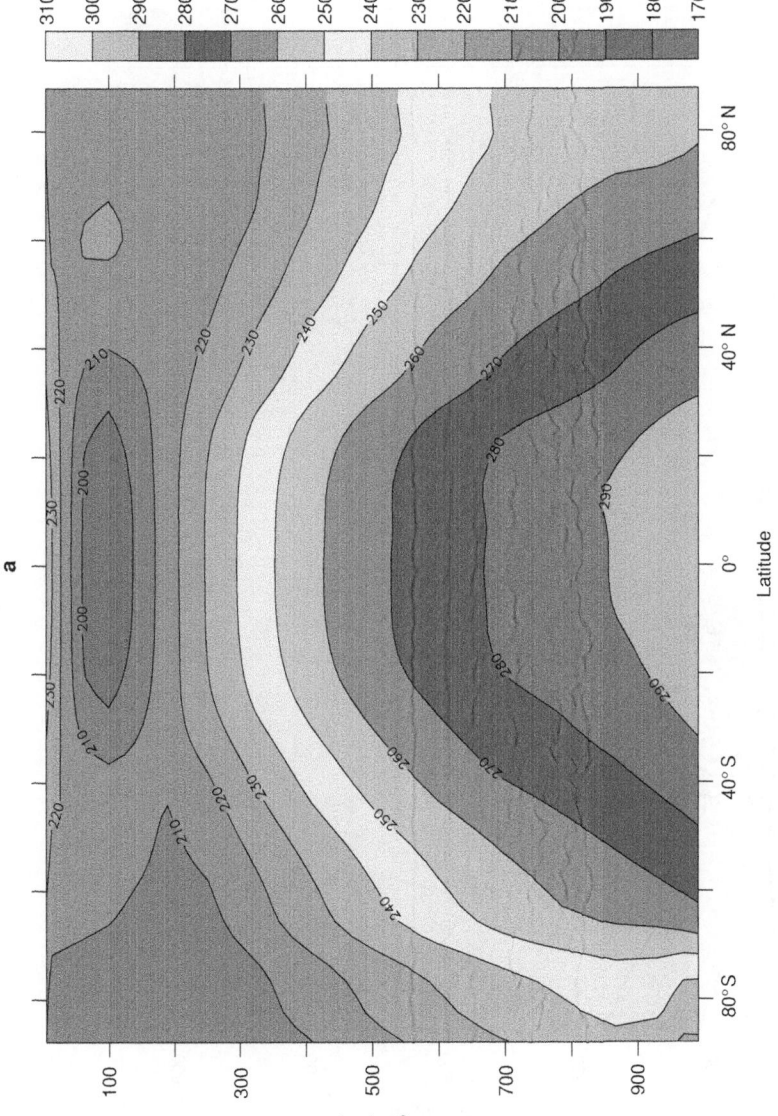

Figure 7.7. Zonal-mean annually averaged atmospheric temperatures in K. (a) CCM3. (b) Difference between CCM+GEA and CCM3. Vertical coordinate is σ.

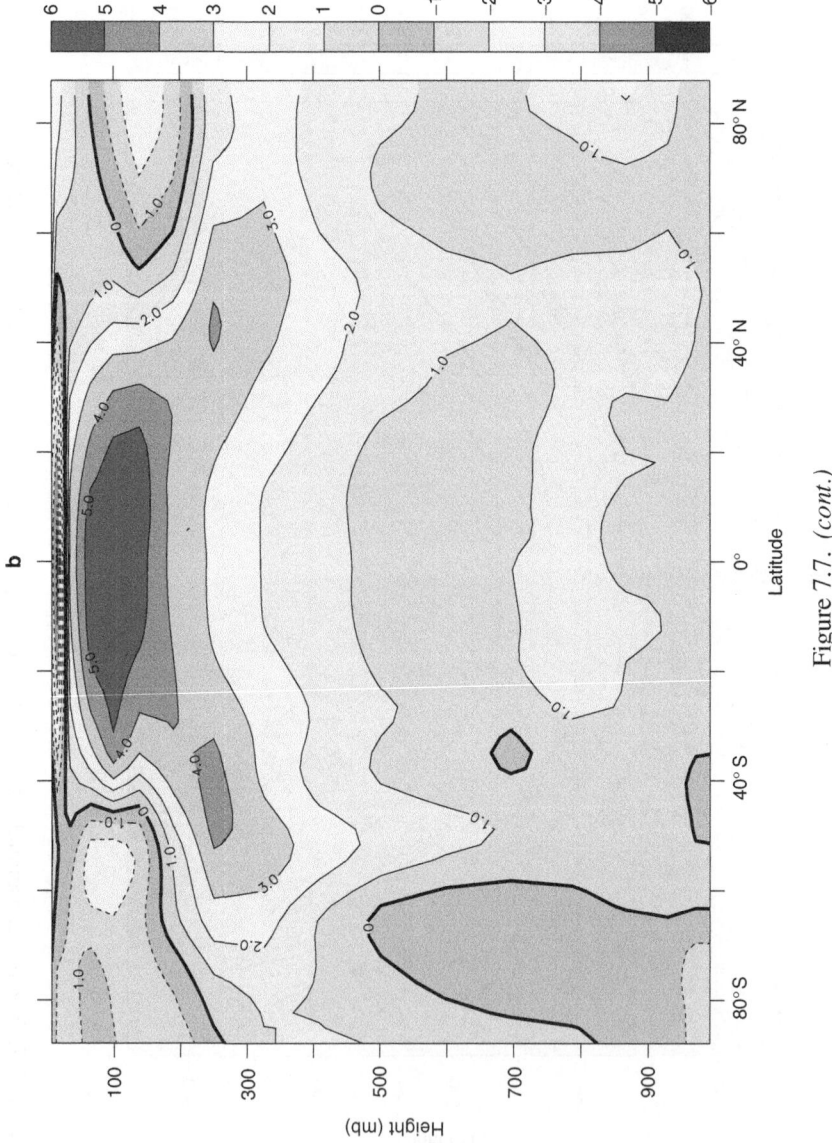

Figure 7.7. (cont.)

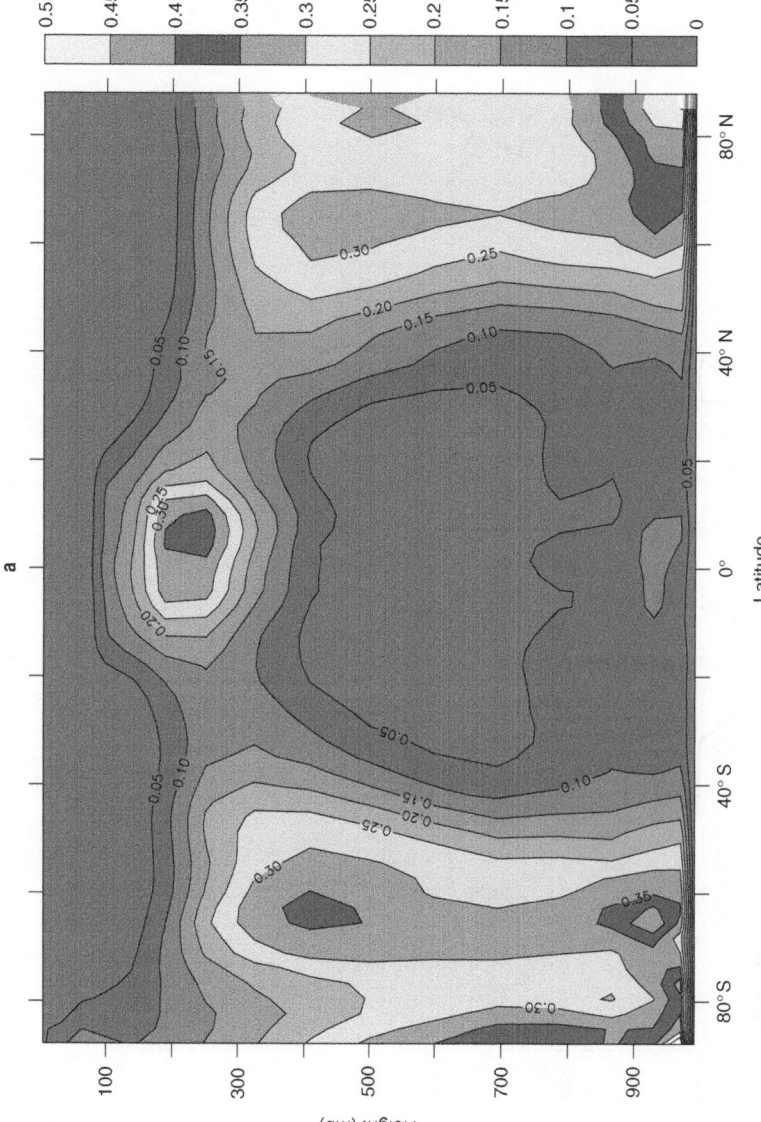

Figure 7.8. Zonal-mean annually averaged fractional cloud amounts. (a) CCM3. (b) Difference between CCM+GEA and CCM3. Vertical coordinate is σ.

201

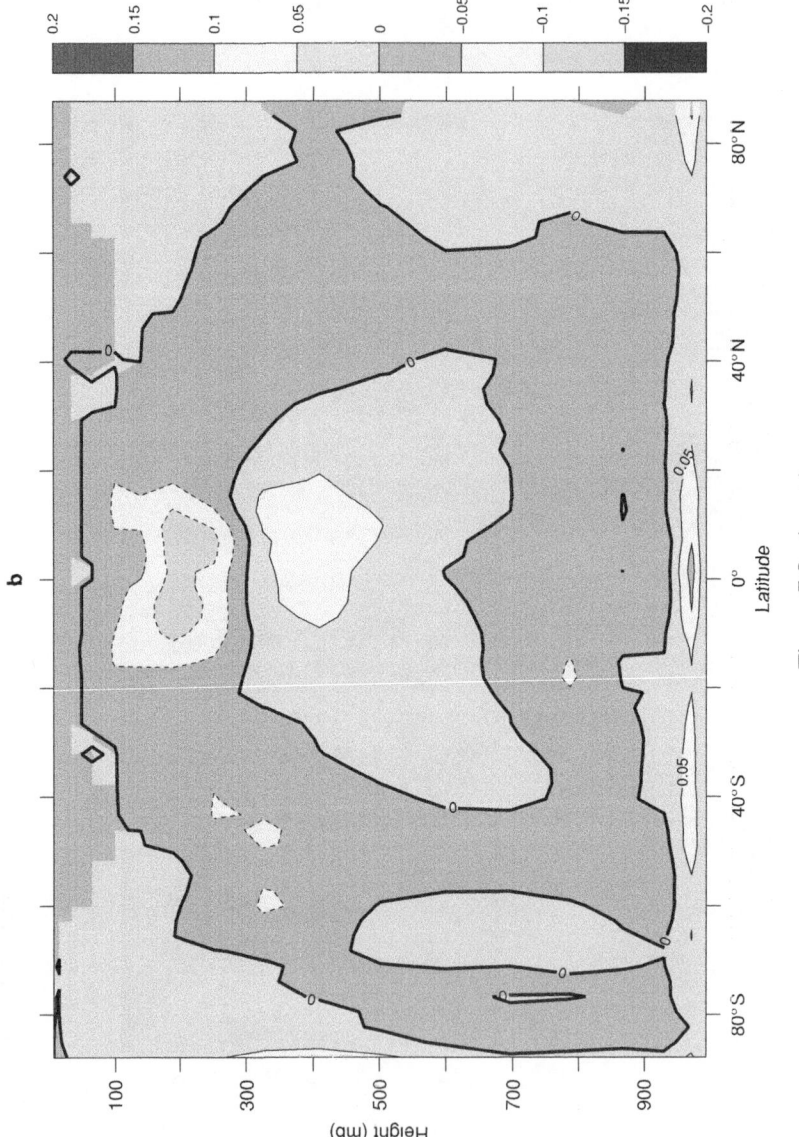

Figure 7.8. *(cont.)*

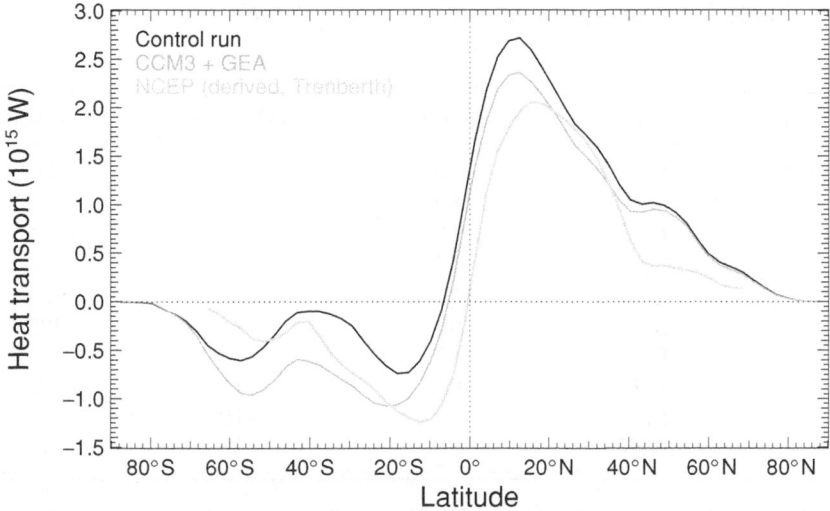

Figure 7.9. Implied ocean heat transport from CCM3 (black line), CCM+GEA (green line), and NCEP re-analysis (blue line, derived from Trenberth and Solomon, 1994).
This figure is available for download in colour from
www.cambridge.org/9780521791328

The atmospheric temperature increases in response to the enhanced absorption. Figure 7.7 shows the temperature profiles from the standard version of CCM and the changes calculated by CCM+GEA. The temperatures from the standard model are generally within 2 K of observed except in the upper tropical troposphere, which is 3 K too cold, and in the polar regions (Hack *et al.*, 1998). The introduction of GEA increases the temperature of the upper tropical troposphere by as much as 5 K near 100 mb, reversing the sign of the model temperature bias. The resulting temperature profiles now show a positive offset of 2–3 K relative to meteorological analyses.

The largest changes in cloud amount occur in the tropical regions as well (Figure 7.8). Cloud amount in the boundary layer increases by 10–15%, and cloud coverage between 300–500 mb increases by 5–10%. High-cloud amount decreases by 10–15% between 300 mb and the tropopause. The change in high-cloud amount results from increased stabilization of the upper atmosphere, which causes convection to detrain moisture at lower altitudes (not shown), and from the reduction in the relative humidity with increased temperature. Cloud coverage in CCM is based upon simple empirical relationships between relative humidity, other meteorological variables, and cloud amount. It follows from these formulae that lower relative humidity leads to lower cloud amounts. The reduction in high cloud and increase in middle-tropospheric cloud amount reduces the systematic biases in CCM tropical cloud coverage relative to satellite-based estimates from ISCCP.

Table 7.3. *Warm-pool energy budgets with GEA (10° N to 10° S,*
140° E to 170° E)

Level	Flux	CCM3+GEA	CSM+GEA	Observed[a]
TOA	S	296	304	309
	S_{clr}	380	380	373
	F	−215	−226	−225
	F_{clr}	−283	−285	−285
Surface	S	177	185	182
	S_{clr}	294	294	282
	F	−45	−45	−49
	LH	−93	−110	−107[b]
	SH	−10	−10	−8[b]
	NET	28	20	18

[a] Ramanathan *et al.* (1995) and Kiehl (1998), except where noted. [b] Zhang and McPhaden (1995).

The change in implied ocean heat transport calculated from the surface-heat budget (Trenberth and Solomon, 1994) is shown in Figure 7.9. The ocean transports from the CCM and CCM+GEA are both comparable to an estimate derived from meteorological analyses constrained by observations. The effect of enhanced absorption is to reduce the magnitude of the northward transport in the northern hemisphere and increase the magnitude of the southward transport in the southern hemisphere. Between approximately 20° S and 20° N, the agreement between the simulated and observationally based heat transports is improved with the introduction of GEA. The sensitivity of the magnitude and even sign of the ocean transports to cloud radiative forcing has been observed in a number of atmospheric GCMs (Gleckler *et al.*, 1995). The implied transport in the southern hemisphere had the wrong sign in many earlier GCMs, and this sign error was eliminated with improvements in simulated cloud-radiative effects. While the transport calculated by the CCM is also affected by the convective parameterization (Hack, 1998), the changes in CCM+GEA are primarily due to changes in the surface fluxes caused by enhanced absorption. These changes support the theoretical inference that the introduction of enhanced shortwave absorption would substantially alter the implied heat transport calculated with global climate models (Kiehl, 1994).

7.4 Response of the coupled climate model to GEA

The annual mean energy budget of the tropical western Pacific from the uncoupled and coupled models with GEA is shown in Table 7.3. While the net energy surface

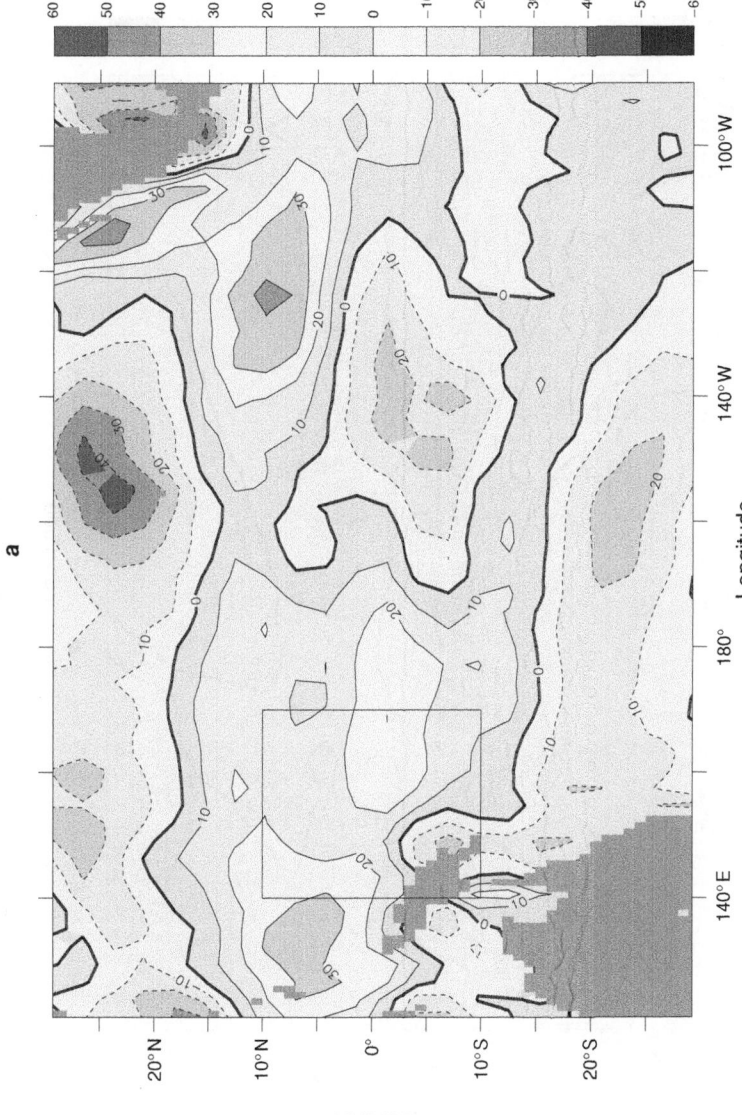

Figure 7.10. Difference in surface fluxes between the coupled integration of CSM+GEA and the atmosphere-only integration of CCM+GEA. Units are W m^{-2}. (a) Latent-heat flux. (b) Net surface shortwave flux.

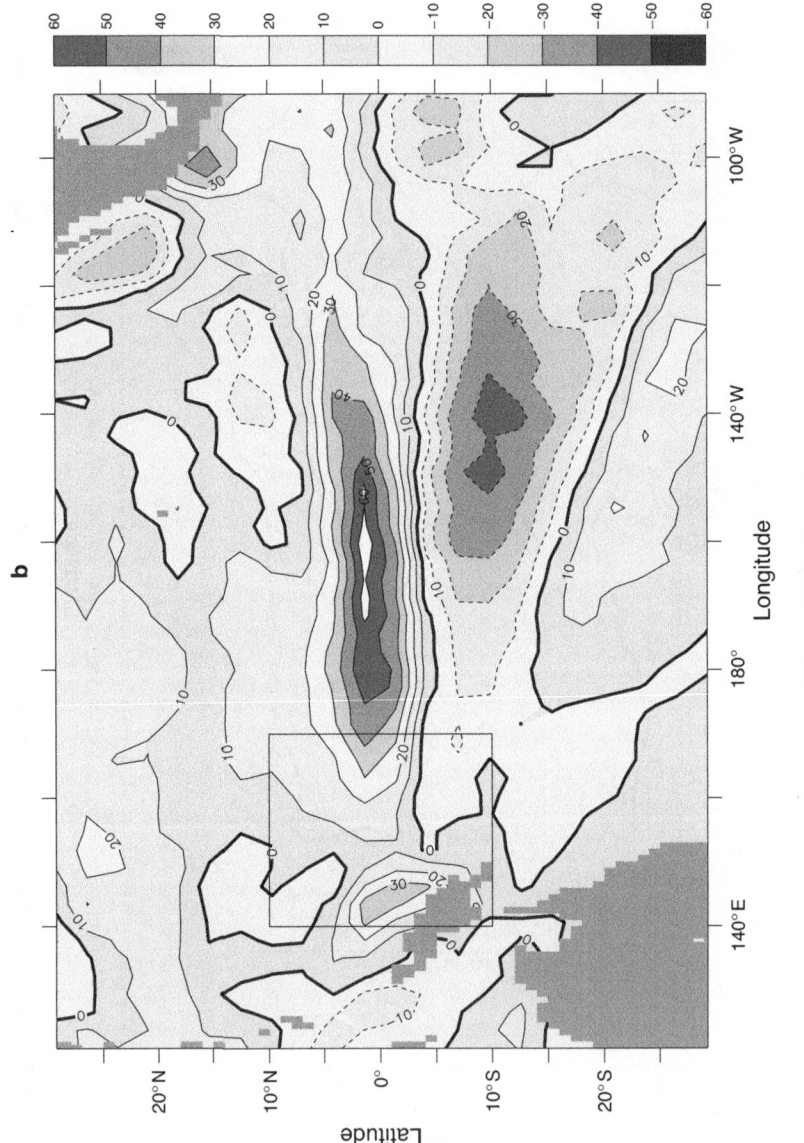

Figure 7.10. (cont.)

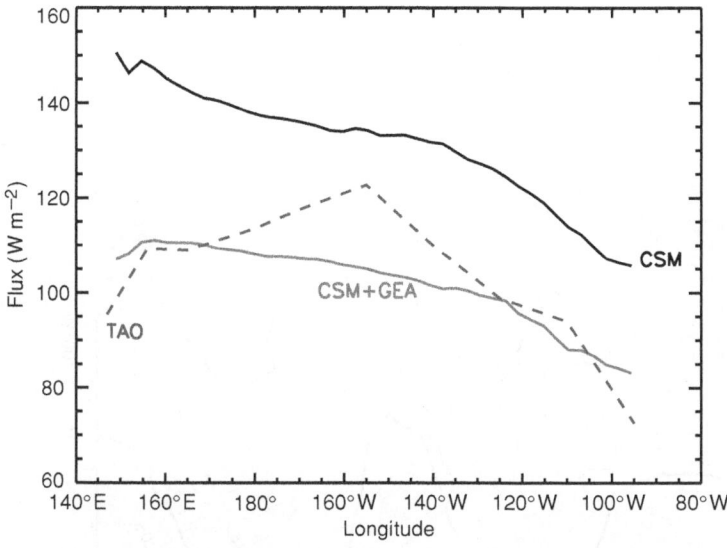

Figure 7.11. Comparison of the meridionally averaged latent-heat flux in the tropical Pacific basin from CSM CSM+GEA, and the TOGA TAO array of buoys (Zhang and McPhaden, 1995).

budget is still overestimated in the uncoupled model relative to observations, the offset has been reduced from 26 W m^{-2} in CCM to 10 W m^{-2} in CCM+GEA. Most of the change in the net heat budget is associated with the reduction in surface insolation. In the coupled model, the latent-heat flux, sensible-heat flux, and the all-sky radiative fluxes at TOA and the surface are each within 5 W m^{-2} of the observations. The net surface-heat budget is within 2 W m^{-2} of observations and satisfies the upper bound imposed by ocean dynamics (Ramanathan *et al.*, 1995). In the transition between the climates simulated by the uncoupled and coupled models, the insolation increases by 9 W m^{-2} and the magnitude of the latent-heat flux increases by 17 W m^{-2}. The changes in these fluxes over the warm pool are consistent with changes in the insolation and latent-heat flux over much of the Pacific basin between CCM+GEA and CSM+GEA (Figure 7.10). Compared to the changes in latent-heat flux between the climates simulated by CCM and CSM (Kiehl, 1998), the increases in latent heat between the uncoupled and coupled models with GEA are generally smaller.

In the coupled model with GEA, the latent heat fluxes are comparable to estimates of the fluxes from the TOGA TAO array (Figure 7.11). The data from the TAO array, a network of tethered buoys in the tropical Pacific, include the state variables necessary to calculate surface heat and momentum fluxes from standard bulk formulae (Zhang and McPhaden, 1995). Over the portion of the warm pool from 150–170° E and over the eastern section of the Pacific from 130–90° W, the flux

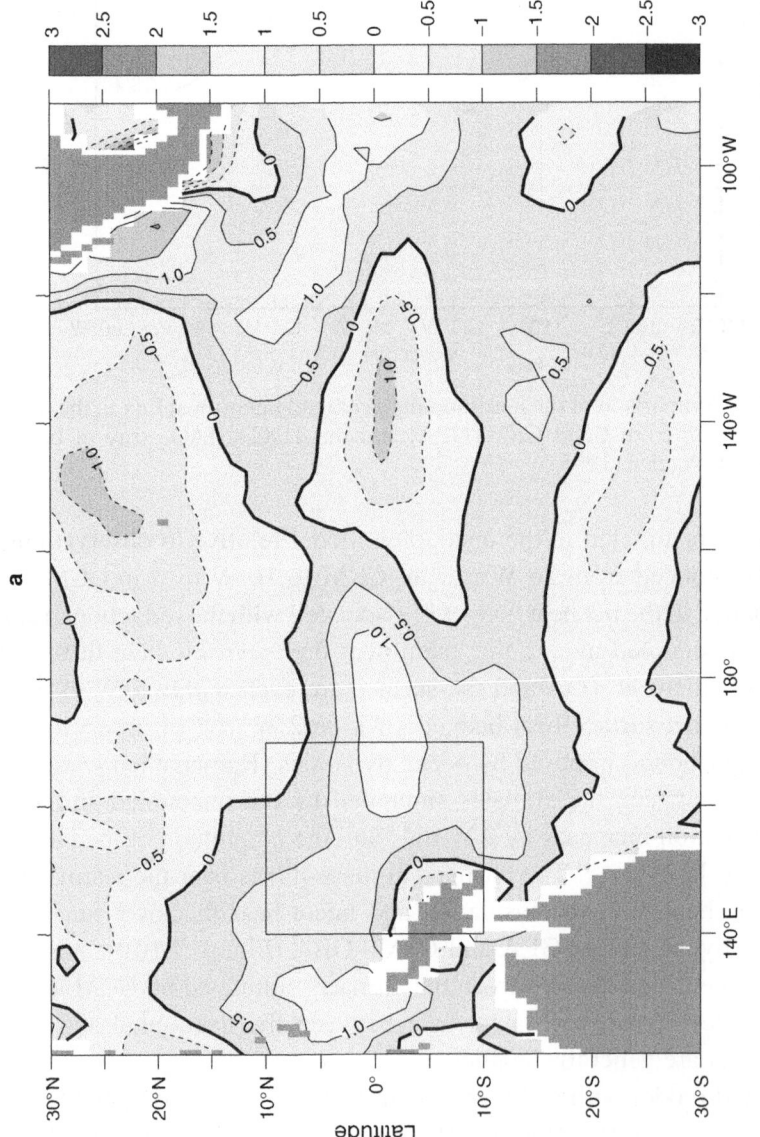

Figure 7.12. Difference in annual mean SST (°C) between the CSM ocean model and the observational estimates (Shea *et al.*, 1992). (a) Uncoupled ocean model forced by fields from CCM+GEA (years 16–20). (b) Coupled ocean model in CSM+GEA (years 1–19).

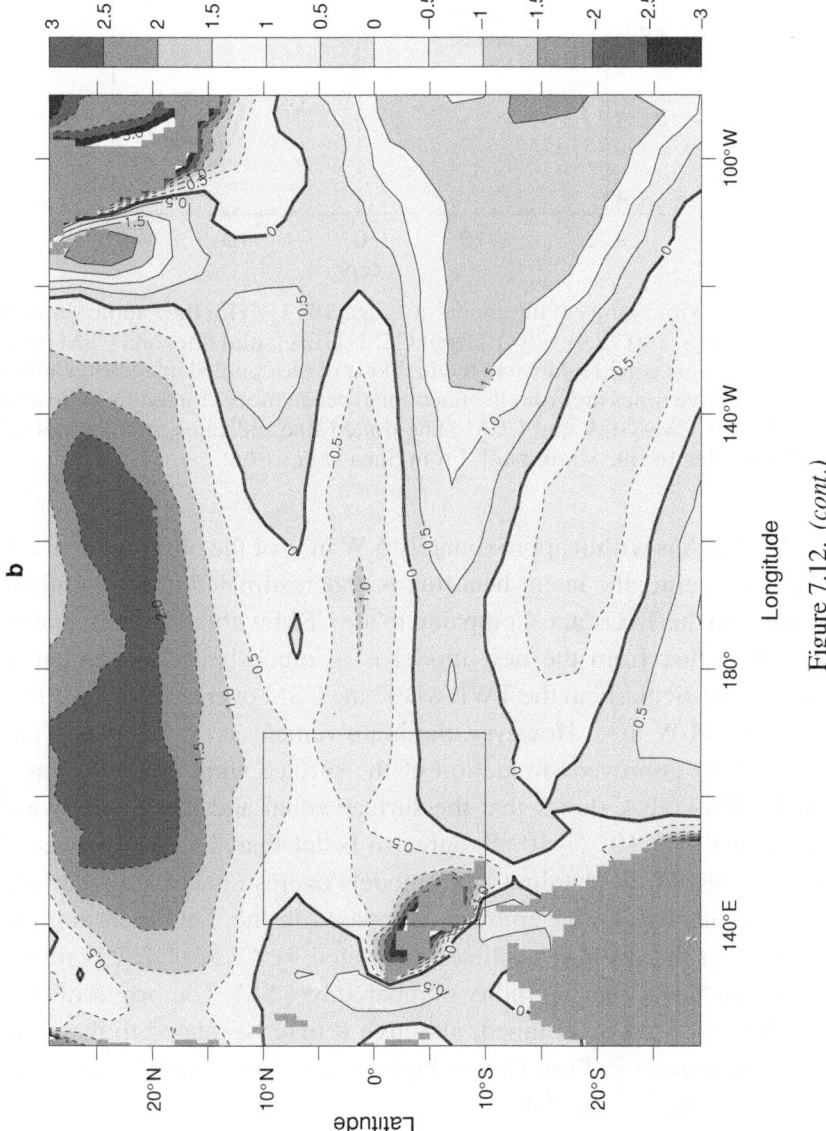

Figure 7.12. (*cont.*)

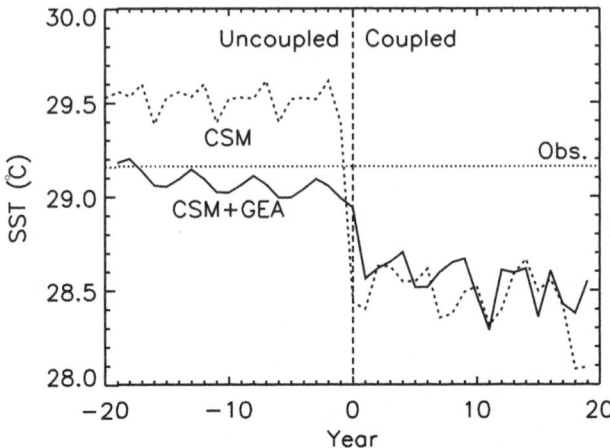

Figure 7.13. Time series of the annual-average SST (°C) for the Pacific warm pool (10° N to 10° S, 140° E to 170° E) from CSM+GEA (solid line) and CSM (dashed line). Time is measured relative to the first year of the coupled integration, and values for negative times are from the uncoupled ocean model forced by atmospheric fields from CCM+GEA and CCM. The dotted line indicates the climatological observed value for the warm pool. From Shea *et al.*, 1992.

from CSM+GEA is within approximately 5 W m^{-2} of the observational estimates. In the central Pacific, the latent-heat flux is underestimated in the simulation with GEA relative to the TAO data. Compared to the CSM without enhanced absorption, the latent-heat flux from the new model is in much better agreement with the observations, particularly in the TWP where the CSM overestimates the magnitude of the flux by 38 W m^{-2}. However, the improvement in the simulated latent flux is not caused by improved simulation of the surface wind fields. Comparison of CSM and CSM+GEA shows that the surface zonal and meridional wind fields, when averaged over 10° N–10° S, agree to better than 1 m s^{-1} between 160° E and 90° W. West of the dateline, both models overestimate the magnitude of the surface easterlies by similar amounts compared to the TAO measurements. The improvement in the latent-heat fluxes simulated by CSM+GEA is related to an increase in the surface-air humidity compared to CSM. The origin of the greater humidity has not been determined, although it may be related to the greater sea-surface temperatures (SST) and hence higher saturation humidities upstream of the TWP in the simulation with GEA.

The SSTs for the Pacific basin from the simulations with enhanced absorption are compared to a climatology of SST (Shea *et al.*, 1992) in Figure 7.12. In the western Pacific, the biases relative to the SST climatology have the same sign but generally lower magnitude than the corresponding biases for the standard climate models. When the ocean model is forced with surface fields from CCM+GEA,

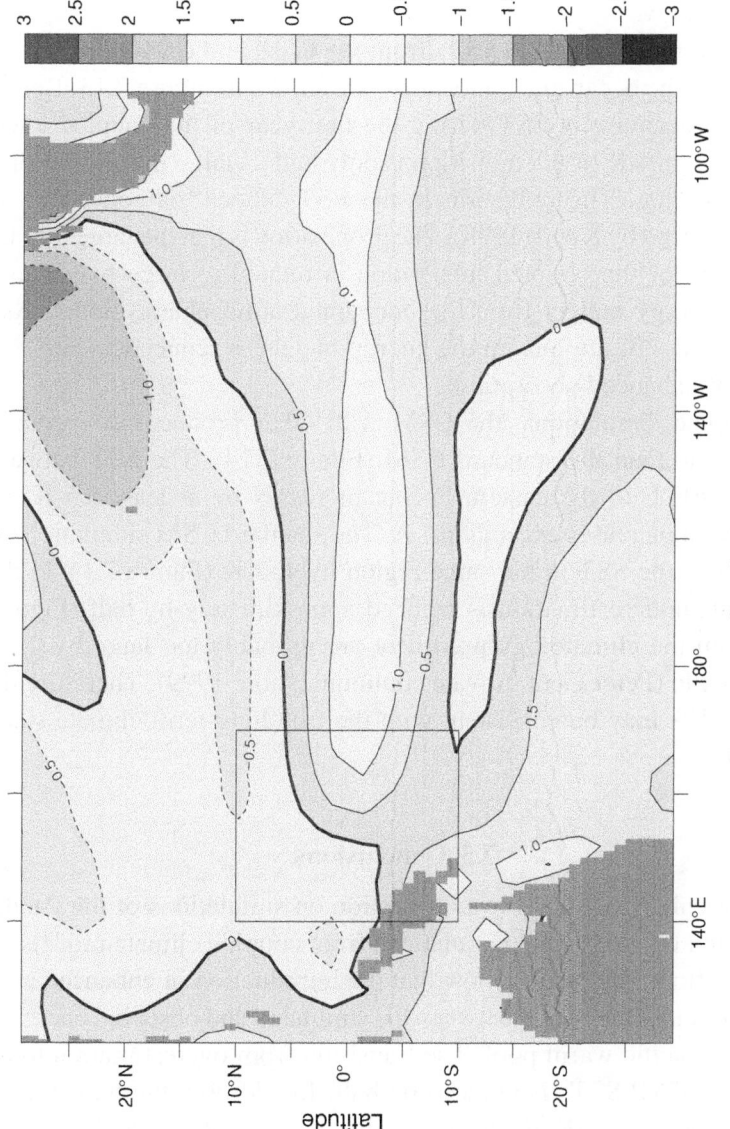

Figure 7.14. Difference between the SST (°C) simulated by CSM+GEA and CSM (years 1–19).

the resulting SSTs are higher than observed in the warm pool (Figure 7.12a). The offset relative to the climatology is caused by the overestimate of the surface-heat budget and the dynamical constraints on advective heat fluxes directed out of the warm pool (Kiehl, 1998). Once the system is fully coupled, the average SST for the TWP from the model with enhanced absorption is approximately 0.7 K lower than the climatology.

Time series for the simulated SSTs from the CSM and CSM+GEA before and after coupling of the ocean and atmosphere are shown in Figure 7.13. Both models exhibit a sharp decrease in SST during the first year of the coupled integration, although the magnitude of the transition is only half as large in the CSM including enhanced absorption. The reduction in the SST "shock" is consistent with the hypothesis advanced by Kiehl (1998). The proposition is that the large SST transient at the beginning of the coupled integration is related to large biases in the net TWP surface-energy budget from the uncoupled atmospheric model. As shown in Tables 7.2 and 7.3, the bias in the energy budget is reduced by 60% with the introduction of enhanced absorption.

In the coupled simulations, the CSM with GEA produces a warmer central equatorial Pacific than the standard CSM (Figure 7.14). The SST between 5° N to 5° S and 170° E to the eastern Pacific increases by at least 0.5 K, with the largest long-term increases exceeding 1 K. The standard CSM simulation yields an underestimate of the SST in the same region by 1–2 K (Figure 7.1). In the CSM with GEA, this underestimation is reduced approximately by half (Figure 7.12). The SSTs from the climatology used here are probably too large by 0.5–1 K in the central Pacific (Peter Gent, private communication, 1999). Therefore the SSTs from CSM+GEA may be consistent with the true long-term climate state of the tropical Pacific.

7.5 Conclusions

The effects of enhanced shortwave absorption on simulations of the Pacific basin have been tested by comparing simulations from coupled climate models with and without absorption. The results show that the introduction of enhanced absorption significantly reduces the biases between the simulated and observed energy budgets of the western Pacific warm pool. The transition from overestimation to underestimation of the TWP SSTs is reduced by half. In addition, the bias in the annual mean SSTs in the central Pacific is reduced by approximately 1 K. In terms of these metrics, the introduction of enhanced absorption has improved the fidelity of the simulation of the TWP. The most important change is the reduction in compensating errors between insolation and latent-heat flux. In the absence of better theoretical understanding and experimental verification of enhanced absorption, these results should be understood as a sensitivity experiment rather than as a definitive simulation.

However, introduction of GEA degrades other aspects of the simulation. First, the variability of the model on intra-seasonal time scales characteristic of the Madden–Julian oscillation (MJO) is reduced relative to the standard model. Since the standard model already exhibits too little variability on these time scales, the model with enhanced absorption is clearly biased. It is possible that the reduction in variability is a feature peculiar to the convective parameterization currently employed in CSM, and future work will include tests of the variability in CSM+GEA with different treatments of convection. Second, the positive SST bias in the eastern sub-tropical Pacific is increased with the introduction of GEA. The standard model underestimates the stratus-cloud cover close to South America, and the resulting excess surface insolation heats the ocean mixed layer to temperatures higher than observed. The bias is increased in the model with enhanced absorption despite the reduction in insolation where clouds are present. The sensitivity of the SST bias in the eastern Pacific and its effects on the simulation of the rest of the Pacific will be examined in future work. The sensitivity could be quantified by improving the stratus-cloud cover in the model through a new cloud parameterization or insertion of satellite cloud amounts in the eastern Pacific.

Resolution of the existence and physical mechanisms of enhanced absorption is critical for further progress in climate modeling. If the enhancement is as large as some studies have suggested, the effects of enhanced absorption on the present-day climate are much larger than those from several climate forcings under more active investigation; for example the direct radiative effects of aerosols. It is essential that more carefully controlled experiments be conducted to test proposed mechanisms for the enhancement of shortwave absorption relative to standard radiative-transfer theory.

Acknowledgments

The author would like to thank James Hack and Jeffrey Kiehl (NCAR) for numerous discussions regarding the NCAR climate models and the effects of enhanced short-wave absorption. Conversations with Frank Bryan and William Large (NCAR) were very helpful in understanding the effects of enhanced solar absorption on ocean circulation. John Bergman (CIRES) kindly provided his shortwave radiative transfer code for generalized cloud overlap. Guang Zhang (SIO) provided the state variables and surface fluxes derived from the TOGA TAO array. Eric Malloney (University of Washington) and Lucrezia Ricciardulli (NCAR) kindly agreed to run their analyses of the variability in the modified climate model. The author would also like to acknowledge invaluable assistance in running the NCAR CSM from Lawrence Buja,

Gokhan Danabasoglu, Brian Kauffman, Jan Morzel, Nancy Norton, Jim Rosinski, and Marianna Vertenstein (NCAR). Esther Brady and Christine Shields (NCAR) assisted in the analysis of the simulated ocean circulation. Computational resources were provided by NCAR.

References

Arking, A. (1996). Absorption of solar energy in the atmosphere: discrepancy between model and observations. *Science* **273**, 779–82.

Bergman, J. W. and H. H. Hendon (1998). Calculating monthly radiative fluxes and heating rates from monthly cloud observations. *J. Atmos. Sci.* **55**, 3471–91.

Bonan, G. B. (1998). The land surface climatology of the NCAR Land Surface Model coupled to the NCAR Community Climate Model. *J. Clim.* **11**, 1307–26.

Boville, B. A. and P. R. Gent (1998). The NCAR Climate System Model, Version One. *J. Clim.* **11**, 1115–1130.

Bryan, F. O., B. G. Kauffman, W. G. Large, and P. R. Gent (1996). The NCAR CSM Flux Coupler. Technical Report NCAR/TN-424+STR, National Center for Atmospheric Research, Boulder, Colorado, 55pp.

Cess, R. D. and M. H. Zhang (1996). Response to "How much solar radiation do clouds absorb?" *Science* **271**, 1133–4.

Cess, R. D., M. H. Zhang, Y. Zhou, X. Jing, and V. Dvortsov (1996). Absorption of solar radiation by clouds: interpretations of satellite, surface, and aircraft measurements. *J. Geophys. Res.* **101**, 23299–309.

Cess, R. D., M. H. Zhang, P. Minnis *et al.* (1995). Absorption of solar radiation by clouds: observations versus models. *Science* **267**, 496–9.

Cess, R. D., M. H. Zhang, F. P. J. Valero *et al.* (1999). Absorption of solar radiation by the cloudy atmosphere. Further interpretations of collocated aircraft measurements. *J. Geophys. Res.* **104**, 2059–66.

Chou, M.-D. and W. Zhao (1997). Estimation and model validation of surface solar radiation and cloud radiative forcing using TOGA COARE measurements. *J. Clim.* **10**, 610–20.

Collins, W. D. (1998). A global signature of enhanced shortwave absorption by clouds. *J. Geophys. Res.* **103**, 31669–79.

Collins, W. D., W. C. Conant, and V. Ramanathan (1994). Earth radiation budget, clouds and climate sensitivity. In *The Chemistry of the Atmosphere: Its Impact on Global Change*, ed. J. G. Calvert, Oxford, Blackwell, pp. 207–216.

Cusack, S., A. Slingo, J. M. Edwards, and M. Wild (1999). The radiative impact of a simple aerosol climatology on the Hadley Centre atmospheric GCM. *Quart. J. Roy. Meteor. Soc.* **124**, 2517–26.

Fouquart, Y., J. C. Buriez, and M. Herman (1990). The influence of clouds on radiation: a climate-modeling perspective. *Rev. Geophys.* **28**, 145–66.

Garratt, J. R. (1994). Incoming shortwave fluxes at the surface – a comparison of GCM results with observations. *J. Clim.* **7**, 72–80.

Gent, P. R., J. Willebrand, T. J. McDougall, and J. C. McWilliams (1995). Parameterizing eddy-induced tracer transports in ocean circulation models. *J. Phys. Oceanogr.* **25**, 463–74.

Gent, P. R., F. O. Bryan, G. Danabasoglu *et al.* (1998). The NCAR Climate System Model global ocean component. *J. Clim.* **11**, 1287–306.

Gleckler, P. J., D. A. Randall, G. Boer *et al.* (1995). Cloud-radiative effects on implied oceanic energy transports as simulated by atmospheric general circulation models. *Geophys. Res. Lett.* **22**, 791–94.

Hack, J. J. (1998). Analysis of the improvement in implied meridional ocean energy transport as simulated by the NCAR CCM3. *J. Clim.* **11**, 1237–44.

Hack, J. J., J. T. Kiehl, and J. W. Hurrell (1998). The hydrologic and thermodynamic characteristics of the NCAR CCM3. *J. Clim.* **11**, 1179–206.

Harrison, E. F., P. Minnis, B. R. Barkstrom *et al.* (1990). Seasonal variation of cloud radiative forcing derived from the Earth Radiation Budget Experiment. *J. Geophys. Res.* **95**, 18687–703.

Ho, C.-H, M.-D. Chou, M. Suarez, and K.-M. Lau (1998). Effect of ice cloud on GCM climate simulations. *Geophys. Res. Lett.* **25**, 71–4.

Jing, X. D. and R. D. Cess (1998). Comparison of atmospheric clear-sky shortwave radiation models to collocated satellite and surface measurements in Canada. *J. Geophys. Res.* **103**, 28817–24.

Kato, S., T. P. Ackerman, E. E. Clothiaux *et al.* (1997). Uncertainties in modeled and measured clear-sky surface shortwave irradiances. *J. Geophys. Res.* **102**, 25881–98.

Kiehl, J. T. (1994). Clouds and their effects on the climate system. *Phys. Today* **47**, 36–42.

Kiehl (1998). Simulation of the tropical Pacific warm pool with the NCAR Climate System Model. *J. Clim.* **11**, 1342–55.

Kiehl, J. T. and K. E. Trenberth (1997). Earth's annual global mean energy budget. *Bull. Amer. Meteor. Soc.* **78**, 197–208.

Kiehl, J. T., J. J. Hack, M. H. Zhang, and R. D. Cess (1995). Sensitivity of a GCM climate to enhanced shortwave cloud absorption. *J. Clim.* **8**, 2200–12.

Kiehl, J. T., J. Hack, G. Bonan *et al.* (1996). Description of the NCAR Community Climate Model (CCM3), Technical Report NCAR/TN-420+STR, Boulder, CO, National Center for Atmospheric Research.

Kiehl, J. T., J. J. Hack, G. B. Bonan *et al.* (1998). The National Center for Atmospheric Research Community Climate Model: CCM3. *J. Clim.* **11**, 1131–49.

King, M. D., L. F. Radke, and P. V. Hobbs (1990). Determination of the spectral absorption of solar radiation by marine stratocumulus clouds from airborne measurements within clouds. *J. Atmos. Sci.* **47**, 894–907.

Li, Z. and L. Moreau (1996). Alteration of atmospheric solar absorption by clouds: simulation and observation. *J. Appl. Meteor.* **35**, 653–70.

Li, Z., H. Barker, and L. Moreau (1995). The variable effect of clouds on atmospheric absorption of solar radiation. *Nature* **376**, 486–90.

Li, Z., A. P. Trishchenko, H. W. Barker, G. L. Stephens, and P. Partain (1999). Analysis of Atmospheric Radiation Measurement (ARM) program's Enhanced Shortwave Experiment (ARESE) multiple data sets for studying cloud absorption. *J. Geophys. Res.* **104**, 19127–34.

Lubin, D., J. P. Chen, P. Pilewskie, V. Ramanathan, and F. P. J. Valero (1996). Microphysical examination of excess cloud absorption in the tropical atmosphere. *J. Geophys. Res.* **101**, 16961–72.

Monin, A. S. (1986). *An introduction to the theory of climate.* Dordrecht, Reidel.

Philander, S. G. (1990). *El Niño, La Niña, and the Southern Oscillation.* New York, NY, Academic Press.

Pilewskie, P. and F. P. J. Valero (1995). Direct observation of excess solar absorption by clouds. *Science* **267**, 1626–9.

Ramanathan, V. and A. M. Vogelmann (1997). Atmospheric greenhouse effect, excess solar absorption and the radiation budget: from the Arrhenius/Langley era to the 1990s. *Ambio* **26**, 38–46.

Ramanathan, V., R. D. Cess, E. F. Harrison *et al.* (1989). Cloud-radiative forcing and climate: results from the earth Radiation Budget Experiment. *Science* **243**, 57–63.

Ramanathan, V., B. Subasilar, G. Zhang *et al.* (1995). Warm pool heat budget and shortwave cloud forcing: a missing physics? *Science* **267**, 499–502.

Shea, D. J., K. E. Trenberth, and R. W. Reynolds (1992). A global monthly sea surface temperature climatology. *J. Clim.* **5**, 987–1001.

Solomon, S., R. W. Portmann, R. W. Sanders *et al.* (1999). On the role of nitrogen dioxide in the absorption of solar radiation. *J. Geophys. Res.* **104**, 12047–58.

Stephens, G. L. and S.-C. Tsay (1991). On the cloud absorption anomaly. *Quart. J. Roy. Meteor. Soc.* **116**, 671–704.

Trenberth, K. E. and A. Solomon (1994). The global heat balance: heat transports in the atmosphere and ocean. *Clim. Dyn.* **10**, 107–34.

Valero, F. P. J., R. D. Cess, M. Zhang *et al.* (1997). Absorption of solar radiation by clouds: interpretation of collocated aircraft measurements. *J. Geophys. Res.* **102**, 29917–27.

Valero, F. P. J., P. Minnis, S. K. Pope *et al.* (2000). The absorption of solar radiation by the atmosphere as determined using consistent satellite, aircraft, and surface data during the ARM Enhanced Shortwave Experiment (ARESE). *J. Geophys. Res.* **105**, 4743–58.

Waliser, D. E., W. D. Collins, and S. P. Anderson (1996). An estimate of the surface shortwave cloud forcing over the western Pacific during TOGA COARE. *Geophys. Res. Lett.* **23**, 519–22.

Weatherly, J. W., B. P. Briegleb, and W. G. Large (1998). Sea ice and polar climate in the NCAR CSM. *J. Clim.* **11**, 1472–86.

Webster, P. J. (1994). The role of hydrological processes in ocean–atmosphere interactions. *Rev. Geophys.* **32**, 427–76.

Webster, P. J. and R. Lukas (1992). TOGA COARE - the Coupled Ocean Atmosphere Response Experiment. *Bull. Amer. Meteor. Soc.* **73**, 1377–416.

Wild, M. and A. Ohmura (1999). The role of clouds and the cloud-free atmosphere in the problem of underestimated absorption of solar radiation in GCM atmospheres. *Phys. Chem. Earth B* **24**, 261–8.

Wild, M., A. Ohmura, H. Gilgen, and E. Roeckner (1995). Validation of general circulation model radiative fluxes using surface observations. *J. Clim.* **8**, 1309–24.

Wiscombe, W. J., R. M. Welch, and W. D. Hall (1984). The effects of very large drops on cloud absorption. Part I: parcel models. *J. Atmos. Sci.* **41**, 1336–55.

Yu, R. C., M. H. Zhang, and R. D. Cess (1999). Analysis of the atmospheric energy budget: a consistency study of available data sets. *J. Geophys. Res.* **104**, 9655–61.

Zhang, G. J. and M. J. McPhaden (1995). The relationship between sea surface temperature and latent heat flux in the equatorial Pacific. *J. Clim.* **8**, 589–605.

Zhang, M. H., R. D. Cess, and X. D. Jing (1997). Concerning the interpretation of enhanced cloud shortwave absorption using monthly-mean ERBE/GEBA measurements. *J. Geophys. Res.* **102**, 25899–905.

Zhang, M. H., W. Y. Lin, and J. T. Kiehl (1998). Bias of atmospheric shortwave absorption in the NCAR Community Climate Models 2 and 3: Comparison with monthly ERBE/GEBA measurements. *J. Geophys. Res.* **103**, 8919–25.

8

Cloud feedbacks

DAVID A. RANDALL

Department of Atmospheric Science, Colorado State University, Fort Collins, CO

MICHAEL E. SCHLESINGER

Department of Atmospheric Sciences, University of Illinois, Urbana, IL

VALENTIN MELESHKO

Main Geophysical Observatory, St Petersburg, Russia

VENER GALIN

Department of Numerical Mathematics, Russian Academy of Sciences, Moscow, Russia

JEAN-JACQUES MORCETTE

European Centre for Medium Range Weather Forecasts, Reading, UK

RICHARD WETHERALD

Geophysical Fluid Dynamics Laboratory, Princeton University, Princeton, NJ

8.1 Introduction

As discussed in Chapter 1, climate feedbacks are an integral aspect of the climate system. This chapter investigates the importance of cloud feedbacks. Although many of the best-known early climate models used prescribed clouds (e.g., Manabe and Bryan, 1969), the importance of potential changes in cloudiness for the problem of climate change has been recognized as a key factor since the 1970s (e.g., Arakawa, 1975; Charney *et al.*, 1979). In particular, it is now widely appreciated that "cloud feedback" is a key source of uncertainty limiting the reliability of simulations of anthropogenic climate change (e.g., Houghton *et al.*, 1990).

Nevertheless the whole concept of cloud feedback continues to be obscure, in part because the term "cloud feedback" is often used without being properly defined at all, and is rarely given a definition precise enough to show how it can be quantitatively measured. Further confusion arises from the fact that there are in fact many types of cloud feedbacks (e.g., Schneider, 1972; Schlesinger, 1985; 1988; 1989; Wielicki *et al.*, 1995). In addition, it is widely perceived that existing atmospheric general circulation models (AGCMs) are incapable of making quantitatively realistic simulations of cloudiness.

Frontiers of Climate Modeling, eds. J. T. Kiehl and V. Ramanathan.
Published by Cambridge University Press. © Cambridge University Press 2006.

The purposes of this chapter are to give a definition of cloud feedback, to discuss some particular types of cloud feedback, and to assess the prospects for simulations of cloud feedback on anthropogenic climate change.

Before embarking on an exploration of the many problems in simulating cloud feedbacks, we would like to point out that there are reasons to believe that these problems can be and are being overcome. The AGCMs which are embedded in climate models are in principle identical to the AGCMs used for global numerical weather prediction (NWP). Miller *et al.* (1999) present comparisons of cloud forecasts performed with the NWP model of the European Centre for Medium Range Weather Forecasts (ECMWF) with cloud observations obtained through the Lidar-in-Space Technology Experiment (LITE; McCormick *et al.*, 1993). The ECMWF model has high spatial resolution and incorporates many state-of-the-art physical parameterizations including the cloud parameterization of Tiedtke (1993). As shown in Figure 8.1, the cloud forecasts are quite successful overall.

In addition, Klein and Jakob (1999) present statistical comparisons of a large number of ECMWF cloud forecasts with cloud observations in extra-tropical cyclones. Figure 8.2 shows that again the cloud forecasts are very successful overall. It would be useful if the study of Klein and Jakob were extended to an analysis of the statistics of cloudiness in long "free runs" of the ECMWF model and/or other AGCMs. This would permit a comparison of the simulated and observed clouds associated with simulated and observed extra-tropical cyclones, respectively. Studies of this type, which relate to cloud feedbacks on synoptic time scales, will no doubt be carried out in the months and years ahead.

These studies demonstrate that GCMs can in fact predict realistic distributions of cloudiness in deterministic forecasts, for which the use of observed initial conditions ensures that the largescale dynamical structures are realistic.

While the results of Miller *et al.* (1999) and Klein and Jakob (1999) do not suffice to show that GCMs can realistically simulate changes in cloudiness that are associated with climate change, they are certainly promising and provide some grounds for optimism. In addition, they illustrate the enormous and as-yet-underexploited utility of NWP for the evaluation of parameterizations.

8.2 The nature of cloud feedback

Clouds affect the Earth system in a variety of ways (Arakawa, 1975). They are produced by and are host to a wide variety of microphysical/hydrological processes, including most fundamentally latent-heat release and precipitation. They strongly affect the flows of both longwave (LW) and shortwave (SW) radiation through the atmosphere. Finally, they are intimately associated with powerful microscale and mesoscale transport processes including deep penetrative convection (e.g., Arakawa and Schubert, 1974) and also the convective turbulence which fills stratiform clouds (e.g., Lilly, 1968), to the extent that a three-dimensional map of cloudiness is tantamount to a three-dimensional map of atmospheric turbulence (at least above the boundary layer). The various cloud processes correspond to tendency terms in the equations describing the evolution of the atmosphere, the oceans, and the land surface. All three types of cloud processes – microphysical/hydrological, radiative, and convective – exert major influences on the climate system.

Because clouds are formed in and affect the atmospheric circulation, the effects of clouds on climate are fundamentally dynamical in character. Clouds couple many processes together, over a very wide range of space and time scales (Arakawa, 1975), giving rise to cloud–climate feedbacks, which are of intense interest in the context of anthropogenic climate change. Many types of cloud–climate feedbacks have been identified (e.g., Senior and Mitchell, 1993; Fowler and Randall, 1994; Yao and Del Genio, 1999; Senior 1999), and probably there are some which have yet to be recognized. A survey is given by Wielicki *et al.* (1995). Since cloud influences involve microphysical and convective processes as well as radiative processes, cloud feedbacks are similarly diverse. For example, Schneider (1972) pointed out that cloud-radiative feedbacks can occur through changes in cloud amount, cloud top-height, and cloud optical properties. Several specific types of cloud feedback are discussed later in this chapter. The various cloud feedbacks can interact with each other through their collective effects on the climate system as a whole.

Any discussion of feedbacks in the climate system (or any other system) must carefully distinguish between internal and external processes. Internal processes are part-and-parcel of the system and so are affected by the state of the system and can interact with each other in potentially complex ways. External processes are unaffected by the state of the system. Only the interactive internal processes can "feed back" to influence the state of the system. The incident top-of-the-atmosphere (TOA) solar radiation is perhaps the most obvious example of a potentially variable process external to the climate system. Cloud processes, on the other hand, are definitely internal to the climate system.[1]

The effects of clouds are often discussed in terms of "cloud forcing," which refers to the effects of clouds on some climate process, most often the TOA radiation. As stressed above, cloud processes are internal to the climate system, and interact strongly with other climate processes. The term "forcing" has a strong connotation of externality. For this reason, we avoid using the term "cloud forcing." The effects of clouds on a process P will be denoted by $C(P)$. For example, the effects of clouds

---

[1] Cess (1976) used an analysis of observations to conclude that cloud feedbacks due to changes in cloud amount are not significant, but further investigation has led him to change his mind about this (fervent personal communications on numerous occasions).

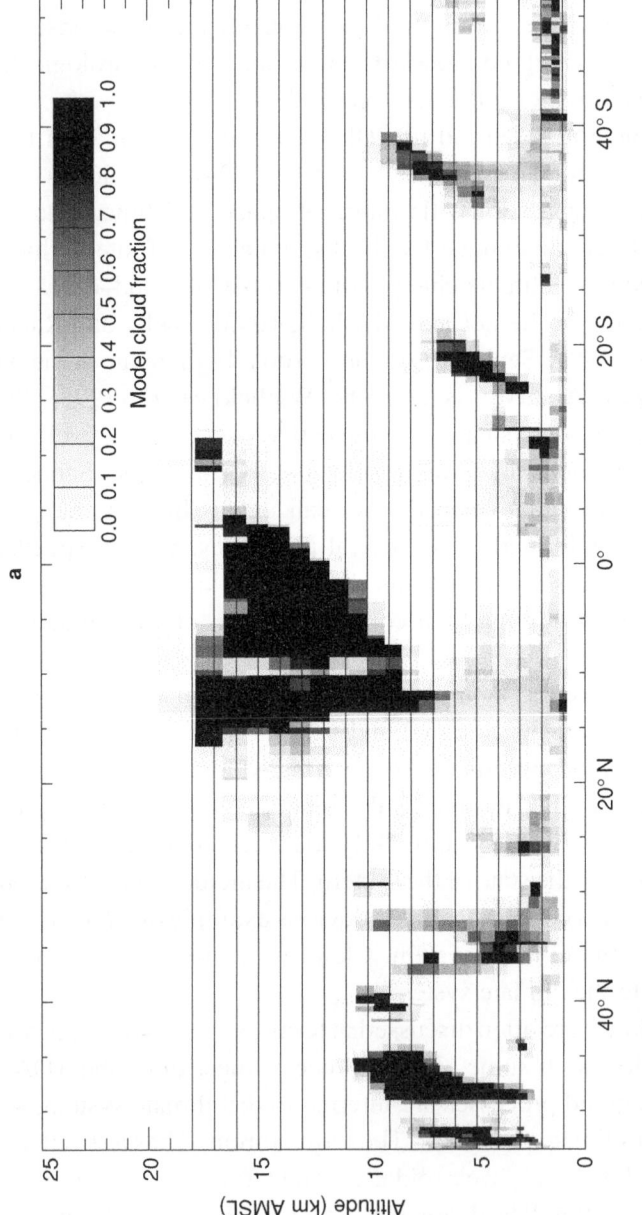

Figure 8.1. Cloud fraction comparison for LITE orbit 124 (September 16, 1994, 1425–1500 universal time coordinated spanning the western Pacific warm pool). The greyscale shading denotes cloud amount, ranging from 0 to 1, plotted as a function of height in km above mean sea level (AMSL), and latitude from 40° S to 40° N. (a) Model results from the ECMWF simulation; (b) the LITE observations, with ECMWF resolution. From Miller *et al.* (1999).

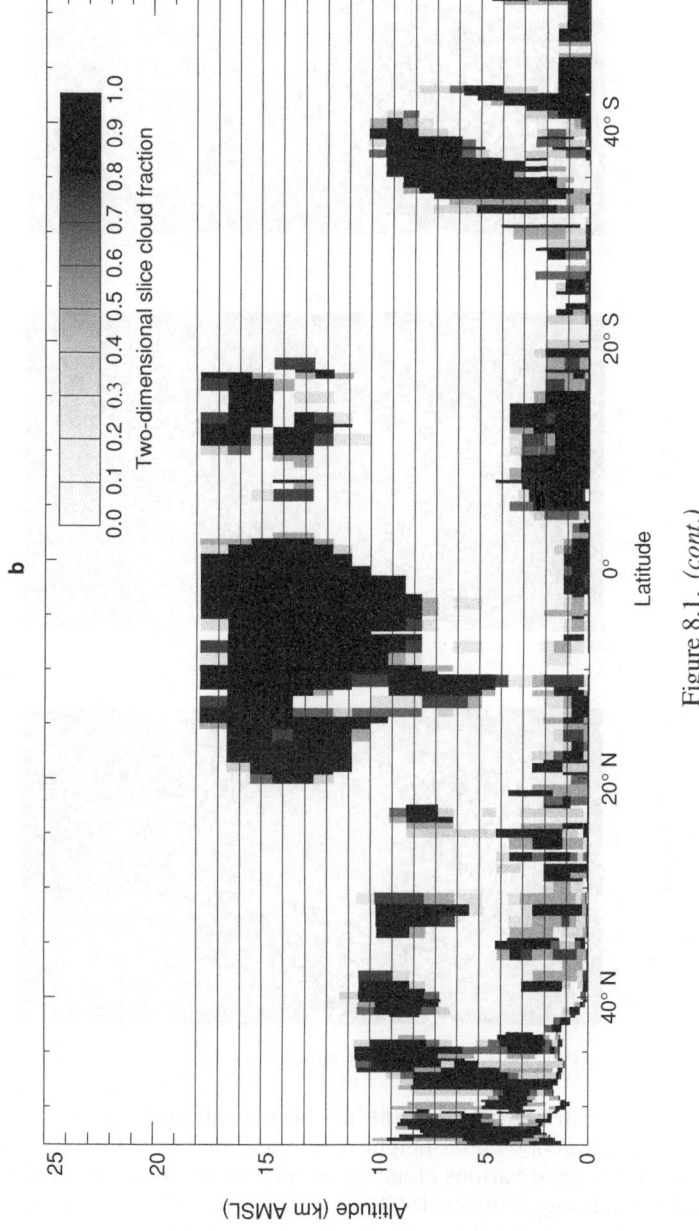

b

Two-dimensional slice cloud fraction

0.0 0.1 0.2 0.3 0.4 0.5 0.6 0.7 0.8 0.9 1.0

Altitude (km AMSL)

Latitude

Figure 8.1. *(cont.)*

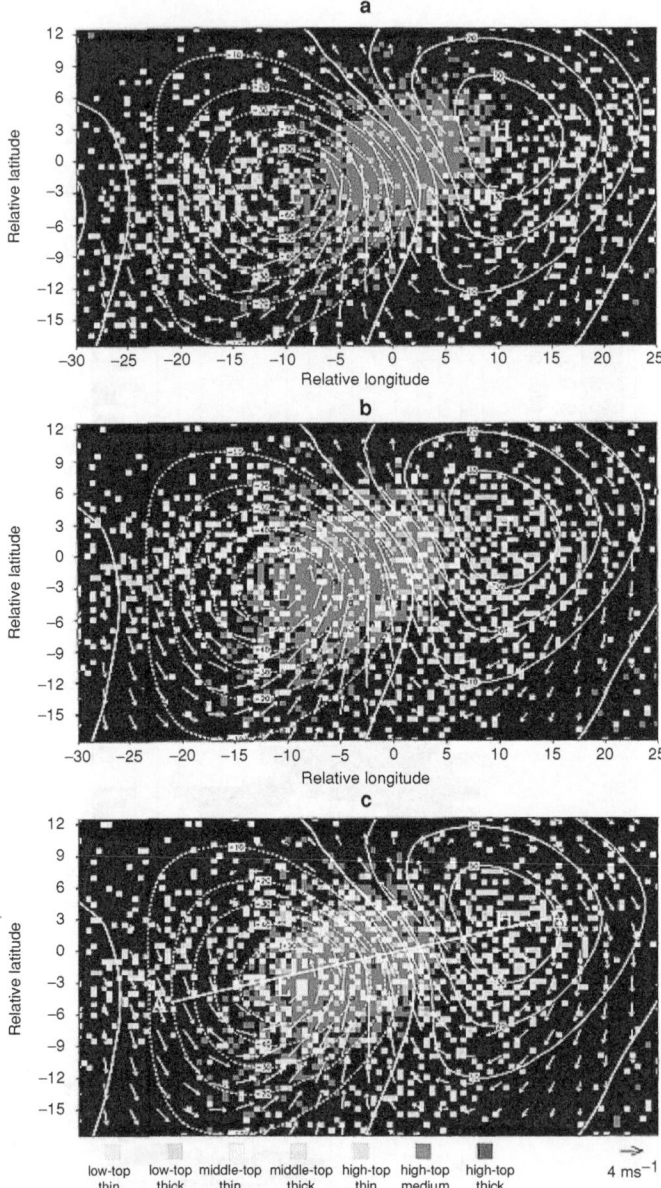

Figure 8.2. (a) Distributions of 1000-hPa horizontal wind (arrows, see scale at bottom right) and geopotential height (contours, interval 10 m) from ECMWF re-analysis (ERA), and various cloud types (pixels) from the International Satellite Cloud Climatology Project (ISCCP) observations. The ordinate (abscissa) of the coordinate system used here corresponds to latitudinal (longitudinal) displacements in degrees from the reference site. Inside each 2.5° by 2.5° grid box of this coordinate system, the presence and relative abundance of a certain cloud type is indicated by plotting a number of randomly scattered pixels

This figure is available for download in colour from
www.cambridge.org/9780521791328

on the outgoing longwave radiation will be represented by $C(R_\infty)$, where R is the net upward longwave radiation (in general a function of height), and the subscript ∞ denotes the "top of the atmosphere." In addition, the symbol C (no parentheses and no argument) will be used to denote cloud influences in general.

The state of the climate system changes with time due to both internal processes and external forcing. For convenience, we refer to these as internal variability and external variability, respectively. Examples of internal variability (sometimes called "free" variability) include baroclinic wave development, tropical intra-seasonal oscillations, and El Niño events. Familiar examples of time-dependent external forcing leading to external variability include the diurnal and seasonal cycles of solar forcing, and volcanic events, each of which can produce large-amplitude fluctuations of the climate system. We are particularly interested in the external variability due to the effects of the external forcings which arise from human activity and from the gradual changes in Earth's orbital parameters.

Each of the various cloud processes mentioned earlier has the potential to change as the climate state evolves due to internal variability and/or externally forced variability. *The change in a cloud process associated with a fluctuation of the climate state represents a cloud–climate feedback.* Climate feedbacks, including cloud–climate feedbacks, thus help to determine the amplitude and character of both internal and external variability. For example, Hall and Manabe (1999) have shown that the internal (unforced) variability of a climate model becomes unrealistically weak when the water-vapor feedback is artificially suppressed. Similarly, negative feedbacks act to reduce internal variability. This suggests that if clouds exert strong positive or negative feedbacks on climate, failure to represent these feedbacks in a model may result in unrealistically strong or weak free fluctuations of the climate system, so that the internal variability of a simulated equilibrium climate state will be unrealistic. This illustrates the important point that cloud–climate feedbacks are at work even when a climate system is in a statistical equilibrium. Cloud–climate feedbacks can thus influence the system even in the absence of climate change.

Figure 8.2. *(cont.)* designated to the cloud species in questions (see legend at bottom). Each pixel represents a 1% increment in cloud fraction; negative values of cloud fraction are not plotted. In this and all following figures, the composite data for all fields represent deviations from background levels estimated by averaging the values for the five-day period entered on the key dates. (b) As in (a), but for cloud data and dynamical fields from the twenty-four-hour ERA forecasts. Clouds in this figure are classified by their physical cloud-top pressure. (c) as in (b) but using emissivity-adjusted cloud-top pressure. From Klein and Jakob, 1999.

Cloud feedback as defined above is measurable, i.e., quantifiable from data, at least in principle; otherwise it would not be a proper subject for scientific study. Feedbacks are perhaps most readily measured through their influences on the internal and external variability of a system, i.e., by watching the system fluctuate. Cloud feedbacks on anthropogenic climate change can be observed by measuring the changes in cloud processes that occur over extended periods of time (decades or longer). Of course, we would like to measure the various cloud–climate feedbacks right now. One approach to doing so is to observe the changes in cloud processes which occur in the context of shorter-time-scale phenomena, such as El Niño events (e.g., Ramanathan and Collins, 1991), the seasonal cycle (e.g., Cess *et al.*, 1997), various quasi-repeatable weather events (e.g., Klein and Jakob, 1999), and the diurnal cycle (e.g., Hendon and Woodberry, 1993). An understanding of how clouds feed back on these relatively "fast" processes can aid us in understanding how clouds feed back in the more ponderous processes of forced and unforced decadal and centennial variability.

Figure 8.3 shows a possible example of an observed (Norris and Leovy, 1994) cloud feedback. The two panels of the figure show observed trends in sea-surface temperature (SST) and stratocumulus cloudiness, over a thirty-year period. Downward trends in SST are correlated with upward trends in stratocumulus amount, and vice versa. There are at least two possible interpretations of these correlated trends in SST and stratocumulus-cloud amount, which do not contradict each other. The first is that a cooling (warming) of the sea favors an increase (decrease) in stratus-cloud amount; this is plausible in light of our understanding of the physics of marine stratus clouds. The second interpretation is that an increase (decrease) in stratus cloud amount favors a decrease (increase) in the SST because the clouds reflect solar radiation which would otherwise be absorbed by the ocean.

In addition to measuring cloud–climate feedbacks, we would like to simulate them. Every climate-change simulation produces a simulation of cloud–climate feedbacks, but we have to ask whether the simulated feedbacks will be found to agree with the observed ones, when suitable observations ultimately become available. Simulations also depict cloud feedbacks on shorter time scales. A demonstration that AGCMs can realistically simulate short-term cloud feedbacks is a positive (though not sufficient) indication that the same models can realistically simulate long-term cloud–climate feedbacks.

8.3 An example of simulated cloud feedbacks

Let R_∞ be the net upward LW radiation at the TOA, S_∞ be the net downward SW radiation at the TOA, and $N \equiv S - R$ be the total radiation at the TOA, positive downward. We also define $(R_\infty)_{clr}$, $(S_\infty)_{clr}$, and $(N_\infty)_{clr}$ as the corresponding "clear-sky"

Stratus Trend (1952 1981)

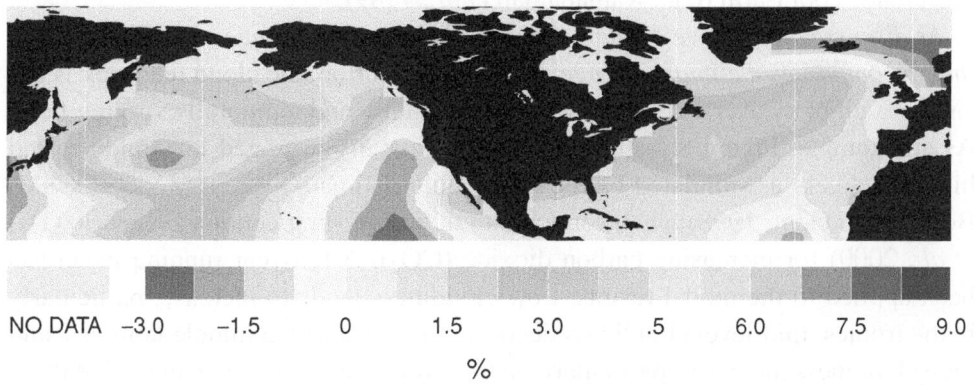

NO DATA −3.0   −1.5   0   1.5   3.0   .5   6.0   7.5   9.0

%

SST Trend (1952-1981)

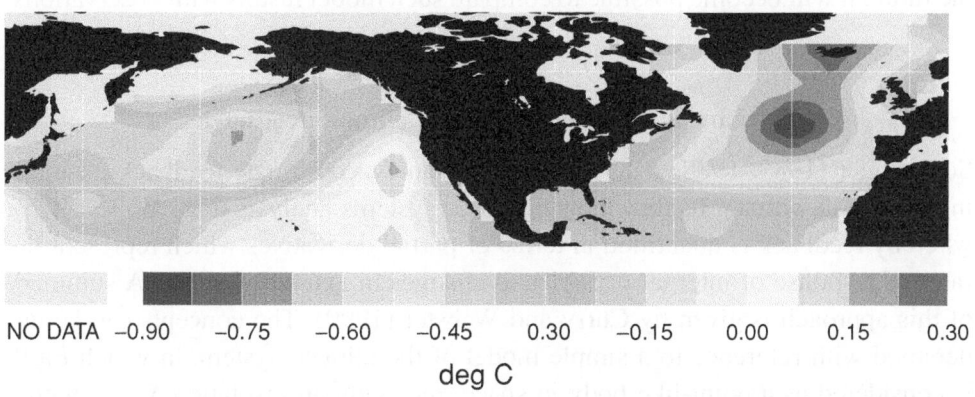

NO DATA −0.90   −0.75   −0.60   −0.45   0.30   −0.15   0.00   0.15   0.30

deg C

Figure 8.3. (a) Observed trends of stratus-cloud amount and (b) sea-surface temperature (SST), for the period 1952–1981. Adapted from Norris and Leovy, 1994. This figure is available for download in colour from www.cambridge.org/9780521791328

fluxes, i.e., the fluxes which would occur if no clouds existed but the system was otherwise unchanged. Using the notation introduced earlier, cloud effects on the LW, SW, and net TOA radiation are then represented by Equations (8.1), (8.2), and (8.3), respectively:

$$\mathcal{C}(R_\infty) \equiv R_\infty - (R_\infty)_{\text{clr}}, \tag{8.1}$$

$$\mathcal{C}(S_\infty) \equiv S_\infty - (S_\infty)_{\text{clr}}, \tag{8.2}$$

$$\mathcal{C}(N_\infty) = \mathcal{C}(S_\infty) - \mathcal{C}(R_\infty). \tag{8.3}$$

With these definitions, we expect $\mathcal{C}(R_\infty) < 0$ and $\mathcal{C}(S_\infty) < 0$. It is observed that in the global mean $C(S_\infty)$ dominates, so that the global mean of the $C(N_\infty)$ is negative, i.e., clouds cool Earth (e.g., Ramanathan *et al.*, 1989).

As discussed above, we define a cloud feedback as the change in a cloud process that accompanies a climate change. With this definition, the cloud feedback is not simply a globally averaged quantity; it is a spatially and temporally varying field. As an example, Figure 8.4 shows changes in zonally averaged low, middle, and high cloudiness as simulated by the Community Climate System Model (CCSM; Boville and Gent, 1998) when forced with a "twenty-first-century" scenario (Dai *et al.*, 2000) for increasing carbon dioxide (CO_2). A ten-year running mean has been applied to the model results. Low cloudiness tends to increase, particularly in the tropics, mid-level cloudiness decreases, particularly in middle latitudes, and high cloudiness increases, particularly in the Arctic. Note, however, that all of these simulated trends are rather weak. Figure 8.5 shows the corresponding simulated trends in $C(S_\infty)$, $-C(R_\infty)$, and $C(N_\infty)$. Again the trends are weak. This model produces only weak cloud–climate feedbacks in terms of the radiation at the TOA. In the future it will become possible to compare such model results with observations.

8.4 Linear systems analysis of climate feedbacks

Schlesinger (1985; 1988; 1989) defined and analyzed climate feedbacks, including the cloud–climate feedback, using linear systems analysis (e.g., Bode, 1975), whereby feedback is quantified in terms of partial derivatives which represent the rates of response of internal variables to changes in external forcing. A summary of this approach is given by Curry and Webster (1999). The concepts can be understood with reference to a simple model of the climate system, in which Earth is considered as a point-like body in space, receiving an insolation $S \downarrow_\infty$, with a planetary albedo α, a bulk emissivity ϵ, and a global mean surface temperature T_S, such that the globally averaged outgoing LW radiation per unit area, R_∞ is given by:

$$R_\infty = \epsilon \sigma T_S^4. \tag{8.4}$$

Here σ is the Stefan–Boltzmann constant. For Earth, $S \downarrow_\infty = 1370$ W m^{-2}, $\alpha = 0.3$, $T_S = 288K$, $\epsilon = 0.6$, and $R_\infty = 240$ W m^{-2}. The radiation budget in equilibrium is expressed by $\pi a^2 S \downarrow_\infty (1 - \alpha) = 4\pi a^2 \epsilon \sigma T_S^4$, where a is the radius of the Earth. After simplifying, we obtain:

$$4\epsilon \sigma T_S^4 = S \downarrow_\infty (1 - \alpha) = S_\infty. \tag{8.5}$$

Here T_S is clearly an internal variable, and $S \downarrow_\infty$ is clearly an external variable.

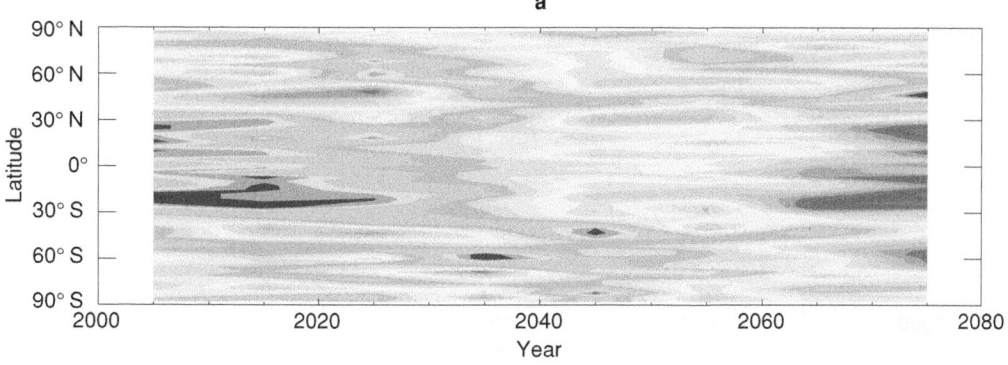

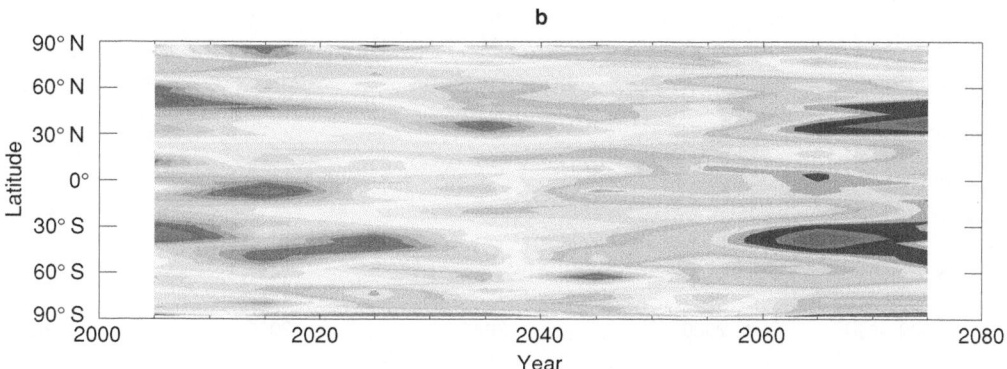

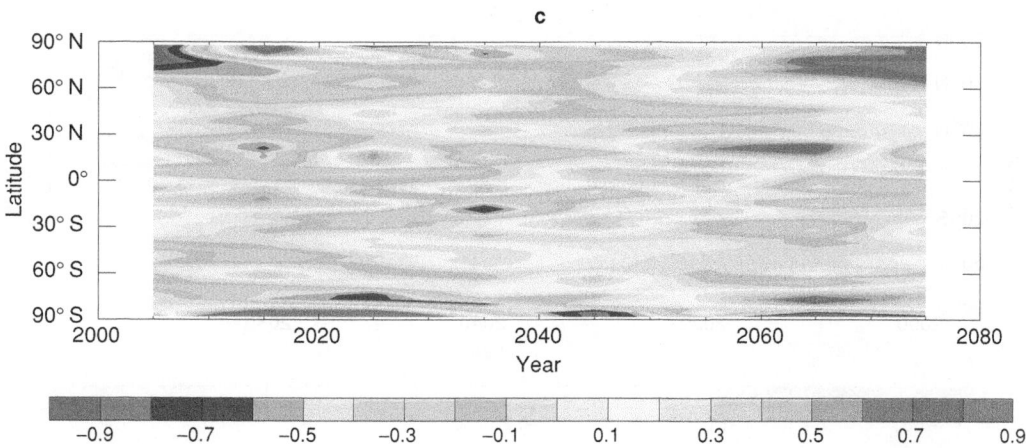

Figure 8.4. Zonally averaged changes in (a) low-, (b) middle-, and (c) high-cloud amount, as simulated by the CCSM. The plots show the departures from the time mean in each latitude (%).

This figure is available for download in colour from
www.cambridge.org/9780521791328

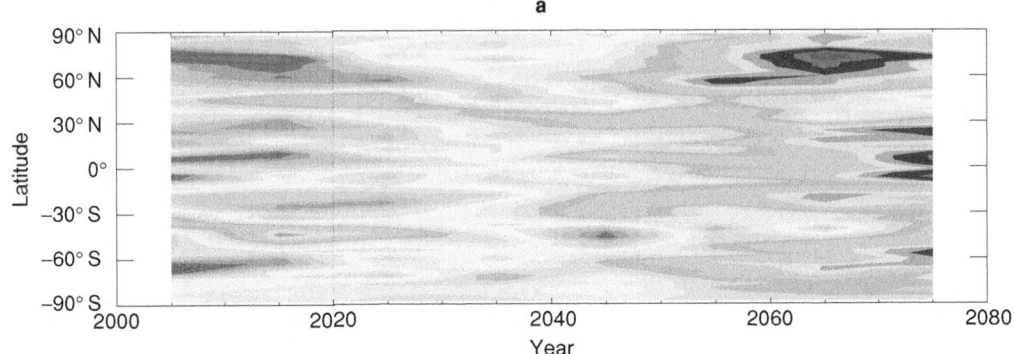

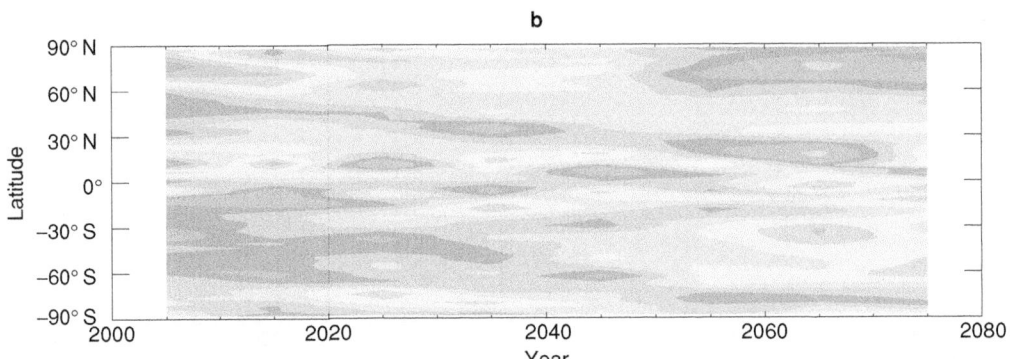

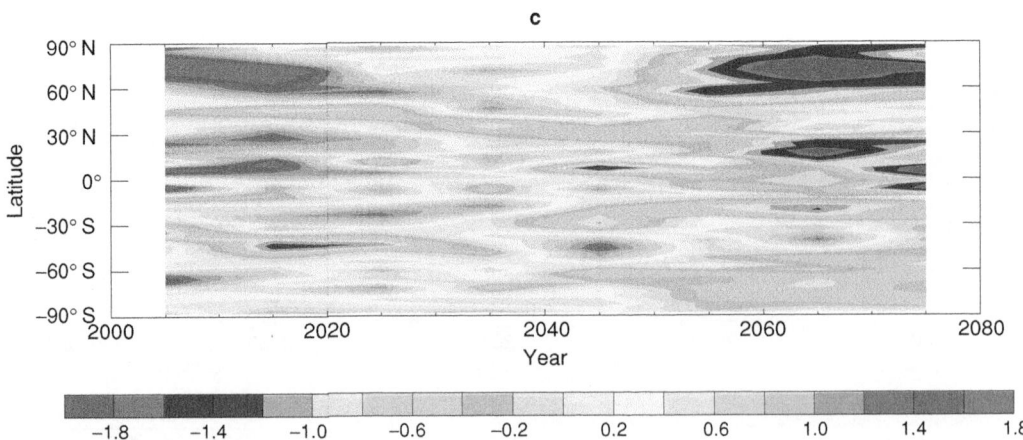

Figure 8.5. As in Figure 8.4, but the trends of (a) $\mathcal{C}(S_\infty)$, (b) $-\mathcal{C}(R_\infty)$, and (c) $\mathcal{C}(N_\infty)$, values in W m^{-2}. Note that $-\mathcal{C}(R_\infty)$ is sometimes called the "longwave cloud forcing."

This figure is available for download in colour from
www.cambridge.org/9780521791328

We linearize about an equilibrium "base" state in which $S\downarrow_\infty = S\downarrow_0$, $\alpha = \alpha_0$, $\epsilon = \epsilon_0$, and $T_S = T_0$, so that the base state satisfies:

$$4\epsilon_0 \sigma T_0^4 = S\downarrow_0 (1 - \alpha_0). \tag{8.6}$$

A linearized version of (8.5) is

$$16\epsilon_0 \sigma T_0^3 \Delta T_S + 4\Delta\epsilon\sigma T_0^4 = \Delta S\downarrow (1 - \alpha_0) - S\downarrow_0 \Delta\alpha. \tag{8.7}$$

Here $\Delta S\downarrow$ is the perturbation to $S\downarrow_\infty$, and ΔT_S, $\Delta\epsilon$, and $\Delta\alpha$ are defined similarly. The left-hand side of Equation (8.7) represents the perturbation of R_∞, and the right-hand side represents the perturbation of $S\downarrow_\infty$. Note that the implied perturbation of N_∞ is zero, because we assume that both the base state and the perturbed state are in equilibrium.

The bulk emissivity can have both an externally modulated component (e.g., due to anthropogenic changes in CO_2) and an internally varying component (e.g., due to changing amounts of cloudiness and/or water vapor), so that we can write

$$\Delta\epsilon = (\Delta\epsilon)_{ext} + (\Delta\epsilon)_{int}. \tag{8.8}$$

Similarly, the planetary albedo can have both an externally modulated component (e.g., due to volcanic events) and an internally varying component (e.g., due to changes in cloudiness and/or snow and ice cover):

$$\Delta\alpha = (\Delta\alpha)_{ext} + (\Delta\alpha)_{int}. \tag{8.9}$$

An external perturbation or "forcing" of the system can arise from one or more of several possible causes: changes in $S\downarrow_\infty$ from its nominal value $S\downarrow_0$, and/or non-zero values of $(\Delta\epsilon)_{ext}$, and/or non-zero values of $(\Delta\alpha)_{ext}$. The response of the system can be measured in terms of the *equilibrium* changes in the various internal variables, which are T_S, ϵ_{int}, and α_{int}.

Now introduce feedbacks. Let $(\Delta\alpha)_{int}$ be defined by:

$$(\Delta\alpha)_{int} = (\Delta\alpha)_{ice,\ clr} - \frac{\Delta[C(S_\infty)]}{S\downarrow_0} \tag{8.10}$$

$$= -I\Delta T_S - A\Delta T_S, \tag{8.11}$$

and $\Delta\epsilon$ by

$$\Delta\epsilon = (\Delta\epsilon)_{CO_2,clr} + (\Delta\epsilon)_{int} \tag{8.12}$$

$$= (\Delta\epsilon)_{CO_2,clr} + (\Delta\epsilon)_{vap,\ clr} + \frac{\Delta[C(R_\infty)]}{\sigma T_0^4} \tag{8.13}$$

$$= (\Delta\epsilon)_{CO_2,clr} - V\Delta T_S - E\Delta T_S. \tag{8.14}$$

where $(\Delta\alpha)_{ice,\ clr} = -I\Delta T_S$ represents the change in the *clear-sky* albedo due to melting ice and snow, $(\Delta\epsilon)_{CO_2,clr}$ represents the externally forced change in the *clear-sky* bulk emittance due to changes in the atmospheric CO_2 concentration, and $(\Delta\epsilon)_{vap,clr} = -V\Delta T_s$ represents the change in the *clear-sky* bulk emittance due to the changing water-vapor content of the atmosphere. The notation $\Delta[C(S_\infty)]$ means the perturbation to $C(S_\infty)$, and $\Delta[C(R_\infty)]$ is defined similarly. The quantities $A, I, E,$ and V are essentially partial derivatives. For example,

$$E\Delta T_S = -\frac{\Delta[C(R_\infty)]}{\sigma T_0^4}, \tag{8.15}$$

implies that

$$E = \frac{-1}{\sigma T_0^4} \frac{\Delta[C(R_\infty)]}{\Delta T_S}, \tag{8.16}$$

i.e., E is a normalized partial derivative of $C(R_\infty)$ with respect to T_S. Similarly,

$$A = \frac{1}{S\downarrow_0} \frac{\Delta C(S_\infty)}{\Delta T_S} \tag{8.17}$$

is a normalized partial derivative of $\Delta C(S_\infty)$ with respect to T_S.

In (8.11) and (8.14) we have assumed that $(\Delta\alpha)_{int}$ and $(\Delta\epsilon)_{int}$ can be written as functions of T_S only. Given the overwhelming complexity of the climate system, this appears to be a rather drastic assumption. It can be rationalized to some extent, however, by thinking of T_S as an "index" of the climate equilibrium state; the linearization used above will be useful to the extent that $(\Delta\alpha)_{int}$ and $(\Delta\epsilon)_{int}$ have one-to-one (i.e., single-valued) relationships with T_S in the neighborhood of the base state about which the linearization is performed. The existence of such one-to-one relationships is plausible, if by no means assured.

It should also be noted that our analysis is linear. Equation (8.7) was derived through linearization of the Planck function. Moreover, we are representing the feedbacks as first-order derivatives of T_S in the vicinity of the base state; we neglect higher-than-first-order derivatives, which is an acceptable approximation if ΔT_S is sufficiently small. These linearized feedbacks are conceptually compatible with the linearization already used to obtain (8.7). Note, however, that the real climate system, when subjected to real perturbations of interest, may or may not behave linearly.

With the use of Equations (8.11) and (8.14), Equation (8.7) can be written as

$$16\epsilon_0\sigma T_0^3 \Delta T_S + 4\sigma T_0^4[(\Delta\epsilon)_{CO_2,clr} - (E+V)\Delta T_S] \tag{8.18}$$

$$= \Delta S\downarrow(1-\alpha_0) + S\downarrow_0[(A+I)\Delta T_S], \tag{8.19}$$

which can be solved for ΔT_S, yielding:

$$\Delta T_S = \frac{G_0[-\sigma T_0^4(\Delta \epsilon)_{CO_2,clr} + \frac{\Delta S\downarrow(1-\alpha_0)}{4}]}{1 - G_0 F}. \tag{8.20}$$

We define G_0 and F as

$$G_0 \equiv \frac{T_0}{4\epsilon_0 \sigma T_0^4} = \frac{T_0}{(1-\alpha_0)S\downarrow_0} = 0.3 \text{ K (W m}^{-2})^{-1}, \tag{8.21}$$

$$F \equiv \sigma T_0^4(E + V) + \frac{S\downarrow_0}{4}(A + I). \tag{8.22}$$

External forcing enters through the numerator of Equation (8.20); in fact, $-4\sigma T_0^4(\Delta \epsilon)_{CO_2,clr} - \Delta S \downarrow (1 - \alpha_0)$ is the change in N_∞ which would occur *instantaneously* if the imposed forcing were suddenly "switched on." Of course, after equilibration the change in N_∞ must be zero. Feedbacks enter through the denominator of (8.20), via the parameter F, which is given, according to (8.22), in terms of A, I, E, and V. We can thus identify A, I, E, and V as "feedback parameters."

Equations (8.16) and (8.17) show how the feedback parameters E and A can be computed in terms of partial derivatives; similar formulae can be given for I and V. Also note that the various feedbacks simply add in (8.22). Positive (negative) values of either $A + I$, or $E + V$ will give positive (negative) feedbacks. The cloud feedback parameters are A and E.

We can imagine a case in which the feedbacks make the denominator on the right-hand side of (8.20) zero, implying an infinite response to a finite forcing. One interpretation would be that the climate system will be unable to achieve equilibrium with the perturbed forcing. A more modest interpretation would be that the linearizations used in the derivation of the model cause the analysis to break down in such a case.

For a CO_2 perturbation in the absence of feedbacks, (8.20) reduces to

$$\Delta T_S = -G_0[(\Delta \epsilon)_{CO_2,clr}\sigma T_0^4] = \frac{-T_0}{4}\left[\frac{(\Delta \epsilon)_{CO_2,clr}}{\epsilon_0}\right]. \tag{8.23}$$

If we instantaneously double CO_2, the effect is to reduce ϵ in such a way that R_∞ instantaneously decreases by about 4 W m^{-2}, thus disrupting the equilibrium. Therefore (8.21) and (8.23) give

$$\Delta T_S = 0.3 \text{ K/(W m}^{-2})(4 \text{ W m}^{-2}) = 1.2 \text{ K}. \tag{8.24}$$

Feedbacks can either increase or decrease this response.

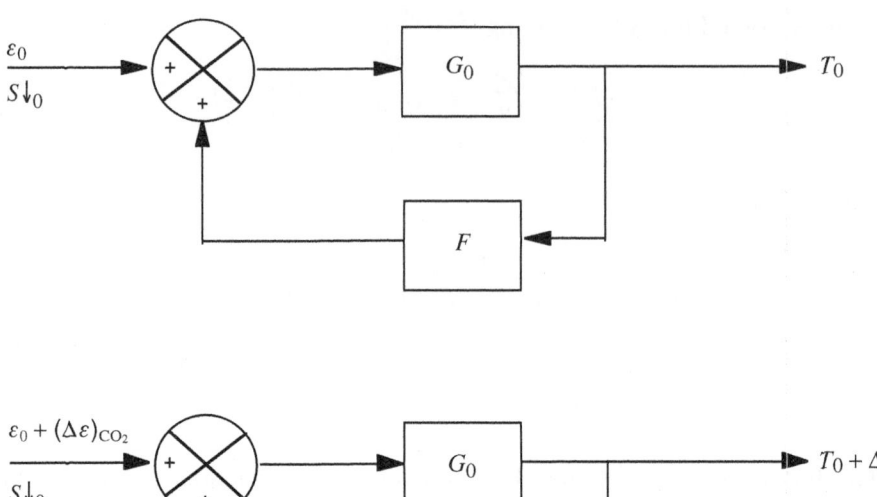

Figure 8.6. "Block diagrams" illustrating the approach of linear systems analysis to the analysis of climate feedbacks. In (a) the external forcing consists of the incident solar radiation, $S\downarrow_0$, and a particular atmospheric composition which gives rise to the bulk emissivity ϵ_0. The response of the system to the forcing is measured in terms of the surface temperature T_S, which is (necessarily) an internal parameter. The parameter G_0 is a "transfer function" which relates the forcing to the response; it is the "gain", which can be defined as the response divided by the forcing, in the absence of feedback. The lower diagram shows a perturbed state of the system. In (b) $(\Delta\epsilon)_{CO_2,clr}$ and ΔT_S represent perturbations to the forcing and the response, respectively. In this example the solar forcing is assumed to be unperturbed. The feedback of the system is represented by F. See text for details. Adapted from Schlesinger, 1985; 1988; 1989.

We now give a graphical interpretation of the preceding discussion. Consider the two "block diagrams" shown in Figure 8.6, which can be considered as an illustration of the linearized model discussed above. The diagram illustrates the logical structure of a system which is subjected to a forcing and produces a response. The system's response to external forcing is determined by the "transfer function," G_0, defined by Equation (8.21), and also called the "zero-feedback gain." Feedbacks are represented by F.

The upper diagram in Figure 8.6 illustrates the base state of the system, in which external forcings given by $S\downarrow_0$ and ϵ_0 give rise to a response T_0. (We write the infrared part of the forcing as ϵ_0 rather than as $\epsilon_0 \sigma T_0^4$ because the latter form would express the forcing partly in terms of an internal parameter, namely T_0.) The base state satisfies:

$$T_0 = G_0 S \downarrow_0 (1 - \alpha_0). \tag{8.25}$$

Note that the feedback parameter, F, does not appear in this equation. The perturbation (relative to the base state) forcing in the lower panel of Figure 8.6 is given by $(\Delta \epsilon)_{CO_2,clr}$; for simplicity, we have assumed that $\Delta S \downarrow = 0$. In the absence of feedback the response of the climate system to the perturbation of the forcing is given by (8.19). When feedback is active (i.e., for $F \neq 0$), ΔT_S becomes

$$\Delta T_S = G_0[-\sigma T_0^4 (\Delta \epsilon)_{CO_2,clr} + F \Delta T_S], \tag{8.26}$$

which gives

$$\Delta T_S = \frac{-G_0 \sigma T_0^4 (\Delta \epsilon)_{CO_2,clr}}{1 - G_0 F}. \tag{8.27}$$

Compare with (8.20). The feedback parameter does appear in (8.27), but it did not appear in (8.25). Nevertheless, the physical processes which give rise to feedback are operating in both the base state and the perturbed state.

The coefficients A and E were introduced by way of linearization; they essentially represent partial derivatives of the albedo and emissivity, respectively, with respect to the surface temperature, which is used here as a sole indicator of the climate state. It is feasible to evaluate these partial derivatives through the use of models; examples are given by Schlesinger (1985; 1988; 1989) and Curry and Webster (1999). It is also possible to evaluate the partial derivatives through the use of data.

The linear systems analysis does not in itself explicitly represent the physical processes responsible for the various feedbacks. The complex physical processes associated with cloud and water-vapor feedbacks are (purportedly) included in detailed climate models, and attempts have been made to incorporate them into simple climate models (e.g., Kelly *et al.*, 1999).

Note, however, that the preceding discussion of cloud feedback in terms of linear systems analysis is formulated entirely in terms of globally averaged quantities. In practice we are interested in analyzing the external forcing, the response, and the feedbacks *as spatially distributed fields*. In fact, an understanding of spatial variations of $C(R_\infty)$, $C(S_\infty)$, and $C(N_\infty)$ and other measures of cloud feedback is necessary because of Earth's wide variety of cloud regimes and the very different circulation regimes in which they occur. It must be expected that the surface temperature and various feedbacks, including cloud feedbacks, would vary in spatially and/or seasonally correlated ways. In addition, we must understand the physical processes at work in producing these spatially distributed fields. Linear systems analysis does not address these processes.

As an example, consider the potential contribution to cloud–climate feedback of a possible change in $C(S_\infty)$ over the eastern North Pacific Ocean, associated with changes in the amount of marine stratocumulus cloudiness there. Among the parameters which are believed to affect marine stratocumulus-cloud amount are the sea-surface temperature, the large-scale vertical motion, the temperature and moisture jumps across the top of the marine layer, the tendencies of marine-layer temperature and moisture due to horizontal advection, and the wind speed and direction including the shear across the top of the marine layer. These quantities are, in turn, affected by various processes in remote locations on Earth; for example, the large-scale vertical motion over the eastern North Pacific Ocean is affected by the strength of the North American monsoon (e.g., Rodwell and Hoskins, 2000).

Despite this complexity, the geographical structure of cloud feedback can be observed, simulated, and otherwise analyzed. For example, the spatial distributions of $C(R_\infty)$, $C(S_\infty)$, and $C(N_\infty)$, and their changes over time, can be studied through both observational and modeling approaches. An example has already been provided in Figure 8.5.

8.5 Can we simulate cloud–climate feedbacks accurately?

In a widely cited analysis, Cess *et al.* (1989, 1990; hereafter C89) presented the results of idealized numerical experiments designed to investigate the role of cloud feedback in climate change. The experiments were performed with more than a dozen AGCMs, and in fact C89 was the first and one of the most successful "intercomparison" activities undertaken by the AGCM community.

Among the most fundamental aspects of climate change is the complex seasonally varying spatial pattern of changes in the SST distribution. A climate forcing, such as the TOA radiation changes associated with increased CO_2, leads, through a complex process with a lag time of decades, to changes in the SST. In order to predict how the SST changes in response to the imposed radiative forcing, climate models must include ocean sub-models. Cess *et al.*, in C89, cleverly elected to "solve the problem backwards" by prescribing the SST change in a simplified manner, and computing the implied TOA radiative forcing. This elegant strategy eliminates the need for ocean sub-models and long integrations. For simplicity, C89 prescribed globally uniform changes in the SST, and made runs with SSTs increased by 2 K, decreased by 2 K, and also fixed at their observed (July) values. For further discussion of the experiment design, see C89.

C89 defined a "climate-sensitivity parameter," λ, as the ratio of ΔT_S to ΔQ, which is defined as the imposed *instantaneous* change in N_∞ which would occur if

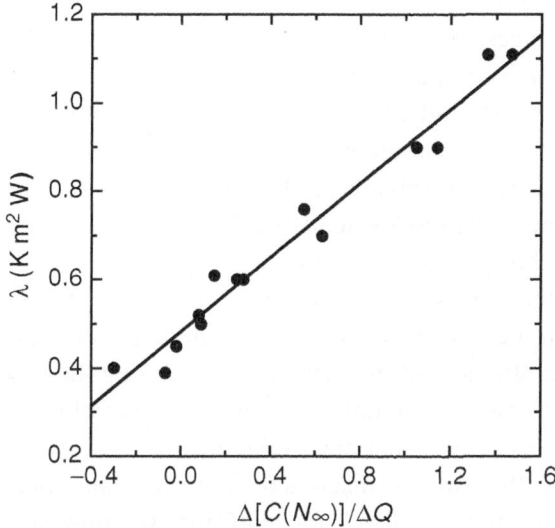

Figure 8.7. The climate-sensitivity parameter λ plotted against the cloud-feedback parameter $\Delta[\mathcal{C}(N_\infty)]/\Delta Q$ produced in 14 GCM simulations. From Cess *et al.*, 1989.

the forcing were suddenly switched on; once again, it is important to keep in mind that in equilibrium $N_\infty = 0$ and, therefore, $\Delta N_\infty = 0$. In the parlance of linear systems analysis, we obtain

$$\lambda = \frac{\Delta T_S}{\Delta Q} = \frac{G_0}{1-f}, \tag{8.28}$$

where $f = G_0 F$. Consistent with the discussion given in the preceding sub-section, C89 defined the cloud feedback in terms of the change in the globally averaged $\mathcal{C}(N_\infty)$, i.e., $\Delta[\mathcal{C}(N_\infty)]$, in response to the imposed SST changes.

Because of the idealized nature of the C89 experiment, the results cannot be used to draw any conclusions about real climate-change scenarios. The idealizations drastically simplify the interpretation of the results, however, and do in fact permit interesting and important conclusions to be drawn. The key findings of C89 can be summarized as follows.

- The climate-sensitivity parameter varied by roughly a factor of three among the models.
- Inter-model differences in the cloud feedback accounted for almost all of the inter-model differences in the climate-sensitivity parameter. This is illustrated in Figure 8.7.
- The "clear-sky" climate-sensitivity parameter, defined by analogy with λ but using the clear-sky TOA radiation in place of the all-sky TOA radiation, agreed very well among the models.

Cess *et al.* (1996) showed that updated versions of the AGCMs exhibit smaller inter-model differences; they note, however, that this does not necessarily indicate that we have arrived at a better understanding of the cloud–climate feedback problem. In a follow-up study, Cess *et al.* (1997) presented results from a further intercomparison based on simulated (and observed) changes in the $C(N_\infty)$ in response to a realistic *seasonal* change, rather than an idealized climate change. They concluded that none of the models realistically reproduced the observed seasonally varying $C(N_\infty)$, although some did better than others.

On the basis of the results of C89 and Cess *et al.* (1996), the question posed by the title of this section, i.e., "Can we accurately simulate cloud–climate feedbacks?," must be answered: "In view of the discrepancies among the model results, at least some of the current models cannot accurately simulate cloud–climate feedbacks, and it remains to be seen whether any of them can."

A limitation of the studies discussed above is that none of them focused on specific cloud–climate-feedback mechanisms. We now consider several such mechanisms.

8.6 A global radiative–convective feedback

The observed globally averaged energy budgets of Earth's surface and the atmosphere are discussed, for example, by Peixóto and Oort (1992). For Earth's surface, the globally averaged net radiative heating approximately balances the globally averaged evaporative cooling.[2] For the atmosphere, the globally averaged net radiative cooling approximately balances the globally averaged latent-heat release.[3] In short, the hydrologic cycle plays a dominant role in the energy budgets of both Earth's surface and the atmosphere.

Radiatively active clouds are themselves products of the hydrologic cycle. The clouds strongly modulate both the net surface radiative heating and the net atmospheric radiative cooling. Clouds reduce the solar radiation absorbed by Earth's surface, but increase the net infrared radiation impinging on the surface. Clouds reduce the solar radiation absorbed by water vapor in the lower troposphere, and they also warm the atmosphere by reducing the infrared emission to space by water vapor and CO_2.

We now explore these ideas more quantitatively, but still in a simplified framework. As discussed above, the globally averaged *atmospheric* energy balance is

---

[2] The surface sensible heat flux plays a relatively small role in the global-mean surface-energy budget, although it can be very important locally.

[3] The surface sensible heat flux plays a relatively small role in the global mean atmospheric energy budget.

approximately expressed by

$$SW_{atm} - LW_{atm} + LP \cong 0. \tag{8.29}$$

Here $SW_{atm} \equiv S_\infty - S_S$ and $LW_{atm} \equiv R_\infty - R_S$ are the globally averaged net SW heating and LW cooling of the atmosphere, respectively, and LP is the globally averaged net latent heating of the atmosphere. Each of these quantities is an energy flux. We can make cloud effects explicit by writing

$$SW_{atm} - LW_{atm} \cong SW_{atmclr} - LW_{atmclr} - C(R_\infty) + C(S_\infty - S_S). \tag{8.30}$$

Here SW_{atmclr} and LW_{atmclr} are the clear-sky net SW heating and LW cooling of the global atmosphere, respectively. We assume here, for simplicity, that $C(S_\infty - S_S)$ is negligible and that $C(R_\infty)$ modulates the longwave cooling of the atmosphere. For example, an increase in cirrus cloudiness will cause $C(R_\infty)$ to become more negative, and this will tend to reduce the net radiative cooling of the atmosphere.

Both precipitation and $C(R_\infty)$ are associated with deep convection, so it is useful to define

$$\beta \equiv \frac{-C(R_\infty)}{LP} \cong \frac{30 \text{ W m}^{-2}}{90 \text{ W m}^{-2}} = \frac{1}{3}. \tag{8.31}$$

Here the observed value of $C(R_\infty)$ is based on the study of Ramanathan *et al.* (1989), and the observed value of LP is based on the globally averaged precipitation rate of 3 mm d^{-1} (e.g., Peixóto and Oort, 1990). This non-dimensional parameter, β, measures the value of $C(R_\infty)$, relative to the rate of latent-heat release. Roughly speaking, the value of β tells how much cirrus cloud is produced per unit of precipitation. Substitution of (8.30) and (8.31) into (8.29) gives

$$LP = \frac{-(SW_{atmclr} - LW_{atmclr})}{1 + \beta}. \tag{8.32}$$

The (positive) numerator of (8.32) is the net clear-sky atmospheric radiative cooling, and shows that as β increases, the globally averaged precipitation rate decreases. In other words, "*the more cirrus is produced per unit precipitation rate, the slower the hydrologic cycle must run.*" In view of the importance of hydrologic processes for the climate system, the value of β is a very important index of the climate state. It increases as the global precipitation efficiency decreases. To see this, consider

$$W = P + C. \tag{8.33}$$

Here W indicates the rate at which water vapor is condensed to form clouds, P is the precipitation rate, and C is the rate of cloud-water formation. The precipitation efficiency can be defined as

$$\eta \equiv \frac{P}{W}. \tag{8.34}$$

From Equation (8.33) we obtain

$$\frac{1}{\eta} = 1 + \frac{C}{P} \sim 1 + \beta. \tag{8.35}$$

Here we have used the fact that $\frac{C}{P}$ is analogous to β. Then we find that

$$\eta \sim \frac{1}{1 + \beta}. \tag{8.36}$$

According to (8.36), as β increases, the precipitation efficiency decreases.

Suppose that the hydrologic cycle undergoes a positive fluctuation, i.e., that the globally averaged rates of precipitation and evaporation increase for some reason. An increase in the speed of the hydrologic cycle[4] will tend to produce an increase in the amount of cirrus-cloud cover, at least in the vicinity of the precipitation event(s). The cirrus can then spread over a large region. From this perspective, we would expect $-\mathcal{C}(R_\infty)$ and LP to increase together, which suggests that β might tend to remain roughly constant.

On the other hand, an increase in $-\mathcal{C}(R_\infty)$ will cause the net atmospheric radiative cooling to decrease. In order to maintain global atmospheric energy balance, the precipitation rate will have to decrease; the initial fluctuation of the hydrologic cycle will, therefore, be damped. From this perspective, we would expect LP to decrease as $-\mathcal{C}(R_\infty)$ increases. This would imply an increase in β.

To the extent that LP decreases as $-\mathcal{C}(R_\infty)$ increases, our hypothesized positive fluctuation of the globally averaged hydrologic cycle will be damped. This is a negative feedback, which can be called a "global radiative–convective feedback" (Fowler and Randall, 1994). This negative feedback will tend to regulate the speed of the hydrologic cycle. It is a cloud feedback, because it involves both cloud formation (principally cirrus formation) and precipitation. This conclusion would be bolstered, rather than weakened, if clouds absorbed a significant amount of solar radiation.

Note, however, that a decrease in LP could lag the increase in $-\mathcal{C}(R_\infty)$ by days or weeks. In contrast, upper-tropospheric cloudiness responds almost immediately to an increase in convective activity. This disparity in time scales between cloud

---

[4] The speed of the hydrologic cycle is measured by the globally averaged rates of evaporation and precipitation, which must balance in equilibrium.

formation and the effects of clouds on the thermal structure of the atmosphere may make it difficult for the system to settle into a steady state.

8.7 The thermostat feedback

Ramanathan and Collins (1991) pointed out that deep convection occurs preferentially over the warmest waters of the western tropical Pacific Ocean, and that the convection produces optically thick cloud masses which reflect much of the locally incident solar radiation back to space. Most of this radiation would otherwise be absorbed by the ocean. Ramanathan and Collins also used data from the Earth Radiation Budget Experiment (ERBE; Barkstrom *et al.*, 1989) to demonstrate that when El Niño causes the SSTs of the western tropical Pacific to decrease, and those of the central and eastern tropical Pacific to increase, the strongest deep convection "follows" the warm water eastward, so that the high-albedo clouds follow the warmest water. Ramanathan and Collins suggested that the tendency of strong SW cloud-radiative forcing to occur over the warm water amounts to a thermostat which tends to limit the warmest SSTs that can occur on Earth to values in the neighborhood of 305 K. It has been speculated that a similar thermostat effect may act globally to regulate the globally averaged surface temperature of Earth.

A key ingredient of the thermostat hypothesis is the idea that there exists a "threshold" SST above which deep convection is favored. As is well known (e.g., Bjerknes, 1969; Graham and Barnett, 1987), the present climate of Earth is in fact characterized by a threshold SST for the onset of deep convection; this temperature is in the neighborhood of 300 or 301 K. When the local SST exceeds this threshold value, deep convection (i.e., reaching the tropical tropopause) is often observed to be widespread; but at colder temperatures, deep convection is observed to be relatively rare.

We can imagine that the threshold SST might be a "universal constant," in the sense that it would be the same on either a warmer or colder Earth. An alternative possibility is that the threshold temperature is a property of the climatic state, increasing when the climate undergoes a general warming, and decreasing when it undergoes a general cooling. We have investigated the universality of the threshold SST using a suite of AGCMs, which have different convection parameterizations, different stratiform cloudiness algorithms, and so on. The models used are listed in Table 8.1, which summarizes some particularly relevant aspects of each model's formulation. We have performed three July runs with each AGCM: a control run in which the SST is not perturbed, a "+2 K" run in which the SST is uniformly increased by 2 K relative to the control run, and a "−2 K" run in which the SST is uniformly decreased by 2 K relative to the control. The experimental design thus follows that of C89.

Table 8.1. *Summary of the AGCMs used in the SST threshold study*

Model	Cumulus parameterization	Stratiform cloud parameterization	Boundary-layer parameterization
CSU	Prognostic Arakawa–Schubert (1974), with multiple cloud bases	Eauliq; Fowler *et al.* (1996)	Explicit boundary layer identitified as the lowest model layer
DNM	Betts–Miller scheme; Betts (1986)	Slingo scheme; Slingo (1987)	Six levels in the boundary layer
ECMWF	Tiedtke (1989)	Tiedtke (1993)	Louis *et al.* (1982)
GFDL	Moist convective adjustment	Largescale saturation	
MGO	Kuo and Arakawa–Schubert, in alternative versions; Meleshko *et al.* (2000)	Cloud amount parameterized in terms of relative humidity; Shneerov *et al.* (1997)	Boundary layer represented the four lowest layers; Shneerov *et al.* (1997)
UIUC	Modified Arakawa–Schubert with prognostic cloud water and diagnostic cloud cover; Oh (1989); Schlesinger *et al.* (1997); Wang and Schlesinger (1999)	Modified Sundqvist, with prognostic cloud water and diagnostic cloud cover; Oh (1989); Schlesinger *et al.* (1997); Wang and Schlesinger (1999)	Boundary layer represented the four lowest layers; Oh (1989); Schlesinger *et al.* (1997); Wang and Schlesinger (1999)

Abbreviations: CSU, Colorado State University; DNM, Department of Numerical Mathematics of the Russian Academy of Sciences (formerly the USSR Academy of Sciences); ECMWF, European Center for Medium Range Weather Forecasts; GFDL, Geophysical Fluid Dynamics Laboratory; MGO, Main Geophysical Observatory; UIUC, University of Illinois at Urbana-Champagne.

The results are summarized in Figure 8.8. The results of the control runs are generally consistent with the observations, indicating the existence of a threshold SST for the current climate, as discussed above. In the $+2$ K runs, the SST threshold has increased by 2 K, and in the -2 K runs it has decreased by 2 K. Inspection of the results for the simulated precipitation rate (not shown) confirms these results.

We conclude that the threshold SST is a function of the climatic state; it is not universal. This is consistent with the findings of Bony *et al.* (1997), Lau *et al.* (1997), Bajuk and Leovy (1998), and Brinkop (2001).

8.8 Cloud feedback, water-vapor feedback, and the tropical general circulation

It is widely but not universally agreed that there exists a positive "water-vapor feedback" which amplifies climate variability. As summarized by Ramanathan (1981), the mechanism of a positive water-vapor feedback is as follows. If an external forcing such as an increased CO_2 concentration tends to produce a warming of the SST, this will tend to cause an increase in surface evaporation, which in turn will lead to an increase in the amount of water vapor in the atmosphere. Because water vapor is a strong greenhouse gas, this increase in the atmospheric water-vapor content will cause an increase in the downwelling infrared radiation at Earth's surface, thus favoring a further increase in the SST and, therefore, a positive feedback on the initial warming.

Surface evaporation can continue only if a mechanism exists to carry water vapor upward away from the surface. Boundary-layer turbulence can do this, but only through the depth of the boundary layer, which is typically less than 1 km. In any case, the air near the boundary-layer top is typically near saturation over the tropical oceans. Further lifting of moisture, beyond the boundary-layer top, is necessary if the total moisture content of the atmospheric column is to increase significantly. The most important mechanism for such further lifting is cumulus convection. It follows that convective clouds play an essential role in the "water-vapor" feedback (Emanuel and Zivkovic-Rothman, 1999).

Lindzen (1990) argued that if convection penetrates to higher levels in a warmer climate, the air near cloud-top will be colder and will therefore have a lower saturation mixing ratio for water vapor. The air detrained from cumulus towers will, therefore, contain less water vapor. He suggested that as a result the upper troposphere will be drier if the surface temperature warms, and on this basis he questioned the sign of the water-vapor feedback. He pointed out that upper-tropospheric water vapor can strongly affect the Earth's radiation budget, even though the amount of water vapor in the upper troposphere is much smaller than that in the lower troposphere. He argued that a drying of the upper troposphere in a warmer climate would,

Cloud feedbacks

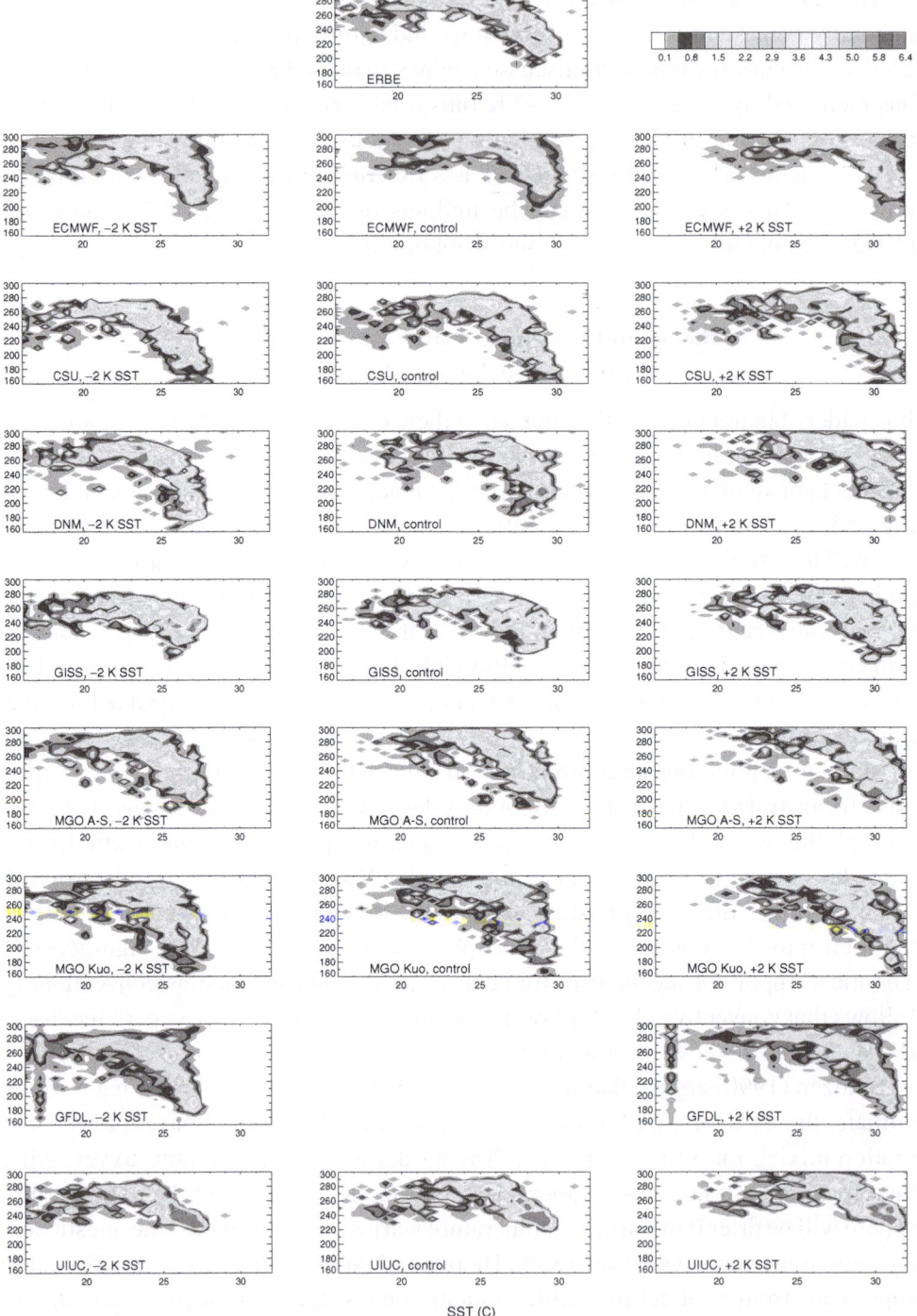

Figure 8.8. Each panel depicts a probability density function. The horizontal axis

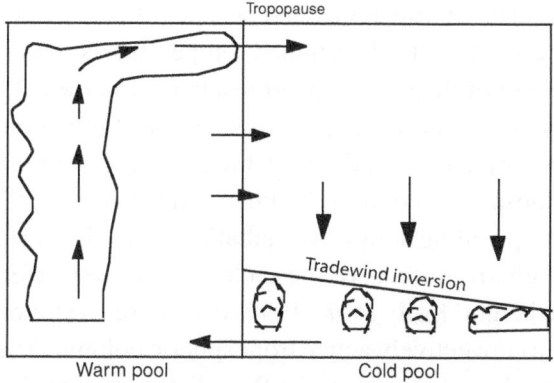

Figure 8.9. A simplified conceptual model of the Walker circulation. The domain is divided into a warm-pool region on the west and a cold-pool region on the east. Deep convection occurs over the warm pool. Only shallow, low-level clouds occur over the cold pool. See text for details.

therefore, significantly reduce the magnitude of a positive water-vapor feedback, and might even give rise to a negative water-vapor feedback. Lindzen's hypothesis of small or negative water-vapor feedback has been met with considerable skepticism (e.g., Held and Soden, 2000), but it has stimulated a lot of useful research by drawing attention to the radiative importance of upper-tropospheric water vapor, and to the importance of deep moist convection for affecting the amount of upper-tropospheric water vapor.

There is an emerging view of the tropical general circulation (Pierrehumbert, 1995; Miller, 1997; Larson *et al.*, 1999; Nilsson and Emanuel, 1999; Sherwood, 1999; Kelly and Randall, 2000) which holds that the magnitude of the mass transport in the Hadley and Walker circulations is driven by clear-sky radiative cooling in the convectively inactive regions of the tropics, rather than by latent-heat release in the convectively active regions. Following Pierrehumbert (1995), we conceptually divide the tropics into "warm-pool" and "cold-pool" regions, as depicted in Figure 8.9. The warm-pool atmosphere is characterized by large-scale rising motion, deep moist convection, and high relative humidities throughout the troposphere. Within the warm-pool atmosphere, strong moist convection ensures that the temperature sounding closely approximates a saturated moist adiabat which

---

Figure 8.8. *(cont.)* is the SST, and the vertical axis is the TOA OLR (W m^{-2}), with smaller values at the bottom and larger values at the top. The center column shows the observations (top) and the AGCM results for the control runs. The left column shows the AGCM results of the −2 K runs, and the right column shows the AGCM results of the +2 K runs.

passes through the surface temperature and pressure. Large-scale dynamical effects impose this same convectively determined temperature sounding on the middle- and upper-troposphere of the cold-pool atmosphere (Charney, 1963).

The deep convection of the warm pool moistens the entire column and produces abundant upper-tropospheric cloudiness which is associated with a net radiative warming of the atmospheric column (Stephens and Webster, 1979), primarily due to the blocking of upwelling longwave radiation from the surface and the lower troposphere. The radiative effects of these clouds can even affect the stratosphere and the tropopause height. Kelly *et al.* (1999) considered a stratosphere in radiative equilibrium above a convectively active tropospheric column. They showed that the cloud-induced reduction of the upward LW radiation across the tropopause leads to a cooling of the stratosphere, which in turn permits deeper penetration of the convective towers and so leads to a higher tropopause.

In the middle and upper troposphere, air flows outward from the warm pool to the cold pool, carrying the water-vapor-mixing-ratio profile impressed by the deep convection of the warm pool. This air then gradually subsides, bringing very dry upper tropospheric air down into the lower and middle troposphere of the cold-pool region (Salathé and Hartmann, 1997). The rate of subsidence in the cold-pool free troposphere (above the tradewind inversion) is determined by a thermodynamic balance between subsidence and radiative cooling, i.e.

$$\omega \frac{\partial \theta}{\partial p} \cong \frac{\theta}{c_p T} Q_R \tag{8.37}$$

where ω is the vertical pressure velocity, p is pressure, θ is the potential temperature, c_p is the specific heat of air at constant pressure, T is temperature, and Q_R is the radiative heating rate, which is typically negative over the cold pool due to dominant emission of infrared radiation by the subsiding air. As discussed above, $\frac{\partial \theta}{\partial p}$ is impressed on the cold pool by the convective physics of the warm pool, in combination with largescale dynamical processes discussed by Charney (1963). Therefore, for a given value of Q_R Equation (8.37) essentially determines the vertical velocity, ω.

Now recall that the radiative cooling rate in the middle and upper troposphere is strongly influenced by the water-vapor content of the air. It follows that the rate of subsidence in the cold pool is controlled by the water-vapor content of the air flowing from the warm pool to the cold pool, and this is largely determined by convective cloud processes at work in the warm pool region.

The total subsiding mass flux over the cold pool is essentially the vertical velocity times the width of the cold pool. To the extent that the width of the cold pool is fixed, the subsiding mass flux over the cold pool is determined by Q_R which, in turn, is determined by the water-vapor content of the subsiding air. Likewise, the total rate of radiative energy loss over the cold pool is Q_R times the width of the cold pool. To first order, the total radiative energy loss from the cold pool must be balanced by latent-heat release over the warm pool. The relative widths of the cold pool and warm pool must adjust so that this overall energy balance is maintained. Further discussion is given by Kelly and Randall (2000).

This perspective on the tropical general circulation emphasizes the roles of clouds and water vapor, and the interactions of radiation with the largescale dynamics. From this point of view, cloud feedbacks play an essential role in the basic dynamics of the tropical circulation, in addition to their roles in climate change.

8.9 A look ahead

The early global atmospheric circulation models of the 1960s and 1970s prescribed the distribution of cloudiness. During those same years, observations of the global distribution of clouds and their effects on the radiation budget and the hydrologic cycle were also very crude or non-existent.

During the 1980s, the importance of clouds for climate achieved near-universal acceptance. At the same time, our observations of cloudiness improved drastically with the advent of the International Cloud Climatology Project (ISCCP; Rossow and Schiffer, 1999) and the ERBE (Barkstrom *et al.*, 1989), as well as the First ISCCP Regional Experiment, FIRE (e.g. Randall *et al.*, 1995).

During the 1990s, the cloud parameterizations used in climate models were drastically improved through the introduction of explicit cloud-water and cloud-ice variables which directly link the simulated hydrologic and radiative processes (e.g. Tiedtke, 1993; Fowler *et al.*, 1996; Del Genio *et al.*, 1996). Our observational capabilities have also improved during the 1990s, through such efforts as ARM (Stokes and Schwartz, 1994), LITE (McCormick *et al.*, 1993), TRMM (Simpson *et al.*, 1996), and CERES (Wielicki *et al.*, 1996). Nevertheless, the global atmospheric models have advanced so rapidly that, as we enter the new century, the models can simulate aspects of global cloudiness (e.g., the seasonally and synoptically and diurnally varying three-dimensional distribution of ice water content) which are beyond our power to observe; in this sense, the models are "ahead of" the observations.

Later in this decade, however, new satellite systems such as CloudSat (Miller and Stephens, 2000; Stephens *et al.*, 2000), PICASSO-CENA (Winker and Wielicki, 1999), and the proposed TRMM follow-on mission (Im *et al.*, 2000) will provide unprecedented data on the vertical structures and meso- and micro-scale structures of cloud systems and their ice and cloud water contents, as well as the global distribution of precipitation. With the advent of these data, the observations may well race ahead of the global models, challenging the modeling community to simulate the newly observed structures and inter-relationships.

This see-saw battle between observations and simulations is bringing about a revolution in our understanding of the role of clouds of all kinds in the climate system, and it will permit both measurement and understanding of the nature and role of cloud feedback on time scales ranging from hours out to a few decades. We will be able to document, and we will try to understand, both transient fluctuations and persistent trends. Will the marine stratocumulus and SST trends shown in Figure 8.3 continue and intensify over the coming decades? Will the hydrologic cycle accelerate while the climate warms, as current climate simulations strongly suggest? Will the upper troposphere moisten or dry? Will the cold pool warm in a perpetual El Niño? We are going to find out, and when we do we will see the changes in cloudiness which accompany these climate shifts. Cloud–climate feedbacks will become manifest in our data. We can't wait.

Acknowledgments

Anthony Del Genio provided the GISS GCM results shown in Section 8.7.

This research has been supported by the National Aeronautics and Space Administration through contract NAS1-98125 and grant NAG1-2081; by the National Science Foundation through grant ATM-9812384; and by the US Department of Energy's ARM Program, through grant number DE-FG03-95ER61968, all to Colorado State University. Support has also been provided by the Russian Fund for Basic Research through Grant 99-06-66274.

Thanks to Graeme Stephens for helpful discussions. Mark Branson and Don Dazlich assisted with some of the plots. Byron Boville and Jim Hack assisted in accessing the results from the Community Climate System Model.

References

Arakawa, A. (1975). Modeling clouds and cloud processes for use in climate models. In *The Physical Basis of Climate and Climate Modelling*. Geneva, ICSU/WMO, GARP Publications, Series No. 16, pp. 181–97.

Arakawa, A. and W. H. Schubert (1974). The interaction of a cumulus cloud ensemble with the large-scale environment, part I. *J. Atmos. Sci.* **31**, 674–701.

Bajuk, L. J. and C. B. Leovy (1998). Seasonal and interannual variations in stratiform and convective clouds over the tropical Pacific and Indian oceans from ship observations. *J. Clim.* **11**, 2922–41.

Barkstrom, B., E. F. Harrison, G. L. Smith *et al.* (1989). Earth Radiation Budget Experiment (ERBE) archival and April 1985 results. *Bull. Amer. Meteor. Soc.* **74**, 591–8.

Betts, A. K. (1986). A new convective adjustment scheme. Part I. Observational and theoretical basis. *Quart. J. Roy. Meteor. Soc.* **112**, 677–91.

Bjerknes, J. (1969). Atmospheric teleconnections from the equatorial Pacific. *Mon. Wea. Rev.* **97**, 163–72.

Bode, H. W. (1975). *Network Analysis and Feedback Amplifier Design*. New York, Krieger.

Bony, S., K. -M. Lau, and Y. C. Sud. (1997). Sea surface temperature and large-scale circulation influences on tropical greenhouse effect and cloud radiative forcing. *J. Clim.* **10**, 2055–77.

Boville, B. A. and P. R. Gent (1998). The NCAR Climate System Model, version one. *J. Clim.* **11**, 1115–30.

Brinkop, S. (2001). Change of convective activity and extreme events in a transient climate change simulation. DLR-Institut für Physik der Atmosphäre, Report No. 142. (Available from DLR-Oberpfaffenhofen, Institut für Physik der Atmosphäre, D-82234 Wessling, Germany.

Cess, R. D. (1976). Climate change: an appraisal of atmospheric feedback mechanisms employing zonal climatology. *J. Atmos. Sci.* **33**, 1831–43.

Cess, R. D., G. L. Potter, J. P. Blanchet *et al.* (1989). Interpretation of cloud–climate feedback as produced by 14 atmospheric general circulation models. *Science* **245**, 513–16.

(1990). Intercomparison and interpretation of climate feedback processes in 19 atmospheric general circulation models. *J. Geophys. Res.* **95**, 16601–15.

Cess, R. D., M. H. Zhang, G. L. Potter *et al.* (1996). Cloud feedback in atmospheric general circulation models: an update. *J. Geophys. Res.* **101**, 12791–4.

(1997). Comparison of atmospheric general circulation models to satellite observations of the seasonal change in cloud-radiative forcing. *J. Geophys. Res.* **102**, 16593–604.

Charney, J. G. (1963). A note on large-scale motion in the tropics. *J. Atmos. Sci.* **20**, 607–9.

Charney, J. G., A. Arakawa, D. J. Baker *et al.* (1979). Carbon dioxide and climate: a scientific assessment. Washington, DC, National Academy Press.

Curry, J. C. and P. J. Webster (1999). *Thermodynamics of Atmospheres and Oceans*. International Geophysics Series 65, London, Academic Press.

Dai, A., T. M. L. Wigley, B. Boville, J. T. Kiehl, and L. Buja (2000). Climates of the 20th and 21st centuries simulated by the NCAR Climate System Model. *J. Clim.* **14**, 485–519.

Del Genio, A. D., M.-S. Yao, W. Kovari, and K. K.-W. Lo (1996). A prognostic cloud water parameterization for global climate models. *J. Clim.* **9**, 270–304.

Emanuel, K. A. and M. Zivkovic-Rothman (1999). Development and evaluation of a convection scheme for use in climate models. *J. Atmos. Sci.* **56**, 1766–82.

Fowler, L. D. and D. A. Randall (1994). A global radiative–convective feedback. *Geophys. Res. Lett.* **21**, 2035–8.

Fowler, L. D., D. A. Randall, and S. A. Rutledge (1996). Liquid and ice cloud microphysics in the CSU General Circulation Model. Part 1: model description and simulated microphysical processes. *J. Clim.* **9**, 489–529.

Graham, N. E. and T. P. Barnett (1987). Observations of sea-surface temperature and convection over tropical oceans. *Science* **238**, 657–9.

Hall, A. and S. Manabe (1999). The role of water vapor feedback in unperturbed climate variability and global warming. *J. Clim.* **12**, 2327–46.

Held, I. and B. J. Soden (2000). Water vapor feedback and global warming. *Ann. Rev. Energy Env.* **25**, 441–75.

Hendon, H. H. and K. Woodberry (1993). The diurnal cycle of tropical convection. *J. Geophys. Res.* **98**, 16623–37.

Houghton, J. T., G. J. Jenkins, and J. J. Ephraums, eds. (1990). *Climate Change. The IPCC Scientific Assessment. World Meteorological Organization / United Nations Environment Programme.* Cambridge, Cambridge University Press.

Im, E., S. L. Durden, G. Sadavy *et al.* (2000). System concept for the next-generation spacebourne precepitation radars, 2000 IEEE conference proceedings 5, pp. 151–8.

Kelly, M. A. and D. A. Randall (2000). The effects of the vertical distribution of water vapor on the strength of the Walker circulation. *J. Clim.* **14**, 3944–64.

Kelly, M. A., D. A. Randall, and G. L. Stephens (1999). A simple radiative–convective model with a hydrologic cycle and interactive clouds. *Quart. J. Roy. Meteor. Soc.* **125**, 837–69.

Klein, S. A. and C. Jakob (1999). Validation and sensitivities of frontal clouds simulated by the ECMWF model. *Mon. Wea. Rev.* **127**, 2514–31.

Larson, K., D. L. Hartmann, and S. A. Klein (1999). Climate sensitivity in a two box model of the tropics. *J. Clim.* **12**, 2359–74.

Lau, K.-M., H.-T. Wu, and S. Bony. (1997). The role of large-scale atmospheric circulation in the relationship between tropical convection and sea surface temperature. *J. Clim.* 381–92.

Lilly, D. K. (1968). Models of cloud-topped mixed layers under a strong inversion. *Quart. J. Roy. Meteor. Soc.* **94**, 292–309.

Lindzen, R. S. (1990). Some coolness concerning global warming. *Bull. Amer. Meteor. Soc.* **71**, 288–99.

Louis, J. F., M. Tiedtke, and J. F. Geleyn (1982). A short history of the operational planetary boundary layer parameterization at ECMWF. Report from Workshop on Planetary Boundary Layer Parameterization, European Centre for Medium Range Weather Forecasts, Reading, UK, November 25–27, 1981, 59–79.

Manabe, S., and K. Bryan (1969). Climate calculation with a combined ocean–atmosphere model. *J. Atmos. Sci.* **26**, 786–9.

McCormick, M. P., D. M. Winker, E. V. Browell *et al.* (1993). Science investigations planned for the Lidar In-Space Technology Experiment (LITE). *Bull. Amer. Meteor. Soc.* **74**, 205–14.

Meleshko, V. P., V. M. Katsov, P. V. Sporysev, S. V. Vavulin, and V. A. Govorkova (2000). Feedbacks in the climate system: cloud, water vapour, and radiation interaction. *Meteorologia and Gidrolagia* **2**, 22–45.

Miller, R. L. (1997). Tropical thermostats and low cloud cover. *J. Clim.* **10**, 409–40.

Miller, S. D., and G. L. Stephens (2000). CloudSat instrument requirements as determined from ECMWF forecasts of global cloudiness. *J. Geophys. Res.* **106**, 17,713–33.

Miller, S. D., G. L. Stephens, and A. C. M. Beljaars (1999). A validation survey of the ECMWF prognostic cloud scheme using LITE. *Geophys. Res. Lett.* **26**, 1417–20.

National Aeronautics and Space Administration (2000). *Understanding Earth System Change: NASA's Earth Science Enterprise Research Strategy 2000–2010.* Available from the Earth Science Enterprise Office of the National Aeronautics and Space Administration (additional bibliographic information to be provided), Washington, DC.

Nilsson, J. and K. A. Emanuel (1999). Equilibrium atmospheres of a two-column radiative-convective model. *Quart. J. Roy. Meteor. Soc.* **125**, 2239–64.

Norris, J. R. and C. B. Leovy (1994). Interannual variability in stratiform cloudiness and sea surface temperature. *J. Clim.* **7**, 1915–25.

Oh, J.-H. (1989). Physically-based general circulation model parameterization of clouds and their radiative interaction. Ph. D. Thesis, Oregon State University, Corvallis.

Peixóto, J. P. and A. H. Oort (1992). *Physics of Climate*. New York, American Institute of Physics.

Pierrehumbert, R. T. (1995). Thermostats, radiator fins, and the runaway greenhouse. *J. Atmos. Sci.* **52**, 1784–806.

Ramanathan, V. (1981). The role of ocean–atmosphere interactions in the CO_2 climate problem. *J. Atmos. Sci.* **38**, 918–30.

Ramanathan, V. and W. Collins (1991). Thermodynamic regulation of ocean warming by cirrus clouds deduced from observations of the 1987 El Niño. *Nature* **351**, 27–32.

Ramanathan, V., R. D. Cess, E. F. Harrison *et al.* (1989). Cloud-radiative forcing and climate: results from the Earth Radiation Budget Experiment. *Science* **243**, 57–63.

Randall, D. A., B. A. Albrecht, S. K. Cox *et al.* (1995). On FIRE at ten. *Adv. Geophys.* eds. A. Berger, R. E. Dickinson, and J. W. Kidson, Washington, DC, American Geophysical Union, **38**, 37–177.

Rodwell, M. J. and B. J. Hoskins (2000). Subtropical anticyclones and summer monsoons. *J. Clim.* **14**, 3192–211.

Rossow, W. B. and R. A. Schiffer (1999). Advances in understanding clouds from ISCCP. *Bull. Amer. Meteor. Soc.* **80**, 2261–87.

Salathé, E. P., Jr. and D. L. Hartmann (1997). A trajectory analysis of tropical upper-tropospheric moisture and convection. *J. Clim.* eds. A. Berger, R. E. Dickinson, and J. W. Kidson, Washington, DC, American Geophysical Union, **10**, 2533–47.

Schlesinger, M. E. (1985). Feedback analysis of results from energy balance and radiativeconvective models. In *The Potential Climatic Effects of Increasing Carbon Dioxide*, eds. M. C. MacCracken and F. M. Luther, Washington, DC, US Department of Energy, DOE/ER-0237, pp. 280–319. (Available from NTIS, Springfield, Virginia.)

  (1988). Quantitative analysis of feedbacks in climate model simulations of CO_2-induced warming. In *Physically-Based Modeling and Simulation of Climate and Climatic Change*, ed. M. E. Schlesinger, Dordrecht, Reidel, NATO Advanced Study Institute Series, pp. 653–736.

  (1989). Quantitative analysis of feedbacks in climate model simulations. In *Understanding Climate Change*, eds. A. Berger, R. E. Dickinson, and J. W. Kidson, Washington, DC, American Geophysical Union, Geophysical Monograph 52, IUGG volume 7, pp. 177–87.

Schlesinger, M. E., N. G. Andronova, B. Entwistle *et al.* (1997). Modeling and simulation of climate and climate change. In *Past and Present Variability of the Solar-Terrestrial System: Measurement, Data Analysis and Theoretical Models*. Proceedings of the International School of Physics "Enrico Fermi" CXXXIII, eds. Cini Castagnoli, G. and A. Provenzale, Amsterdam, IOS Press, pp. 389–429.

Schneider, S. H. (1972). Cloudiness as a global climatic feedback mechanism: the effects on radiation balance and surface temperature of variations in cloudiness. *J. Atmos. Sci.* **29**, 1413–22.

Senior, C. A. (1999). Comparison of mechanisms of cloud–climate feedbacks in GCMs. *J. Clim.* 12, 1480–9.

Senior, C. A. and J. F. B. Mitchell (1993). Carbon dioxide and climate: the impact of cloud parameterization. *J. Clim.* **6**, 393–418.

Sherwood, S. C. (1999). Feedbacks in a simple prognostic tropical climate model. *J. Atmos. Sci.* **56**, 2178–200.

Shneerov, B. E., V. P. Meleshko, A. P. Sokolov *et al.* (1997). MGO global atmospheric general circulation model coupled to mixed layer ocean. *Proc. Voeikov Main Geophys. Observ.* **544**, 3–122.

Simpson, J., C. Kummerow, W.-K. Tao, and R. F. Adler (1996). On the Tropical Rainfall
    Measuring Mission (TRMM). *Meteor. Atmos. Phys.* **60**, 19–36.
Slingo, J. (1987). The development and verification of a cloud prediction scheme for the
    ECMWF model. *Quart. J. Roy. Meteor. Soc.* **133**, 899–927.
Stephens, G. L. and P. J. Webster (1979). Sensitivity of radiative forcing to variable cloud
    and moisture. *J. Atmos. Sci.* **36**, 1542–56.
Stephens, G. L., D. G. Vane, R. J. Boain *et al.* (2002). The CloudSat mission and the
    A-train. *Bull. Amer. Meteor. Soc.* **83**, 1771–90.
Stokes, G. M. and S. E. Schwartz (1994). The Atmospheric Radiation Measurement
    (ARM) Program: programmatic background and design of the Cloud and Radiation
    Test Bed. *Bull. Amer. Meteor. Soc.* **75**, 1201–21.
Tiedtke, M. (1993). Representation of clouds in large-scale models. *Mon. Wea. Rev.* **121**,
    3040–61.
Wang, W. and M. E. Schlesinger (1999). The dependence on convection parameterization
    of the tropical intraseasonal oscillation simulated by the UIUC 11-layer atmospheric
    GCM. *J. Clim.* **12**, 1423–57.
Wielicki, B. A., R. D. Cess, M. D. King, D. A. Randall, and E. F. Harrison (1995).
    Mission to Planet Earth: role of clouds and radiation in climate. *Bull. Amer. Meteor.
    Soc.* **76**, 2125–53.
Wielicki, B. A., B. R. Barkstrom, E. F. Harrison *et al.* (1996). Clouds and the Earth's
    Radiant Energy System (CERES): an Earth Observing System experiment. *Bull.
    Amer. Meteor. Soc.* **77**, 853–68.
Winker, D. M. and B. A. Wielicki (1999). The PICASSO-CENA Mission. In *Sensors,
    Systems, and Next-Generation Satellites III*, eds. H. Fujisada and J. B. Lurie,
    (Europto Series) Proceedings of SPIE 3870, pp. 26–36.
Yao, M.-S., and A. D. Del Genio (1999). Effects of cloud parameterization on the
    simulation of climate changes in the GISS GCM. *J. Clim.* **12**, 761–79.

9

Water-vapor feedback

DAVID H. RIND

NASA Goddard Institute of Space Studies, New York, NY

9.1 Introduction

As mentioned in the previous chapter, water vapor is an important feedback factor in the climate system. The present chapter considers the observational evidence and modeling for this feedback factor. In 1989, commenting on an article in which Raval and Ramanathan (1989) showed from ERBE data that warmer sea-surface temperatures (SSTs) were associated with greater water-vapor absorption of outgoing radiation, Bob Cess (1989) noted that if the relationship held for climate change as well, the water-vapor feedback would amplify doubled CO_2 warming by 60%. What have we learned in the years since then?

As described in Chapter 8, the water-vapor feedback is defined as the response of the atmospheric moisture to changes in climate forced by some other means. The canonical point of view is that when the oceans are forced to warm, say by increasing atmospheric CO_2, they will evaporate more moisture into the atmosphere. This effect should occur because the atmospheric water-vapor holding capacity (the saturation vapor pressure) increases exponentially with increasing temperature (via the Claussius–Clapeyron relationship).

Increasing water vapor in the atmosphere will lead to further warming of the surface, as water vapor is itself a greenhouse gas. It absorbs outgoing terrestrial radiation in a broadband continuum from 8–12 m, through rotational bands in the far-infrared (FIR) (> 20 m), and via vibrational–rotational bands near 6.3 m (Harries, 1996). More water vapor thus increases the greenhouse capacity of the atmosphere, radiating more energy back to the surface, where it can lead to further increases in evaporation and even more water vapor, an iterative positive feedback.

Frontiers of Climate Modeling, eds. J. T. Kiehl and V. Ramanathan.
Published by Cambridge University Press. © Cambridge University Press 2006.

Actually, what we define as the water-vapor feedback on the climate system has three components. An increase in the column average water vapor increases the atmospheric greenhouse effect, as indicated above. Furthermore, if water vapor extends to greater heights and colder temperatures in the atmosphere, its greenhouse effect is amplified, for it would now radiate to space at these colder temperatures, losing less energy in the process (outgoing radiation being proportional to temperature to the fourth power). However, counterbalancing this last tendency, with increased water vapor, the temperature aloft should increase (i.e., the atmospheric lapse rate should decrease), since more water vapor means more condensation and warmer air rising to higher altitudes – so the air at any given level, now at a warmer temperature, would radiate more energy to space. In an assessment of their general circulation model (GCM) results, Hansen *et al.* (1984) concluded that for doubled atmospheric CO_2, itself inducing direct greenhouse warming of some 1.25 °C, the increased atmospheric water vapor (of some 33%) would provide an additional 1.7 °C warming; the increased vertical distribution of water vapor would lead to another 0.9 °C warming; while the decreased atmospheric lapse rate would lead to a cooling of some 1.2 °C. Adding the three effects together indicated that the water-vapor feedback was totaling 1.4 °C, or more than the direct CO_2 warming.

This result is more than the 60% amplification indicated by Cess (1989), and part of the reason is that we tend to think of the water-vapor feedback, and other feedbacks, as operating in isolation. Given that all feedbacks result from the total temperature changes, anything which alters the temperature also alters the magnitude of the feedback. For example, Rind *et al.* (1997) found in the same GCM used by Hansen *et al.* (1984) that if sea ice was not allowed to change, the global warming was reduced by some 35%, and the water-vapor feedback was reduced by a similar percentage. In the highly coupled system, the water-vapor feedback depends on the sea-ice and cloud-cover feedbacks; its total magnitude will vary with the response of the entire system. It therefore is not necessarily a single value; for example, it may well differ in a cooling climate, where sea ice has greater room to expand and can interact more strongly with solar radiation. In the following discussion we will often refer to the water-vapor feedback as if it were an isolated, single-valued property, but this broader perspective is important to keep in mind.

9.2 Radiative considerations

Over the past decade, the predominant consideration in understanding the appropriate magnitude of the water-vapor feedback has been the realization that the change in emissivity (the effectiveness in absorbing thermal radiation) is approximately proportional to the fractional change in water vapor at any level (Shine and Sinha, 1991). Thus, although most of the water vapor is in the lower atmosphere (water vapor has a scale height for exponential decrease of only a few km), the change of water vapor in the upper troposphere can contribute significantly to the overall greenhouse effect. The importance of this realization is that while water vapor in the boundary layer would most likely increase as climate warms, there is less certainty associated with water vapor at higher levels.

The statement above indicated that the proportionality factor is only approximate. In a one-dimensional model study, for a given fractional change in absolute humidity, the lower tropospheric response, dominated by continuum effects, was 1.5 to 2 times greater than that of the upper troposphere, where FIR (rotational) effects dominate (Sinha and Allen, 1994). Thus even with little change in the upper troposphere, the water-vapor feedback would still be positive; it could only be negated by a strong negative change in upper-tropospheric water vapor, as suggested by Lindzen (1990). Even these comparisons are only approximate; as noted by Sinha and Harries (1997), the continuum absorption parameters and FIR line strengths are not well validated under atmospheric conditions. Improvements in water-vapor spectroscopy are still needed, as indicated by the increase in its shortwave absorption that has been recently discussed (Chagas *et al.*, 2001).

The importance of water-vapor emissions from dry regions is not limited to the upper troposphere. Spencer and Braswell (1997) noted that since humidity fluctuations at 10% relative humidity have three time as much impact on outgoing longwave radiation as do fluctuations around 90% relative humidity, the dry regions in the tropics and sub-tropics become especially important, and about one-half of the region from 30 °N to 30 °S has relative humidity less than 20%. Pierrehumbert (1999) has calculated that a four-times change in sub-tropical relative humidity could change global mean temperature by 5–6 °C. The sub-tropics can have such a large effect because dynamical constraints force the tropics/sub-tropics to have relative uniform temperature. As the coriolis force becomes small, the radius of deformation becomes large – strong pressure gradients cannot exist, and neither can strong temperature gradients. Therefore, the radiative losses incurred in the sub-tropics influence the temperatures throughout the region. Pierrehumbert (1995) has popularized the term "dry radiator fins" to illustrate the importance of the sub-tropical longwave emissions in this regard. So in addition to what happens in the upper troposphere, the water-vapor response in the dry regions of the tropics becomes of great importance to the total water-vapor feedback.

9.3 Tropical/sub-tropical moisture budgets

To know how the moisture budget in these regions will change with climate, it would be useful to know how they are maintained today. Unfortunately, the sparcity and imprecision of our observations has left room for considerable differences in opinion, both for the upper troposphere and the sub-tropics. For example, Lindzen (1990) conceived of the tropical upper-tropospheric moisture budget as being a balance between water vapor detrained from deep tropical clouds near the tropopause, and drying resulting from the compensatory subsidence accompanying the deep convection. Betts (1990) pointed out that the detrainment occurs at various levels, as has recently been emphasized by observations of the vertical moisture response to convection over the oceans (Soden, 2000), and re-evaporation of precipitation above the freezing level is probably a more important moistening process. The distinctions are important; if Lindzen's balance is the dominant one, and if convection goes to higher and colder levels as climate starts to warm, then detrainment would be occurring at colder conditions (with a lower saturation vapor pressure), with less moisture available to moisten the upper troposphere, providing a negative water-vapor feedback. In this assessment, the increased moisture from lower levels would all rain out in the ascending supersaturated convective column. The precipitation efficiency then becomes an issue (Renno *et al.*, 1994). In contrast, with re-evaporation of the (increased) precipitation the dominant moistening process, increased water vapor at low levels would likely result in increases aloft.

However, of course the tropics is not a one-dimensional column, and it has become apparent that the moisture sources in the upper troposphere are not necessarily local. An often-repeated point of view is that the boundary-layer (tradewind) inversion in the sub-tropics prevents moisture evaporated from the oceans from getting into the free troposphere. Instead, the sub-tropical low level moisture is advected toward the equator, where it rises in convective or generally ascending air, and is transported by various processes to the drier regions. The dynamical transporting mechanisms could be the largescale Hadley and Walker circulations (Sun and Lindzen, 1993; Gershunov *et al.*, 1998); with these mechanisms, the moisture transport to the dry regions is associated with radiatively-driven subsidence, and moisture content decreases away from cloud and convective regions (Salathé and Hartmann, 1997; Soden, 1998). Hence decreasing sub-tropical humidity in a warming climate could be brought about by an increase in subsidence associated with these largescale circulations, bringing more dry air downward, or by a contraction of the convective zone (say removal of the west Pacific warm pool) (Pierrehumbert, 1998).

Sherwood (1996) and Pierrehumbert (1998) have emphasized the potential importance of subsidence drying being balanced primarily by lateral mixing of moist plumes advected from convective regions by transient eddies. The vertical transport of moisture by eddies from the boundary layer to the free troposphere has not generally been considered important, with the boundary-layer inversion viewed as preventing communication between these levels. In reality, our view of the sub-tropics has often been quite idealized; there are a number of regions in the sub-tropical zone, primarily over the continents and ocean convective areas, where no consistent boundary-layer inversion exists. An early study of the sub-tropical North Atlantic (Holland and Rasmusson, 1973) found that vertical eddy transports of moisture produced a large divergence out of the boundary layer, and moistening extending upward for several hundred mb. It was unclear whether this was due to transport through the boundary-layer inversion, or occurred where the boundary-layer inversion broke down. Sherwood (1996) also found that vertical turbulent transports provided moisture as high as the 500 mb level in arid tropical regions; with the gradients so large, just occasional erosions of the inversion, due to gravity waves or intense surface fluxes from surface wind gusts, would be sufficient. Zhu *et al.* (2000), using ECMWF re-analysis winds in conjunction with Microwave Limb Sounding (MLS) water-vapor observations, calculated that transient vertical advection was important for moistening the sub-tropical upper troposphere, although the horizontal scale of the phenomena is uncertain. Bates and Jackson (2001) emphasize the role of equatorially propagating mid-latitude transient waves in moistening the sub-tropical upper troposphere via vertical advection. We will see below that our GCM studies also emphasize the importance of transient-eddy vertical-moisture transports. To the extent that eddy transports, either horizontal or vertical, are important in the sub-tropical and upper-troposphere moisture budgets, unless the eddies weaken dramatically or their propagation changes as climate warms, they will be available to mix moisture from the increasingly moist lower troposphere/convective areas to the radiatively important dry regions.

The situation is less contentious for the extra-tropical upper troposphere, although it has also not had nearly as much attention. Enhanced upper-tropospheric humidity and greenhouse effect appear in the MLS data over the storm tracks in the North Pacific and North Atlantic. Strong baroclinic activity, and a large number of associated deep convective clouds, both help transport water vapor to the highest reaches of the troposphere (Hu and Liu, 1998). The results are consistent with aircraft observations showing that the southern-hemisphere upper troposphere during winter is two to four times drier than the northern hemisphere from the mid latitudes to the pole. Air that is rising and heading poleward encounters colder air in the southern hemisphere from Antarctica, with greater condensational drying (Kelly *et al.*, 1991). Again, unless eddies weaken dramatically in a warming climate, they will apparently be available to transport increased low-level moisture to higher altitudes in the extra-tropics.

9.4 Moisture trends

What is actually happening as far as moisture trends are concerned? Our sensing techniques are better suited to evaluate total precipitable water or lower-tropospheric trends. Radiosonde data for the last 21 years in the western hemisphere north of the equator show generally increasing precipitable water from the surface to 500 mb, at the rate of 3–7% per decade; changes in the upper portion of the region (500–700 mbar) are as large or larger than those in the middle or lower layers (Ross and Elliot, 1996). In the eastern hemisphere, over China, trends in both the (warming) surface temperature and precipitable water were highly correlated, with a positive greenhouse effect (Zhou and Eskridge, 1997). However, in the tropics, using TIROS Operational Vertical Sounder (TOVS) radiances calibrated with radiosondes, Schroeder and McGuirk (1998a) found that there has been an average drying of some 3% over the last 16 years, occurring in oceanic sub-tropical high- and desert-land areas. This conclusion has been debated, due to the use of changing radiosonde data for calibration purposes (Ross and Gaffen, 1998; Schroeder and McGuirk, 1998b). Considering upper-tropospheric humidity (above 500 mbar), Bates and Jackson (2001) used High-resolution Infrared Radiation Sounder (HIRS) data to conclude that the deep tropics had become more moist during the last several decades, while the sub-tropics dried, with only a small moistening for the region as a whole; they related the results primarily to an increased frequency of El Niños. It is also important to realize that tropical temperatures in the middle and upper troposphere may have actually cooled somewhat, or warmed minimally over this time period (Angell, 1988; Christy *et al.*, 1998). This trend is in contrast to the projections from climate models responding to greenhouse-gas increases, which show unambiguous increases in tropical upper-troposphere temperature and moisture (e.g., Rind, 1998a). However, not included in those climate model projections were (observed) decreases in stratospheric ozone or increases in stratospheric water vapor, both of which have been shown to help cool or minimize the warming especially near the tropopause (Bengtsson *et al.*, 1999; Forster and Shine, 1999; Oinas *et al.*, 2001). To the extent that these additional effects have limited the projected warming in the tropical upper troposphere, the moisture response may well be dominated by episodic events such as El Niño.

The sparcity of such trend studies results from the lack of long-term well-calibrated moisture observations. In lieu of this data, researchers have focused on inter-annual, or geographical variations, to assess how water vapor responds to increased SSTs. Starting with the work referred to above of Raval and Ramanathan (1989), various researchers have found that on the local scale, increased sea-surface temperatures are associated with greater tropospheric water vapor and an increased greenhouse capacity (e.g., Stephens and Greenwald, 1991; Rind *et al.*, 1991; Inadmar and Ramanathan, 1994; Sun and Oort, 1995; and also Chapter 5). Responding to the challenge that local effects can be misleading, and that one should average over largescale systems (Lindzen *et al.*, 1995), researchers then explored a wide range of space and time scales (Soden and Fu, 1995; Soden, 1997; Yang and Tung, 1998; Charboureau *et al.*, 1998; Inadmar and Ramanathan, 1998), with similar results. The vast predominance of evidence is that increased sea-surface and tropospheric temperatures are associated with increased moisture at all levels, although sometimes at a rate less than fixed relative humidity (Sun and Oort, 1995). However, it is still not evident that such studies are analogs for climate change; Gutzler (1996) pointed out that on the inter-annual time scale, during El Niño the western Pacific specific-humidity anomaly is negative in the boundary layer. Yet over the last 20 years, with an upward trend in El Niño, the overall trend in moisture (and western Pacific SST) is positive, indicating that the inter-annual and trend responses can differ. The results relating surface-warming and moisture increases throughout the troposphere so far can be thought of as highly suggestive but not conclusive.

Trends in stratospheric water vapor have recently become of great interest, as it appears as if water vapor in this region has been increasing by about 1% per year (Oltmans *et al.*, 2000; Rosenlof *et al.*, 2001). It is larger than would be implied by methane increases, and may be due to added water vapor in the upper troposphere, or more water-vapor advection through the sub-tropics, hence avoiding the tropical "cold trap" associated with minimum tropical tropopause temperatures (Rosenlof *et al.*, 2001). It is apparently not due to increased tropical tropopause temperatures, for they appear to be decreasing (e.g., Zhou *et al.*, 2001). Whatever the reason, Shindell (2001) has calculated that it may have contributed $\sim 24\%$ to global warming over the past two decades.

9.5 General circulation models

To this point in time, then, our assessment of the likely water-vapor feedback to climate change comes from general circulation models (GCMs). The IPCC (1995) issued the following statement concerning the GCM results. "Feedback from the redistribution of water vapour remains a substantial uncertainty in climate models. . .Much of the current debate has been addressing feedback from the tropical upper troposphere, where the feedback appears likely to be positive. However, this is not yet convincingly established; much further evaluation of climate models with regard to observed processes is needed." Bates and Jackson (1997) found significant differences between observed and modeled upper-tropospheric humidity (UTH) seasonal and inter-annual variations in models run for AMIP I. By IPCC (2001), the conclusion, though hedged, was somewhat more positive. "Models are capable of simulating the moist and very dry regions observed in the tropics and sub-tropics and how they evolve with the seasons and from year to year. While reassuring, this does not provide a check of the feedbacks, although the balance of evidence favours a positive clear-sky water vapour feedback of the magnitude comparable to that found in simulations."

In the past decade, there have been numerous comparisons of GCM-produced moisture distributions with those obtained from remote-sensing observations, primarily UTH, obtained from brightness temperature observations, or total precipitable water (TPW) from microwave observations. In general, the TPW values in models look appropriate (Soden and Bretherton, 1994: Gaffen *et al.*, 1997). Comparisons with model UTH values often indicate horizontal gradients that are too weak or too strong (Schmetz and van de Berg, 1994; Soden and Bretherton, 1994; Salathé *et al.*, 1995; Harries, 1997). It appears that the problem is associated with the strength of the Hadley circulation; for example, if the circulation is too weak, the gradient between the tropics and sub-tropics will be too weak, a situation common to most models. Comparisons with the GISS model did not indicate this particular problem; tropical UTH was a little too moist, but the sub-tropical dry values were appropriate, as is now found in the newer models (IPCC, 2001). If the largescale circulations are not well simulated for the current climate, there would be less confidence in their change as climate warms.

Sun and Held (1996) found in a comparison between radiosonde data and the GFDL model that the model showed too high a correlation between tropical moisture and temperature changes at all levels in the troposphere. The implication was that certain parameterizations, such as the convection scheme, were allowing low-level perturbations to extend too high into the atmosphere. However, Sun *et al.* (2001), investigating the results of numerous models run for the Atmospheric Model Intercomparison Project (AMIP), concluded that it was a common model problem, independent of particular convection scheme, and seemed to affect even re-analysis data (see also Chen *et al.*, 1998). They claimed this result arose even when they restricted their sampling of GCM results to the radiosonde locations, a conclusion that disagrees with several others (Soden and Lanzanthe, 1996; Bauer, 1998). If models were really to exaggerate the connection between low and high levels, they might convert low-level moisture increases to high-level increases too efficiently. That in itself would not necessarily imply an overestimate of the water-vapor feedback, as noted by Sun *et al.* (2001), which is a function of the magnitude of change at all levels.

One way to compare the model's water-vapor feedback is to assess how its clear-sky radiative forcing changes with a change in surface temperature. The simplest approach is to look at how the outgoing radiation at the top of the atmosphere (F) changes with surface temperature (T_s). Sun and Held (1996) calculated from inter-annual variations that the clear-sky value for the tropics of $dF/dT_s = 3.2$ W m^{-2} K^{-1}; for blackbody radiation at the tropical temperatures of 300 K, the Stefan–Boltzman relationship indicates that the value of $4F/T_s$ would be about 6.2 W m^{-2} K^{-1}, the difference being the water-vapor feedback. In contrast, the GFDL model value was 2.4 W m^{-2} K^{-1}, close to that for an atmosphere with constant relative humidity. Sun and Held concluded that if this result was relevant for global-warming studies, then the model would overestimate global warming by 15% (which can be compared to the factor-of-three uncertainty still prevalent in IPCC estimates of climate sensitivity).

Considering the atmosphere more directly, the clear-sky forcing, G_a, is the difference between the longwave radiation emitted from Earth's surface and the longwave radiation leaving the top of the atmosphere. Because the surface emission increases with temperature, so will the trapping; hence G_a is often normalized by the surface emission (the normalized value then being g_a). The observed value of $G_a = 146$ W m^{-2}; clouds increase the value by about 33 W m^{-2} (Raval and Ramanathan, 1989). Observations of inter-annual/geographical variations in surface temperature show that $dG_a/dT_s = 3.2$–3.5 W m^{-2} K^{-1} globally (Sun and Held, 1996; Inadmar and Ramanathan, 1998), with values of 5.5–6 W m^{-2} K^{-1} in the tropics (Inadmar and Ramanathan, 1998). The corresponding observed value of $dg_a/dT_s = 3.3 \times 10^{-3}$ K^{-1} (Raval and Ramanathan, 1989). Using the GFDL 2% solar constant increase experiment, Raval and Ramanathan (1989) obtained a value for dG_a/dT_s of 3.7 W m^{-2} K^{-1} (hence again a 15% overestimate), and a value in the NCAR model using latitudinal variations of 3.1 W m^{-2} K^{-1}, also close to the observed value (see also other studies verifying the apparent accuracy of the GFDL water-vapor response: Soden and Fu, 1995; Soden, 1997; Yang and Tung, 1998). Hall and Manabe (1999) found that the GFDL model can only reproduce the observed variability with its water-vapor feedback; without it, the inter-annual variations are too small. If the observed inter-annual variations of outgoing radiation are relevant to long-term climate-change studies, it would appear as if the GCM water-vapor feedback is of the approximate appropriate magnitude, despite its apparent excessive water-vapor vertical correlation. (The cloud-cover variation might well be affected by such a discrepancy).

The results for an individual model are important because of the overall similarity of the water-vapor feedback in different GCMs. In a landmark study, Cess *et al.* (1990) compared the values of dF/dT_s in various GCMs whose sea-surface temperatures were perturbed by ±2 °C. For global temperatures, the blackbody value of dF/dT_s would be 3.3 W m^{-2} K^{-1}. The model results were very similar to one another, with a mean value of 2.34 W m^{-2} K^{-1} for clear-sky conditions (2.46 W m^{-2} K^{-1} for all-sky conditions), and a standard deviation of only 0.20. Hence, while the question of whether GCMs provide proper water-vapor estimates for climate change is still an open one, they seem to be responding very similarly. Therefore, in the next section, we will take a close look at the water-vapor feedback in one particular GCM.

9.6 The GISS GCM upper troposphere and sub-tropical water-vapor budgets

The first question to address is how does this GCM produce its water vapor in the dry regions of the troposphere. As indicated above, comparison with observed UTH values indicates that the GISS model is appropriately dry in these regions; actually, it appears somewhat too dry in the extra-tropical upper troposphere. An assessment of how the water vapor is produced will be important when considering the water-vapor response to various climate changes.

The version of the GISS model used in this analysis is Model II (e.g., Rind and Lerner, 1996; Hansen *et al.*, 1998), at $4° \times 5°$ horizontal resolution and nine levels. The relevant model characteristics are covered in those publications; in particular, the GISS moist-convective scheme includes two convective plumes, one non-entraining, a downdraft associated with re-evaporated moisture, and compensatory subsidence. We will briefly consider how the budget assessment changes when using finer horizontal and vertical resolution; as models are continually improving their resolution, that might suggest how models might react in producing water-vapor feedbacks in the coming years.

In the following discussion, various processes are referred to, and it is important to clarify their meaning in the GCM. "Moist convection" changes the moisture balance as the result of detrained moisture plus the effects of compensatory subsidence (similar to the definition used by Lindzen, 1990). "Dry convection" results from superadiabatic processes, and its mixing of moisture is primarily in the lowest layers. "Largescale condensation" includes the condensation and re-evaporation of detrained water in an anvil cloud; in the tropics these are obviously related to convection, but in the GCM they are part of the grid-scale saturation process computationally separate from the convection parameterization. The "mean circulation" includes the transports, both vertical and horizontal, of the time-averaged zonally invariant dynamical motions. "Eddies" represent all transports not included in the mean circulation definition; this means all longitudinally varying effects, including actual eddies plus longitudinal circulation cells such as the Walker cell, as well as all transient phenomena.

The effect of these processes on the moisture budget for the current climate is shown in Figure 9.1, with global averages provided in Table 9.1 (the "control").

Figure 9.1. Annual global moistening by various processes i.e., eddies, mean circulation, largescale condensation (LSC), moist convection (MC), and dry convection (DC), in the control run as a function of altitude (a) and a function of latitude (b) for the upper troposphere (at altitudes above the 550 mb pressure level). Values shown in the boxed legend are global averages.

a

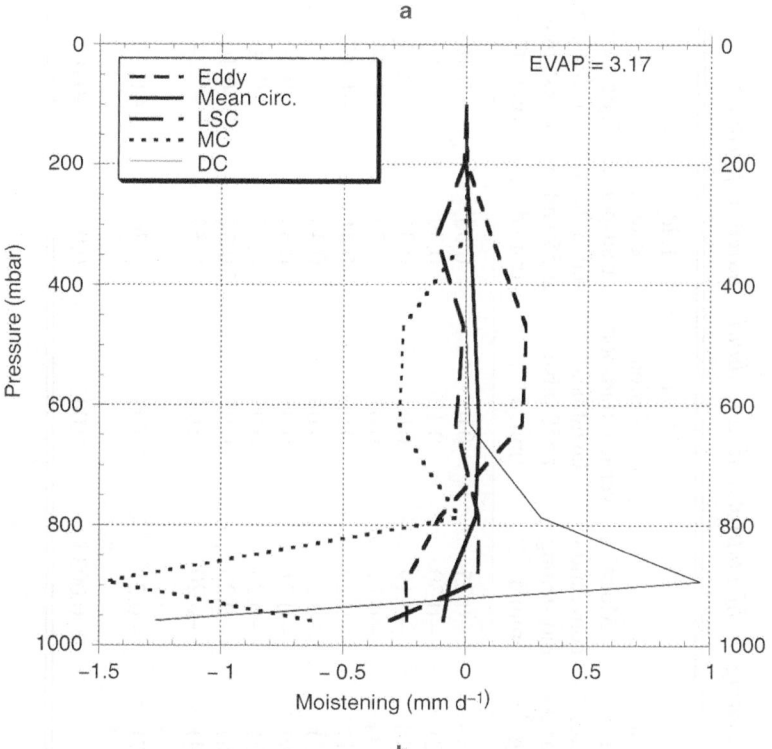

b

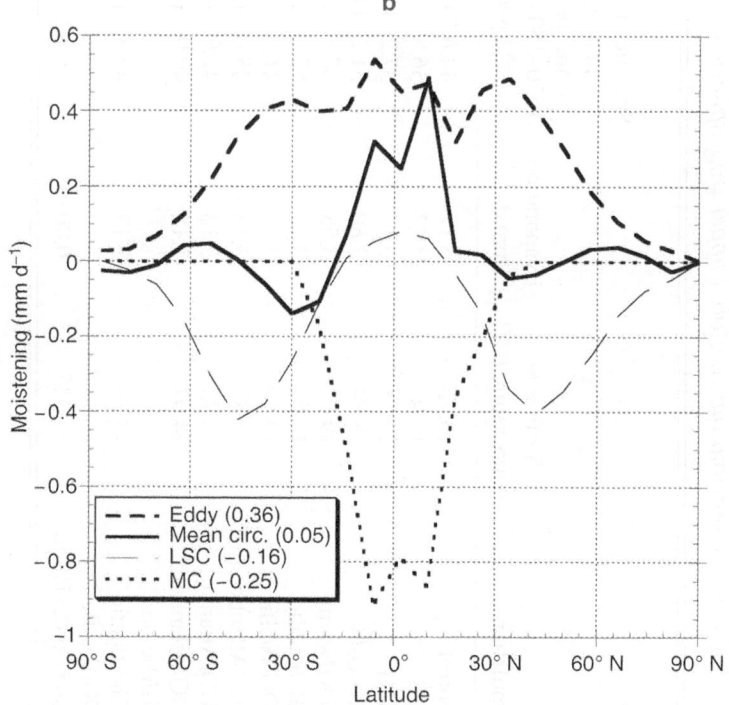

Table 9.1. *Contributions to global upper tropospheric moistening in the control and latitudinal gradient experiments*

Simulation	Surface air temperature (°C)	Evaporation (mm d⁻¹)	Specific humidity in upper troposphere, (550–150 mbar) (10^{-5} kg kg⁻¹)	Moist convection moistening (mm d⁻¹)	Mean circulation vertical transport latent heat at 550 mbar (mm d⁻¹)	Eddy vertical transport latent heat at 550 mbar (mm d⁻¹)	Large-scale condensate moistening (mm d⁻¹)
Control	13.75	3.17	44.6 (6)	−0.25	0.05	0.36	−0.16
A IG	13.95	3.22	56.8 (2)	−0.30	0.10	0.39	−0.19
B DG	13.58	3.12	37.2 (10)	−0.19	0.03	0.31	−0.14
C IG/cold	8.84	2.80	31.2 (11)	−0.21	0.07	0.29	−0.14
D DG/warm	19.04	3.58	67.7 (1)	−0.29	0.06	0.41	−0.17
H IG/Pacific	14.09	3.22	52.2 (3)	−0.23	0.08	0.40	−0.24
F DG/Pacific	13.69	3.15	41.6 (9)	−0.21	0.05	0.34	−0.18
G IG/Atlantic	13.85	3.16	45.6 (5)	−0.23	0.05	0.37	−0.20
E DG/Atlantic	13.75	3.13	42.9 (8)	−0.20	0.05	0.34	−0.19
J DG/Atlantic IG/Pacific	14.07	3.18	49.9 (4)	−0.20	0.07	0.39	−0.24
I IG/Atlantic DG/Pacific	13.80	3.17	43.3 (7)	−0.21	0.05	0.36	−0.20
Standard deviation	0.025	0.0079	0.12	0.003	0.002	0.0035	−0.004

Considering first the vertical distribution of effects (Figure 9.1a), throughout most of the free troposphere, eddies and the mean circulation are moistening the atmosphere, acting on the strong vertical gradient of specific humidity. Moist convection and largescale condensation are providing drying, either through subsidence of drier air, or condensation of vapor to rain. Dry convection is moistening the top of the boundary layer, but is unimportant at higher levels. The latitudinal variation of effects for the upper troposphere (above 550 mbar), given in Figure 9.1b, shows that moist convection is drying most of the tropics, while the other processes are producing moistening. In the extra-tropics, eddy moistening is balanced by the largescale condensation.

Given the discussion in Section 9.3, the most surprising feature is the predominance of eddy vertical advection effects in moistening the upper troposphere, a result previously obtained in different versions of this model by Del Genio (1991) and Del Genio *et al.* (1994). It is clearly the dominant moistening process globally. It is also obviously a significant process throughout the tropics and sub-tropics where, as indicated in the explanation of the "eddy" contribution, some of it may be due to longitudinal circulation effects, such as the Walker circulation. Significant upward vertical moisture transport occurs in the west Pacific, as well as over South America and Africa, with downward transport over the east Pacific. We calculate that this amounts to about one third of the total "eddy" effect in the tropics. By comparison with waves in the extra-tropics, we estimate that the actual wave transports are producing another 60% of the "eddy" value, with 10% associated with transient aspects of the mean circulation. Hence the model moistening of the upper troposphere is associated with actual eddy processes, even from 30° N to 30° S.

The significance of such wave vertical transports raises the question of whether the model's results are the product of some obvious deficiencies. For example, does the model actually produce an appropriate tradewind inversion? Examination of the model's zonal wind and potential temperature structures indicates that the inversion appears realistic. Does the model have excessive eddy energy? Comparison with the values of Oort (1983) indicates that the model's values do look appropriate; if anything, the eddy energy is slightly too small. The GISS model uses the quadratic upstream scheme (Prather, 1986) for moisture advection in three dimensions, known for its highly non-diffusive nature, so it would also not appear to be an obvious numerical problem.

Could the results be the product of coarse resolution? To investigate this issue, we compared the standard model with simulations at $4° \times 5°$ horizontal resolution and 48 levels, and $2° \times 2.5°$ horizontal resolution and 32 levels. The results indicate that the same basic relationships still hold (not shown, but see the contributions to sub-tropical moistening given in Table 9.7). Eddy moistening is somewhat larger in the extra-tropics at the finer horizontal resolution, perhaps due to fronts being better resolved, while in the tropics, with less convective mass flux, there is less convective drying. With finer vertical resolution, tropical drying by moist convection and moistening by largescale condensation (including the evaporation of anvil cloud moisture) are increased by a factor of two or more. Apparently the level of detrainment and subsequent condensation, as well as re-evaporation of moisture, is strongly dependent on resolution, which can affect the stability and relative-humidity profiles. This appears to be the degree of variability that should be expected, at the very least, between different GCMs.

9.7 The GISS GCM response to specified SST changes

We next explore the water-vapor response to specified SST changes. In contrast to the experiments of Cess *et al.* (1990), here we specify changes in the latitudinal gradient of SSTs, as well as changes to the mean temperature. A complete description of these experiments is found in Rind (1998b); the resulting surface air-temperature perturbations are shown in Figure 9.2a. Experiment A has an increased SST gradient, and no change in global mean temperature. Experiment B has a decreased SST gradient, again with no change in global mean temperature. Experiment C is like experiment A except it is colder by some 5 °C. Experiment D is like B except it is warmer by 5 °C. These experiments thus allow us to determine how water vapor responds to changes in gradients that affect atmospheric dynamics, as well as changes in mean temperature that affect the amplitude of the hydrologic cycle (Rind, 1998b).

The results are given in Table 9.1 and in Figure 9.3. Table 9.1 also shows the results for experiments in which the gradient was increased (IG) or decreased (DG) in a manner similar to that of experiments A and B but only in the Atlantic or Pacific (see Rind *et al.*, 2001). In allowing us to indicate the effects of changes in the different ocean basins separately, this is relevant for phenomena like El Niños or La Niñas, which alter the gradient primarily in the Pacific; or climate-change experiments that show decreased future gradients in the Pacific, with increases in the Atlantic (e.g., Russell and Rind, 1999). The largest amount of upper-tropospheric water vapor arises in experiment D, with the next most in experiment A. This is true at all levels globally (Figure 9.3a); in the tropics, experiment A actually has more upper-tropospheric moisture at the lowest latitudes (Figure 9.3b), where its SST is commensurate with the value in D (Figure 9.2a). Either warmer temperatures, or warmer tropical temperatures enhance upper-tropospheric moisture in the model. In both cases, the predominant global effect is produced by increases in eddy vertical moisture transports (Table 9.1).

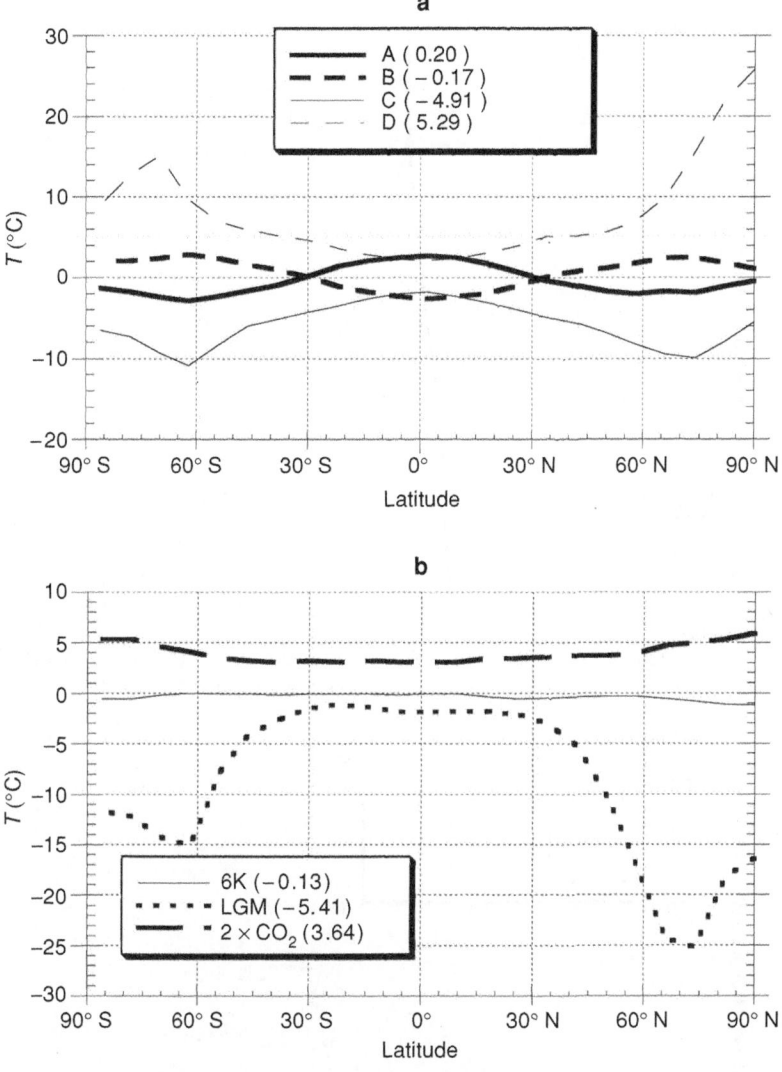

Figure 9.2. Annual surface air-temperature change as a function of latitude for the altered-gradient/mean-temperature experiments (a) and climate simulations (b). Global anomalies are indicated in parenthesis.

Shown in Figure 9.4a are the differences in the upper-tropospheric moistening mechanisms between experiments A and B, in which only the gradients are different. With the increased gradient, the mean circulation is providing much stronger vertical advection, moistening the tropical region. Table 9.2 gives the magnitudes of the relevant dynamical processes in these different runs; the Hadley circulation is indeed much stronger in experiment A than in experiment B, associated with the

Water-vapor feedback

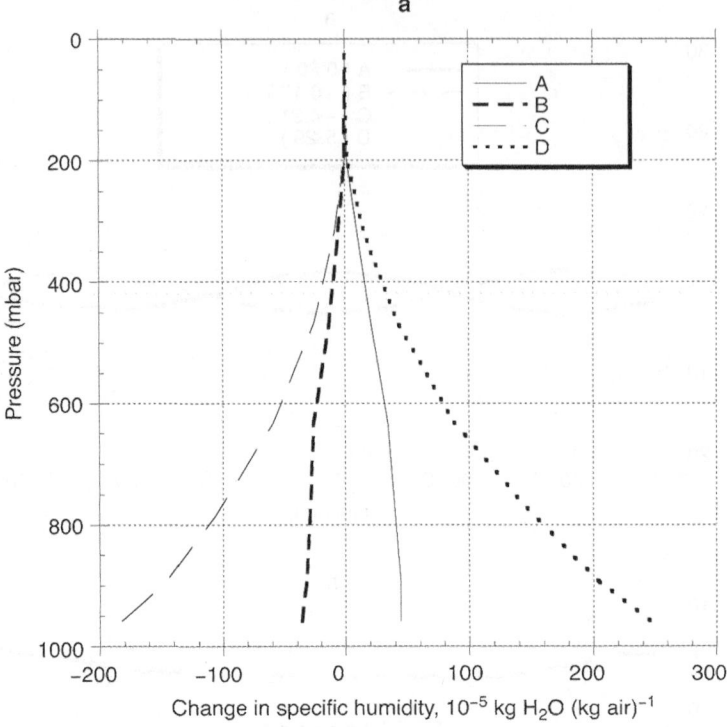

a

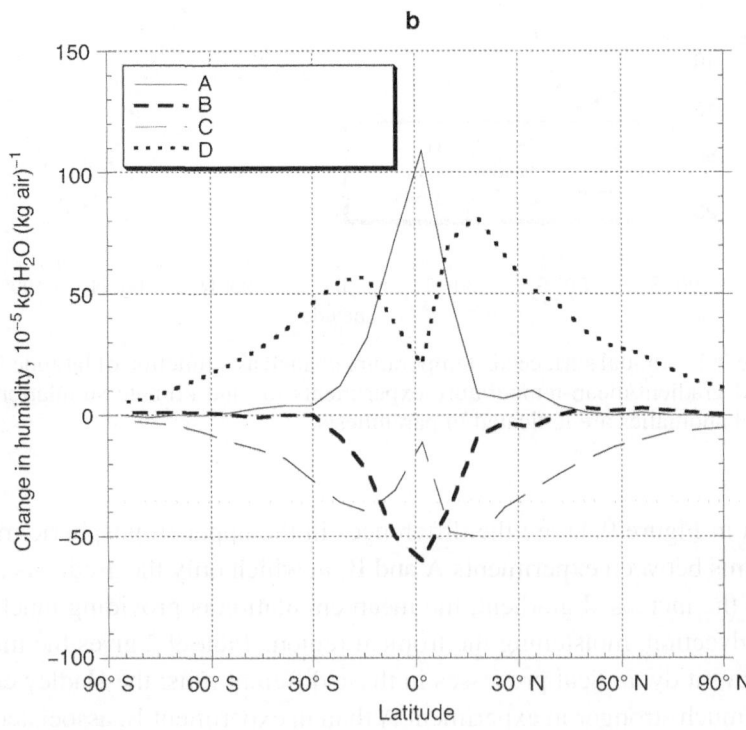

b

larger SST gradient and gradient in latent-heat release. While this might be expected to produce drier conditions in the sub-tropics, a tendency that does exist for the mean circulation alone (Figure 9.4a), the net effect, including horizontal moisture transports, is that the sub-tropics are slightly wetter as well (Figure 9.3b). Moist convection is acting as a drying mechanism from 30° N to 30° S (Figure 9.1b), and with reduced moist convective mass fluxes due to the increased sub-tropical subsidence in A, there is less moist convective drying (hence moistening) in the sub-tropics. Eddies are still playing a moistening role, and eddy energy is somewhat greater in experiment A, although the Walker circulation (as estimated from the difference in vertical velocity maxima between the west and east Pacific) is somewhat smaller (Table 9.2).

Given in Figure 9.4b are the differences in transport mechanisms between the two sets of temperature-change experiments that have the same general gradient; A, C and D, B. The results are similar in the two comparisons, although the difference in tropical temperature gradients produces some offset near the equator – with reduced tropical gradients, warmer climates produce less moistening differences right at the equator. Again the mean circulation is producing moistening at low latitudes in the warmer climates with warmer tropical SSTs (experiments A and D), although in this case the Hadley circulation is actually less intense in each of the warmer runs (Table 9.2). Apparently the vertical moisture gradient predominates over the change in dynamical velocities, presumably as long as the dynamics change is not too large. Eddy vertical transports are also important at all latitudes, and produce more moistening in the warmer climate even when the eddy energy is weaker (as in D versus B).

Tables 9.1 and 9.2 show rankings (numbers in parenthesis) for upper-tropospheric moisture and total atmospheric water vapor. The rankings for each of these quantities are in the same order – more water vapor globally, dominated by values at low levels, also means more water vapor in the upper troposphere. In these experiments that result is unequivocal. The results also show that increased gradients in the Pacific are most effective in producing global and upper-tropospheric moisture increases.

We can also calculate the radiative effects in these experiments, relative to the discussion in Section 9.5. Given in Table 9.3 are the longwave radiation values (total sky) leaving the surface, and also exiting the model top, the greenhouse effect (G) and the normalized greenhouse effect (g). The experiments have little change

---

Figure 9.3. Annual global specific humidity change from the control simulation for the gradient/mean temperature change experiments as a function of altitude (a), and upper-tropospheric specific humidity changes as a function of latitude (b).

a

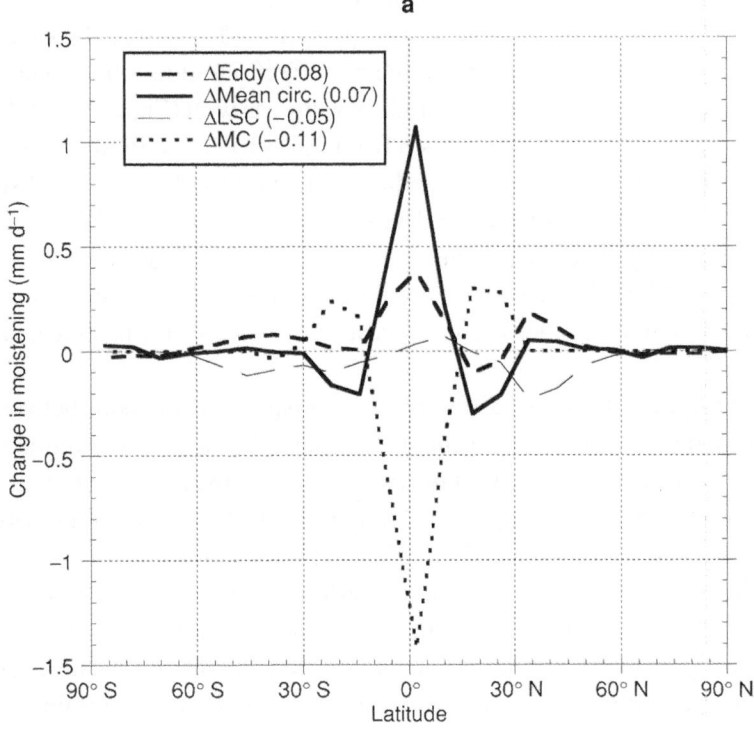

b

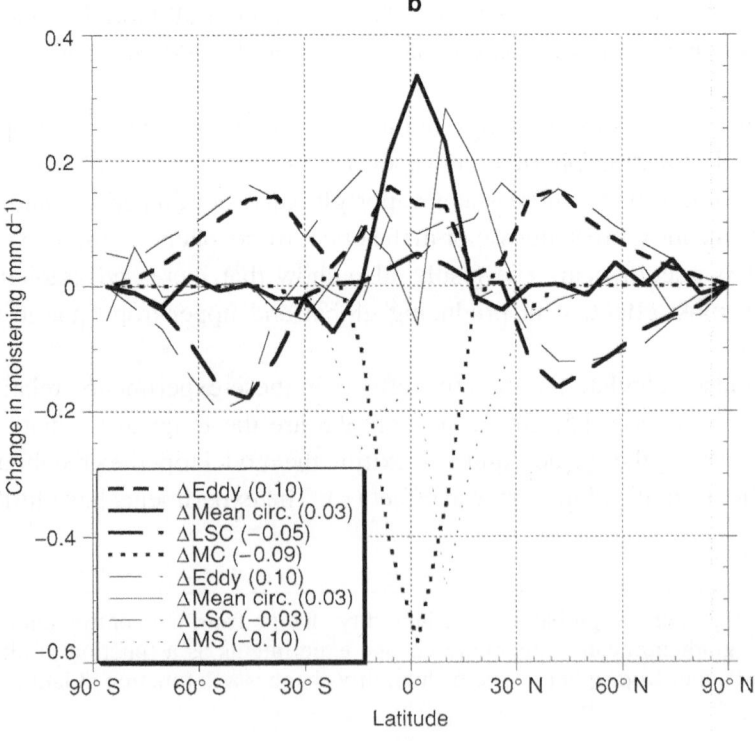

in either total or high-level cloud cover; most of the changes in greenhouse capacity are due to the water-vapor changes (a 1% change in high-level cloud cover in the model appears to produce a change in g of <0.002).

The control-run values compare favorably with estimated longwave radiation leaving the surface of 398 W m^{-2}, leaving the top of the atmosphere of 237 W m^{-2}, and a G value of 161 W m^{-2} (see Chapter 5). The normalized greenhouse effect is largest in experiment D and lowest in experiment C (the warmest and coldest experiments, respectively); the smallness of the absolute change in this parameter can be appreciated by noting that these experiments differ in global mean temperature by more than 10 °C (Table 9.3). The variations in G and g among the different experiments now do not always match the upper-tropospheric water-vapor ordering, which in many experiments differ by less than one standard deviation (and have cloud-cover differences as well). From the entire suite of experiments we can calculate the sensitivity of G and g to the global surface air-temperature changes; values are given in the footnote in the table. The value of dG/dT_s for all-sky conditions is slightly larger than the clear-sky value calculated for observed inter-annual variations (Section 9.5), while the dg/dT_s value is also somewhat larger. We will compare these values to those which arise in actual climate-change experiments in the next section.

9.8 The GISS GCM response in climate change experiments

We now apply a similar analysis to climate-change experiments purporting to represent real (or at least modeled) climate states. We use three such situations: the 6000 BP (before present) climate (6K), in which there was greater solar radiation during northern hemisphere summer due to orbital variations; the climate of the last glacial maximum (LGM, *c.* 21 000 BP); and a modeled doubled CO_2 climate ($2 \times CO_2$). The surface air-temperature changes in these three experiments are shown in Figure 9.2b. The SSTs were assumed to be the same as today's for 6K, were specified from CLIMAP (1981) values for the LGM (which also included large ice sheets), and were calculated in the GISS model with doubled atmospheric CO_2 using a q-flux mixed-layer ocean.

Figure 9.4. Change in upper-tropospheric moistening by different processes between experiments (A–B) with a latitudinal gradient increase (and no global temperature change) (a), and between experiments with a global mean warming (without a low-sub-tropical latitude gradient change) (b). Bold values in the legend of the bottom figure represent experiments A–C, with the lighter lines representing experiments D–B.

Table 9.2. *Dynamical processes contributing to upper tropospheric moistening in the latitudinal gradient experiments*

Simulation	Surface air temperature (°C)	Atmospheric water vapor (mm)		Moist convective mass flux at 550 mbar (10^9 kg s^{-1})	Dec–Feb Hadley cell peak (10^9 kg s^{-1})	Jun–Aug Hadley cell peak (10^9 kg s^{-1})	Vertical velocity peak difference between west Pacific and east Pacific (10^{-4} m s^{-1})	Average eddy kinetic energy 10^{17} J
Control	13.75	22.9	(6)	1529	−158	194	119	1646
A IG	13.95	25.2	(2)	1326	−183	199	96	1882
B DG	13.58	21.3	(10)	1840	−139	188	125	1502
C IG/Cold	8.84	17.3	(11)	1384	−205	226	123	1847
D DG/Warm	19.04	30.5	(1)	1767	−128	209	105	1470
H IG/Pacific	14.09	24.5	(3)	1383	−150	216	181	2026
F DG/Pacific	13.69	22.3	(9)	1583	−162	196	119	1673
G IG/Atlantic	13.85	23.3	(5)	1447	−172	191	110	1690
E DG/Atlantic	13.75	22.7	(8)	1543	−156	182	92	1746
J DG/Atlantic IG/Pacific	14.07	24.0	(4)	1439	−158	219	203	2138
I IG/Atlantic DG/Pacific	13.80	22.7	(7)	1552	−146	192	144	1757
Standard deviation	0.025	0.05		6.83	4.80	12.1	11.9	16.34

Table 9.3. *Radiative processes in the different latitudinal gradient experiments*

Simulation	Surface air temperature (°C)	Specific humidity in upper troposphere 550–150 mbar (10^{-5} kg kg^{-1})	Longwave radiation leaving surface (10^9 kg s^{-1})	Longwave radiation at model top (W m^{-2})	Greenhouse effect, G^* (W m^{-2})	Normal greenhouse effect, g^{**}	Total cloud cover (%)	High cloud cover (%)
Control	13.75	44.6 (6)	396	232.3	163.7	0.4134	56.4	14.0
A IG	13.95	56.8 (2)	398	233.5	164.5	0.4133	55.8	13.7
B DG	13.58	37.2 (10)	394.5	231.1	163.4	0.4142	57.1	13.5
C IG/Cold	8.84	31.2 (11)	371.6	226	145.6	0.3918	55.7	11.9
D DG/Warm	19.04	67.7 (1)	423.2	238.8	184.4	0.4357	57.1	15.8
H IG/Pacific	14.09	52.2 (3)	397.9	233.1	164.8	0.4142	56.1	13.1
F DG/Pacific	13.69	41.6 (9)	395.3	231.2	164.1	0.4151	56.9	14.0
G IG/Atlantic	13.85	45.6 (5)	396.9	232.1	164.8	0.4152	56.3	13.8
E DG/Atlantic	13.75	42.9 (8)	395.9	231.2	164.7	0.4160	56.6	13.7
J DG/Atlantic IG/Pacific	14.07	49.9 (4)	396.1	231.2	164.9	0.4163	57.0	13.7
I IG/Atlantic DG/Pacific	13.80	43.3 (7)	397.9	232.7	165.2	0.4152	56.2	12.9
Standard deviation	0.025	0.12	0.12	0.12	0.14	0.0006	0.07	0.09

*dG/DT_s = 3.80 W m^{-2} K^{-1}; **dg/dT_s = 4.3 × 10^{-3} K^{-1}.

The resulting specific humidity values are shown in Figure 9.5, with the contributions to upper-tropospheric moistening given in Table 9.4. As was the result for the
previous experiments, the warmest climate ($2 \times CO_2$) had the most moisture, and
the coldest climate the least, and this was true at every altitude and latitude. The 6K
climate, which ended up slightly cooler, had slightly reduced upper-tropospheric
moisture. Globally, the moisture differences were associated with changed vertical
moisture transport by eddies (Table 9.4; note that the $2 \times CO_2$ climate has its own
control run, whose values are presented as well). This occurred despite the fact that
the LGM climate had much greater eddy energy than the $2 \times CO_2$ run (Table 9.5),
emphasizing again that the increased moisture gradient from low levels dominated
the resulting process. Considering the individual runs relative to the control run,
in the 6K climate the (slightly) intensified Hadley cell (Table 9.5) led to some in-
creases/decreases in vertical transport by the mean circulation (Figure 9.6a), with
little net global effect. The LGM climate (Figure 9.6b) had reduced moistening
of the tropical upper troposphere by the mean circulation (not only were moisture
values and vertical gradients reduced, so was the mean circulation itself, Table 9.5),
and the colder climate also had reduced moist convective drying (associated with
reduced moist convection). In the extra-tropics, where eddy energy was considerably larger, eddy vertical moisture transports were nevertheless reduced. For the
doubled CO_2 climate (Figure 9.6c), eddies and the mean circulation moistened the
upper troposphere at various latitudes (despite slightly weaker mean circulation and
eddies), while moist convection and largescale condensation produced drying. The
results in Figure 9.6c are similar in nature to those shown for the climate changes
associated with little latitudinal gradient change in Figure 9.4b (and the doubled
CO_2 run had very little gradient change in the tropics, as indicated in Figure 9.2b),
despite the fact that in the earlier case the SSTs were specified, and here they
were calculated. Apparently, the model responds qualitatively similarly regardless
of how the SSTs are forced (the specified SST experiments in effect were forcing
the gradient change by implicitly altering the ocean heat transports, as discussed in
Rind, 1998b).

The radiative properties for these experiments are given in Table 9.6. Using these
results we again calculate the derivative with respect to temperature of the total-
sky greenhouse and normalized greenhouse effects. In comparison to the results in
Table 9.3, these derivatives are larger. Part of the reason might be that in the earlier
experiments, the warmest climate had a reduced tropical SST gradient, while the
coldest climate had an increased value (Figure 9.2a), while in the latest experiments
that did not occur. As shown from experiments A and B, those SST gradient changes
would have worked against the water-vapor feedback being generated by the global
temperature changes, and hence minimized the model sensitivity. The larger values
here again emphasize that the value of these greenhouse-effect derivatives will

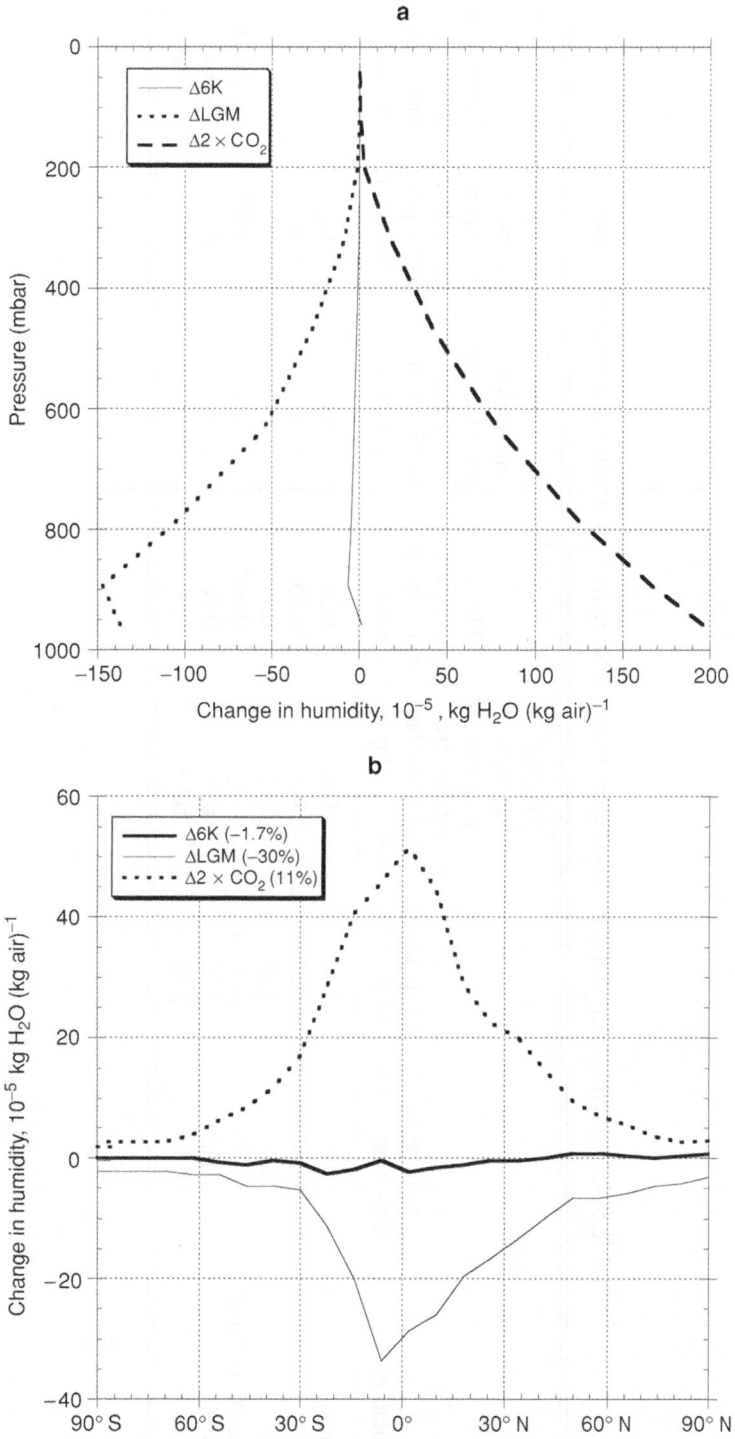

Figure 9.5. As in Figure 9.3 except for the modeled climate simulations.

Table 9.4. *Contributions to upper-tropospheric moistening in the climate-change experiments*

Simulation	Surface air temperature (°C)	Evaporation (mm d^{-1})	Specific humidity in upper troposphere 550–150 mbar (10^{-5} kg kg^{-1})	Moist convection moistening (mm d^{-1})	Mean circulation vertical transport latent heat at 550 mbar (mm d^{-1})	Eddy vertical transport latent heat at 550 mbar (mm d^{-1})	Large-scale condensate moistening (mm d^{-1})
Control	13.75	3.17	45.6	−0.25	0.05	0.36	−0.16
6K	13.62	3.20	44.8	−0.25	0.05	0.36	−0.16
LGM	8.34	2.86	31.8	−0.17	0.05	0.31	−0.18
Standard deviation	0.025	0.0079	0.12	0.003	0.002	0.0035	0.004
Control	13.88	3.22	45.9	−0.26	0.06	0.36	−0.15
2×CO$_2$	17.39	3.44	68.9	−0.30	0.08	0.42	−0.19

Table 9.5. *Dynamical processes contributing to upper-tropospheric moistening in the climate-change experiments*

Simulation	Surface air temperature (°C)	Atmospheric water vapor (mm)	Moist convective mass flux at 550 mbar (10^9 kg s^{-1})	Dec–Feb Hadley cell peak (10^9 kg s^{-1})	Jun–Aug Hadley cell peak (10^9 kg s^{-1})	Vertical velocity peak difference between west Pacific and east Pacific (10^{-4} m s^{-1})	Average eddy kinetic energy 10^{17} J
Control	13.75	22.9	1529	−158	194	131	1646
LGM	8.34	18.6	1462	−130	146	93	2068
6K	13.62	23.0	1535	−166	199	127	1646
Standard deviation	0.025	0.05	6.83	4.80	12.1	11.9	16.34
Control	13.88	23.4	1530	−182	196	123	1660
2×CO$_2$	17.39	29.7	1543	−170	181	101	1648

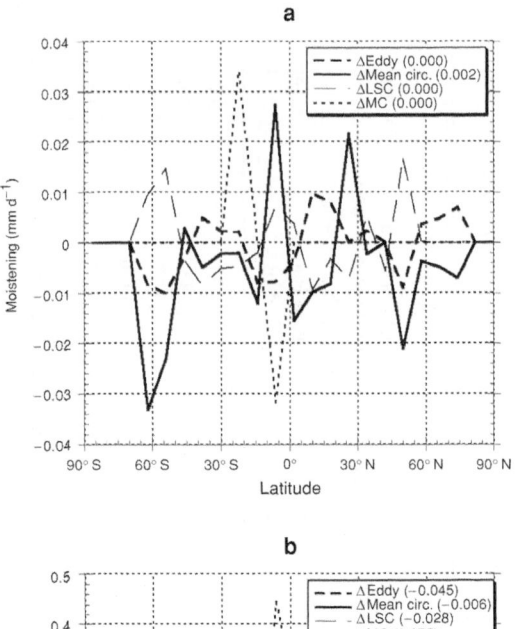

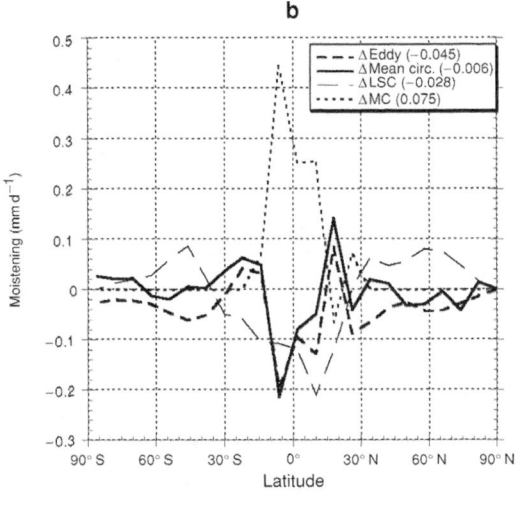

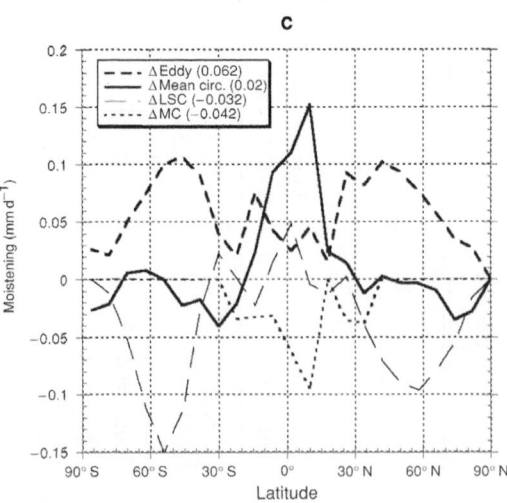

Figure 9.6. As in Figure 9.4 except for the modeled climate simulations: 6K (a), LGM (b), 2 CO$_2$ (c).

Table 9.6. *Radiative processes in the climate-change experiments*

Simulation	Surface air temperature (°C)	Specific humidity in upper troposphere 550–150 mbar (10^{-5} kg kg^{-1})	Longwave radiation leaving surface (W m^{-2})	Longwave radiation at model top (W m^{-2})	Greenhouse effect, G* (W m^{-2})	Normalized greenhouse effect, g**	Total cloud cover (%)	High cloud cover (%)
Control	13.75	45.6	397	233	164	0.4131	56.6	13.8
6K	13.62	44.8	396	235	161	0.4066	56.7	13.8
LGM	8.34	31.8	373	232	141	0.3780	53	12.2
Stand Dev	0.025	0.12	0.12	0.12	0.14	0.0006	0.07	0.09
Control	13.88	45.9	397	233	164	0.4131	56.8	14.3
2×CO$_2$	17.39	68.9	415	234	181	0.4361	56.6	16.7

*$dG/DT_s = 4.25$ for LGM, 4.84 for $2 \times CO_2$; **$dg/dT_s = 6.6 \times 10^{-3}$ for LGM and $2 \times CO_2$ simulations.

Table 9.7. *Sub-tropical (20° N to 30° S) upper-troposphere moisture balance (mm d^{-1})*

	Control	6K	LGM	$(2 \times CO_2)$	$4 \times 5, 48L$	$2 \times 2.5, 32L$
Evaporation	3.92	3.99	3.84	4.33	3.62	3.82
Specific humidity (mm)	5.2	5.0	3.9	7.7	5.2	3.6
Largescale condensation	−0.15	−0.16	−0.19	−0.16	0.13	−0.01
Moist Convection	−0.17	−0.17	−0.16	−0.18	−0.38	−0.24
Vertical transport by eddies	0.41	0.40	0.39	0.47	0.40	0.35
Vertical transport by mean circulation	−0.03	−0.03	0.01	−0.04	−0.06	−0.10
Horizontal transport by eddies	−0.05	−0.04	−0.06	−0.06	0	−0.03
Horizontal transport by mean circulation	0.02	0.01	0	0.10	0.02	0.01

vary with both the mean temperature and the low latitude gradient; thus using inter-annual, geographical, or latitudinal surface-temperature/moisture variations might not be suitable analogs for climate changes.

9.9 Discussion

The GISS model, and apparently most GCMs, effectively translate a warmer sea surface into increased moisture at all levels, regardless of whatever dynamical changes occur in accompaniment. The essence of this mechanism was deduced by Del Genio *et al.* (1991; 1994) in which they pointed out that due to the Claussius–Clapeyron relationship, increasing temperatures will augment the saturation vapor pressure more at warmer temperatures, hence at lower altitudes, steepening the vertical moisture gradient. When acting upon this stronger gradient, dynamical processes, such as mean latitudinal and longitudinal circulation cells and eddies will transport more moisture vertically. Throughout a variety of climate-change experiments we see this process operating in the model.

As noted earlier, this result is somewhat at variance with the theoretical discussions in the field concerning the tropical and sub-tropical moisture budgets, in which the moisture response is related to variations in the dynamical processes (e.g., Section 9.3). To emphasize this point, we present in Table 9.7 the upper-tropospheric sub-tropical moisture budget in the control run (at different resolutions) and for the climate-change experiments presented in Section 9.8. Expectations are that a decrease in the Hadley cell intensity will increase mid- and upper-tropospheric moisture due to reductions in subsidence of dry air from above. The sub-tropical upper-tropospheric specific humidity is considerably less in the modeled LGM despite its weaker Hadley circulation (Table 9.5). A weaker meridional circulation in the $2 \times CO_2$ climate is associated with increased moisture, but the weaker mean circulation is moistening the sub-tropics via horizontal transport. It is actually producing drying from vertical transport, for the vertical gradient of moisture has become larger in the warmer climate, and in acting on this increased gradient even the reduced subsidence is capable of producing vertical moisture transport divergence. The vertical transport by longitudinal variations (e.g., eddies plus longitudinal circulation cells) dominates in all climates and at all model resolutions. In contrast, the horizontal transport by eddies, suggested as being a major player in providing moisture to the sub-tropics (Section 9.3) has little net effect, and is in fact a slight drying mechanism; while there is eddy transport from convective tropical regions into the sub-tropics, there is actually greater eddy transport poleward from the sub-tropics, producing a net, small, divergence which changes little among the experiments.

Arguments concerning the effect of convection appear irrelevant to these results, since convection (plus largescale condensation) is a drying process in the model, and with increased convection in a warmer climate, the (combined) effect is even more negative. Nevertheless, moisture increases anyway. Throughout the tropics, the model parameterizations do satisfy the basic concept that moisture detrains at various levels, and the "largescale condensation" by itself (specifically the reevaporation of cloud moisture) is a significant component in helping to balance convective drying when the vertical resolution is sufficiently large ($4° \times 5°$, 48-layer model [$4 \times 5, 48L$]). In that sense it qualitatively agrees with expectations (although the vertical advection by the mean circulation and eddies are still important), and reproduces on the long-term time scale the processes seen on the diurnal time scale in SAGE II and ISCCP observations (Liao and Rind, 1997).

The modeling results and theoretical discussions are so dichotomous that they suggest two very different interpretations. At one extreme, one can assume that the model parameterizations are so crude that they have little relevance to the real-world situation; or, at the other extreme, one can conclude that the theoretical discussions are occurring in a vacuum, with little reference to the real world as simulated by the models. The first interpretation would seem at odds with studies that have indicated the models seem capable of reproducing the approximate magnitude of the water vapor feedback, at least for inter-annual or geographical variations in SSTs. If the models are producing water-vapor responses via the wrong mechanisms, then their suitability for climate-change depictions is highly suspect, and only observed trends are relevant (even inter-annual variations might be inappropriate because they do not necessarily involve the same dynamical-mechanism changes as would occur in an actual climate change).

If the second interpretation is correct, then we are left with (uncertain) trend observations and models as the only way of understanding the water-vapor feedback, and no amount of theorizing with current understanding would be relevant to estimating the future water-vapor response (and climate sensitivity). To see how difficult it will be to determine trends from available observations, the most ubiquitous of the observing methods involves determination of UTH from the brightness temperatures. We calculated the model UTH integrated over the 200–550 mbar levels, hence at pressures appropriate to the weighting functions used for UTH observations. Even though the experiments here differed in global mean temperature by 10 °C, zonal UTH differences were generally on the order of a few percent. Considering that models may act toward conserving relative humidity somewhat more than seems to occur in the real world, this result may be somewhat of an underestimate. Nevertheless, with an order of magnitude smaller temperature change likely during the next few decades, we might consider this an upper estimate. Peak latitudinal changes as deduced by Bates and Jackson (2001) from ENSO-dominated decadal variation over the past 20 years are of this magnitude, with confidence levels generally on the order of 50%.

9.10 Concluding remarks

The water-vapor feedback has the potential to amplify anthropogenic greenhouse-gas climate forcing by 100%, and could potentially convert moderate climate forcings into highly deleterious climate responses. All available observational studies indicate that the water-vapor feedback, as indicated by GCMs, is at least qualitatively correct, and may be of the proper order of magnitude. The major caveat in this regard is that the conclusions are based on inter-annual or geographic variations in water vapor and SSTs, and the suitability of these studies as analogs for climate change cannot be proven. A disconcerting result is that while most GCMs act similarly as far as the water-vapor response is concerned, detailed analysis of how one GCM is producing its upper-tropospheric water-vapor budget and feedback is at variance with theoretical expectations of how the real world is doing it. Diminishing the distance between the modeled and theorized processes should be a top priority for those attempting to better understand the water-vapor feedback and climate sensitivity.

Acknowledgments

The GISS modeling studies are supported by the NASA Climate Modeling Program.

References

Angell, J. K. (1988). Variations and trends in tropospheric and stratospheric global temperatures, 1958–87. *J. Clim.* **1**, 1296–313.

Bates, J. J. and D. L. Jackson (1997). A comparison of water vapor observations with AMIP I simulations. *J. Geophys. Res.* **102**, 21837–52.

(2001). Trends in upper-tropospheric humidity. *Geophys. Res. Lett.* **28**, 1695–8.

Bauer, M. (1998). Upper tropospheric humidity in the GISS GCM: mechanisms regulating interannual change. Masters Thesis, Columbia University, NY.

Bengtsson, L., E. Roeckner, and J. Stendel (1999). Why is the global warming proceeding much slower than expected? *J. Geophys. Res.* **104**, 3865–76.

Betts, A. K. (1990). Greenhouse warming and the tropical water budget. *Bull. Amer. Meteor. Soc.* **71**, 1464–5.

Cess, R. D. (1989). Gauging water-vapour feedback. *Nature* **342**, 736–7.

Cess, R. D., G. L. Potter, J. P. Blanchet *et al.* (1990). Intercomparison and interpretation of climate feedback processes in 19 atmospheric general circulation models. *J. Geophys. Res.* **95**, 16601–15.

Chagas, J. C., D. A. Newnham, K. M. Smith, and K. P. Shine (2001). Effects of improvements in near-infrared water vapour line intensities on short-wave atmospheric absorption. *Geophys. Res. Lett.* **28**, 2401–4.

Charboureau, J.-P., A. Chedin, and N. A. Scott (1998). Relationship between sea surface temperature, vertical dynamics, and the vertical distribution of atmospheric water vapor inferred from TOVS observations. *J. Geophys. Res.* **103**, 23173–80.

Chen, M., R. B. Rood, and W. G. Read (1998). Upper tropospheric water vapor from GEOS reanalysis and UARS MLS observation. *J. Geophys. Res.* **103**, 19587–94.

Christy, J. R., R. W. Spencer, and E. S. Lobl (1998). Analysis of the merging procedure for the MSU daily temperature time series. *J. Clim.* **11**, 2016–41.

CLIMAP (1981). Seasonal reconstructions of the Earth's surface at the last glacial maximum. Geological Society of America, Map and Chart Series 36.

Del Genio, A. D. (1991). Simulations of the effect of a warmer climate on atmospheric humidity. *Nature* **351**, 382–5.

Del Genio, A. D., W. Kovari, Jr., and M.-S. Yao (1994). Climatic implications of the seasonal variation of upper tropospheric water vapor. *Geophys. Res. Lett.* **21**, 2701–4.

Forster, P. M. de F. and K. Shine (1999). Stratospheric water vapour changes as a possible contributor to observed stratospheric cooling. *Geophys. Res. Lett.* **26**, 3309–12.

Gaffen, D. J., R. D. Rosen, D. A. Salstein, and J. S. Boyle (1997). Evaluation of tropospheric water vapor simulations from the Atmospheric Model Intercomparison Project. *J. Clim.* **10**, 1648–61.

Gershunov, A., J. Michaelsen, and C. Gautier (1998). Large-scale coupling between the tropical greenhouse effect and latent heat flux via atmospheric dynamics. *J. Geophys. Res.* **103**, 6017–31.

Gutzler, D. (1996). Low-frequency ocean–atmosphere variability across the tropical western Pacific. *J. Atmos. Sci.* **53**, 2773–85.

Hall, A. and S. Manabe (1999). The role of water vapor feedback in unperturbed climate variability and global warming. *J. Clim.* **12**, 2327–46.

Hansen, J. E., A. Lacis, D. Rind *et al.* (1984). Climate sensitivity: analysis of feedback mechanisms. In *Climate Processes and Climate Sensitivity,* Geophysical Monograph 29, eds. J. E. Hansen and T. Takahashi, Washington, DC, AGU, pp. 130–63.

Hansen, J. E., M. Sato, R. Ruedy *et al.* (1998). Forcings and chaos in interannual to decadal climate change. *J. Geophys. Res.* **102**, 25679–720.

Harries, J. E. (1996). The greenhouse effect: a view from space. *Quart. J. Roy. Meteor. Soc.* **122**, 799–818.

(1997). Atmospheric radiation and atmospheric humidity. *Quart. J. Roy. Meteor. Soc.* **123**, 2173–86.

Holland, J. Z. and E. M. Rasmusson (1973). Measurements of the atmospheric mass, energy and momentum budgets over a 500 km square of tropical ocean. *Mon. Wea. Rev.* **101**, 44–53.

Hu, H. and W. T. Liu (1998). The impact of upper tropospheric humidity from Microwave Limb Sounder on the midlatitude greenhouse effect. *Geophys. Res. Lett.* **25**, 3151–4.

Inadmar, A. K. and V. Ramanathan (1994). Physics of greenhouse effect and convection in warm oceans. *J. Clim.* **7**, 715–31.

(1998). Tropical and global scale interactions among water vapor, atmospheric greenhouse effect, and surface temperature. *J. Geophys. Res.* **103**, 32177–94.

IPCC (1995). *Climate Change 1995*. J. T. Houghton, L. G. Meira Filho, B. A. Callander, N. Harris, A. Kattenberg, and K. Maskell, eds., Cambridge, Cambridge University Press.

(2001). *Climate Change 2001*. Cambridge, Cambridge University Press.

Kelly, K. K., A. F. Tuck, and T. Davies (1991). Winter asymmetry of upper tropospheric water between the northern and southern hemispheres. *Nature* **353**, 244–47.

Liao, X. and D. Rind (1997). Upper tropospheric/lower stratospheric water vapor and tropospheric deep convection. *J. Geophys. Res.* **102**, 19543–58.

Lindzen, R. S. (1990). Some coolness regarding global warming. *Bull. Amer. Meteor. Soc.* **71**, 288–99.

Lindzen, R. S., B. Kirtman, D. Kirk-Davidoff, and E. K. Schneider (1995). Seasonal surrogate for climate. *J. Clim.* **8**, 1681–4.

Oinas, V., A. A. Lacis, D. Rind, D. T. Shindell, and J. E. Hansen (2001). Radiative cooling by stratospheric water vapor: big differences in GCM results. *Geophys. Res. Lett.* **28**, 2791–4.

Oltmans, S. J., H. Vomel, D. J. Hofmann, K. H. Rosenlof, and D. Kley (2000). The increase in stratospheric water vapor from balloonborne, frostpoint hygrometer measurements at Washington, DC and Boulder, Colorado. *Geophys. Res. Lett.* **27**, 3453–6.

Oort, A. H. (1983). NOAA Paper 14, Washington, DC, US Department of Commerce.

Pierrehumbert, R. T. (1995). Thermostats, radiator fins, and the local runaway greenhouse. *J. Atmos. Sci.* **52**, 1784–1806.

(1998). Lateral mixing as a source of subtropical water vapor. *Geophys. Res. Lett.* **25**, 151–54.

(1999). Subtropical water vapor as a mediator of rapid global climate change. In *Roles of High and Low Latitudes in Millennial-Scale Global Climate Change*, Geographical Monograph 112, Washington, DC, American Geophysical Union.

Prather, M. (1986). Numerical advection by conservation of second-order moments. *J. Geophys. Res.* **91**, 6671–81.

Raval, A. and V. Ramanathan (1989). Observational determination of the greenhouse effect. *Nature* **342**, 758–62.

Renno, N. O., K. A. Emanuel, and P. H. Stone (1994). Radiative–convective model with an explicit hydrologic cycle. I. Formulation and sensitivity to model parameters. *J. Geophys. Res.* **99**, 14429–41.

Rind, D. (1998a). Just add water vapor. *Science* **281**, 1152–3.

(1998b). Latitudinal temperature gradient and climate. *J. Geophys. Res.* **103**, 5943–71.

Rind, D. and J. Lerner (1996). The use of on-line tracers as a diagnostic tool in GCM model development. *J. Geophys. Res.* **101**, 12667–83.

Rind, D., R. Healy, C. Parkinson, and D. Martinson (1997). The role of sea ice in $2\times CO_2$ climate model sensitivity. Part I. The total influence of sea ice thickness and extent. *J. Clim.* **8**, 449–63.

Rind, D., M. Chandler, J. Lerner, D. G. Martinson, and X. Yuan (2001). Climate response to basin-specific changes in latitudinal temperature gradients and implications for sea-ice variability. *J. Geophys. Res.* **106**, 20161–73.

Rind, D., E. W. Chiou, W. Chu *et al.* (1991). Positive water vapour feedback in climate models confirmed by satellite data. *Nature,* **349**, 500–3.

Rosenlof, K. H., S. Oltmans, D. Kley *et al.* (2001). Stratospheric water vapor increases over the past half-century. *Geophys. Res. Lett.* **28**, 1195–8.

Ross, R. J. and W. P. Elliot (1996). Tropospheric water vapor climatology and trends over North America: 1973–93. *J. Clim.* **9**, 3561–74.

Ross, R. J. and D. J. Gaffen (1998). Comment on widespread tropical atmospheric drying from 1979 to 1995 by Steven R. Schroeder and James P. McGuirk. *Geophys. Res. Lett.* **25**, 4357–8.

Russell, G. and D. Rind (1999). Atmosphere–ocean response to CO_2 transient increase in the GISS coupled model. *J. Clim.* **12**, 531–9.

Salathé, E. P. and D. L. Hartmann (1997). A trajectory analysis of tropical upper-tropospheric moisture and convection. *J. Clim.* **10**, 2533–47.

Salathé, E. P., D. Chesters, and Y. C. Sud (1995). Evaluation of the upper-tropospheric moisture climatology in a general circulation model using TOVS radiance observations. *J. Clim.* **8**, 2404–14.

Schmetz, J. and L. van de Berg (1994). Upper tropospheric humidity observations from Meteosat compared with short-term forecast fields. *Geophys. Res. Lett.* **21**, 573–6.

Schroeder, S. R. and J. P. McGuirk (1998a). Widespread tropical atmospheric drying from 1979 to 1995. *Geophys. Res. Lett.* **25**, 1301–4.

(1998b). "Reply". *Geophys. Res. Lett.* **25**, 4359–60.

Sherwood, S. C. (1996). Maintenance of the free-tropospheric water vapor distribution. Part II: simulation by large-scale advection. *J. Clim.* **9**, 2919–34.

Shindell, D. T. (2001). Climate and ozone response to increased stratospheric water vapor. *Geophys. Res. Lett.* **28**, 1551–4.

Shine, K. P. and A. Sinha (1991). Sensitivity of the Earth's climate to height-dependent changes in water vapour mixing ratio. *Nature* **354**, 382–4.

Sinha, A. and M. R. Allen (1994). Climate sensitivity and tropical moisture distribution. *J. Geophys. Res.* **99**, 3707–16.

Sinha, A. and J. E. Harries (1997). The Earth's clear-sky radiation budget and water vapor absorption in the far infrared. *J. Clim.* **10**, 1601–14.

Soden, B. J. (1997). Variations in the tropical greenhouse effect during El Niño. *J. Clim.* **10**, 1050–5.

(1998). Tracking upper tropospheric water vapor radiances: a satellite perspective. *J. Geophys. Res.* **103**, 17069–81.

(2000). The diurnal cycle of convection, clouds, and water vapor in the tropical upper troposphere. *Geophys. Res. Lett.* **27**, 2173–6.

Soden, B. J. and F. P. Bretherton (1994). Evaluation of water vapor distribution in general circulation models using satellite observations. *J. Geophys. Res.* **99**, 1187–210.

Soden, B. J. and R. Fu (1995). A satellite analysis of deep convection, upper-tropospheric humidity, and the greenhouse effect. *J. Clim.* **8**, 2333–51.

Soden, B. J. and J. R. Lanzanthe (1996). An assessment of satellite and radiosonde climatologies of upper-tropospheric water vapor. *J. Clim.* **9**, 1235–50.

Spencer, R. W. and W. D. Braswell (1997). How dry is the tropical free troposphere? Implications for global warming theory. *Bull. Amer. Meteor. Soc.* **78**, 1097–106.

Stephens, G. L. and T. J. Greenwald (1991). The Earth's radiation budget and its relation to atmospheric hydrology, 1. Observations of the clear sky greenhouse effect. *J. Geophys. Res.* **96**, 15311–24.

Sun, D.-Z. and I. M. Held (1996). A comparison of modeled and observed relationships between interannual variations of water vapor and temperature. *J. Clim.* **9**, 665–75.

Sun, D.-Z. and R. Lindzen (1993). Distribution of tropical tropospheric water vapor. *J. Atmos. Sci.* **50**, 1644–60.

Sun, D.-Z. and A. H. Oort (1995). Humidity–temperature relationships in the tropical troposphere. *J. Clim.* **8**, 1974–87.

Sun, D.-Z., C. Covey, and R. S. Lindzen (2001). Vertical correlations of water vapor in GCMs. *Geophys. Res. Lett.* **28**, 259–62.

Yang, H. and K. K. Tung (1998). Water vapor, surface temperature and the greenhouse effect – a statistical analysis of tropical-mean data. *J. Clim.* **11**, 2666–97.

Zhou, P. and R. E. Eskridge (1997). Atmospheric water vapor over China. *J. Clim.* **10**, 2533–47.

Zhou, X.-L., M. A. Geller, and M. Zhang (2001). Cooling trend of the tropical cold point tropopause temperatures and its implications. *J. Geophys. Res.* **106**, 1511–22.

Zhu, Y., R. E. Newell, and W. G. Read (2000). Factors controlling upper-troposphere water vapor. *J. Clim.* **13**, 836–48.

10

Water-vapor observations

Geophysical Fluid Dynamics Laboratory, Princeton Forrestal Campus, Princeton, NJ

10.1 Introduction

Water vapor is a key climate variable, serving to link a variety of complex and poorly understood processes. Although comprising less than 1% of the atmospheric mass, water vapor is the dominant gaseous absorber of thermal radiation. The disproportional importance of water vapor stems from the fact that it is the only atmospheric constituent to possess a permanent dipole moment. This feature, combined with the asymmetrical arrangement of mass in the water-vapor molecule, leads to a rich and complex distribution of absorption lines through the electromagnetic spectrum. Radiative absorption by water vapor not only plays a key role in determining the atmosphere's "greenhouse effect" but, because the concentration of water vapor depends strongly on the surface temperature, it also comprises the largest known feedback mechanism for amplifying global warming (ICCP, 1990). Current estimates are that radiative feedback by water vapor increases the climatic sensitivity to carbon dioxide by roughly a factor of two when considered in isolation from other feedbacks, and by as much as a factor of three or more when the interactions with other feedbacks are considered. In addition to its radiative effects, the strong dipole moment of water vapor is also responsible for the large latent heat associated with its phase transitions which, in turn, provides much of the energy for driving the atmosphere's largescale circulation.

The fundamental importance of water vapor in Earth's climate underscores the need for an accurate understanding of its distribution and variation. Accordingly, the task of observing water vapor has received considerable attention over the last several decades. However, water vapor is highly variable in both space and time, and its mass concentration in the troposphere varies by over three orders of

Frontiers of Climate Modeling, eds. J. T. Kiehl and V. Ramanathan.
Published by Cambridge University Press. © Cambridge University Press 2006.

magnitude. Both of these characteristics hinder its observational determination. To further complicate matters, water vapor, unlike atmospheric temperature or pressure fields, is not dynamically constrained. Therefore, data-assimilation methods which supplement the observational data with model-generated fields offer little additional information above that contained in the original observations. Consequently, describing the distribution of water vapor with sufficient detail fully to understand the processes which lead to its variability, represents a substantial challenge – yet such observations are necessary to develop confidence in our ability to predict changes in water vapor associated with global warming. Indeed, it is the fundamental importance of water vapor in the climate system combined with the challenge of monitoring its distribution and variation that has made water-vapor measurements a priority for current and future observing systems (National Research Council, 1999).

Fortunately, a wide variety of instruments exist for measuring water vapor including surface, *in situ* and space-borne sensors. However the variations in water vapor relevant to the issues of climate and climate change necessitate observations with global or near-global coverage. Therefore this chapter focuses on two classes of observing systems capable of such coverage – the global radiosonde network and satellite-based remote sensing. The following sections provide an overview of the observing capabilities of these systems, summarizes their strengths and weaknesses, and highlights future opportunities for improvements in our monitoring capabilities.

10.2 The global radiosonde network

Historically, the main source of information on atmospheric water vapor has been the global radiosonde network. These measurements, designed to provide data in support of operational weather forecasts, are typically performed twice daily at over 800 stations around the globe. Initiated following World War II, these observations provide the longest record of water-vapor measurements. Not surprisingly, much of our original understanding of the distribution and variations of water vapor were based on radiosonde observations (Rasmusson, 1967; Piexoto and Oort, 1983; Gaffen *et al.*, 1991; Piexoto and Oort, 1996). Figure 10.1 depicts the distribution of water vapor obtained from the global radiosonde network plotted as a function of height and latitude (Piexoto and Oort, 1996). The concentrations of water vapor can be expressed in terms of its mass mixing ratio or the relative humidity, offering two very different perspectives on the moisture distribution. The former is largely determined by the atmospheric temperature, as evidenced by its strong vertical and meridional gradients, reflecting the nearly exponential dependence of the saturation vapor pressure of water on temperature.

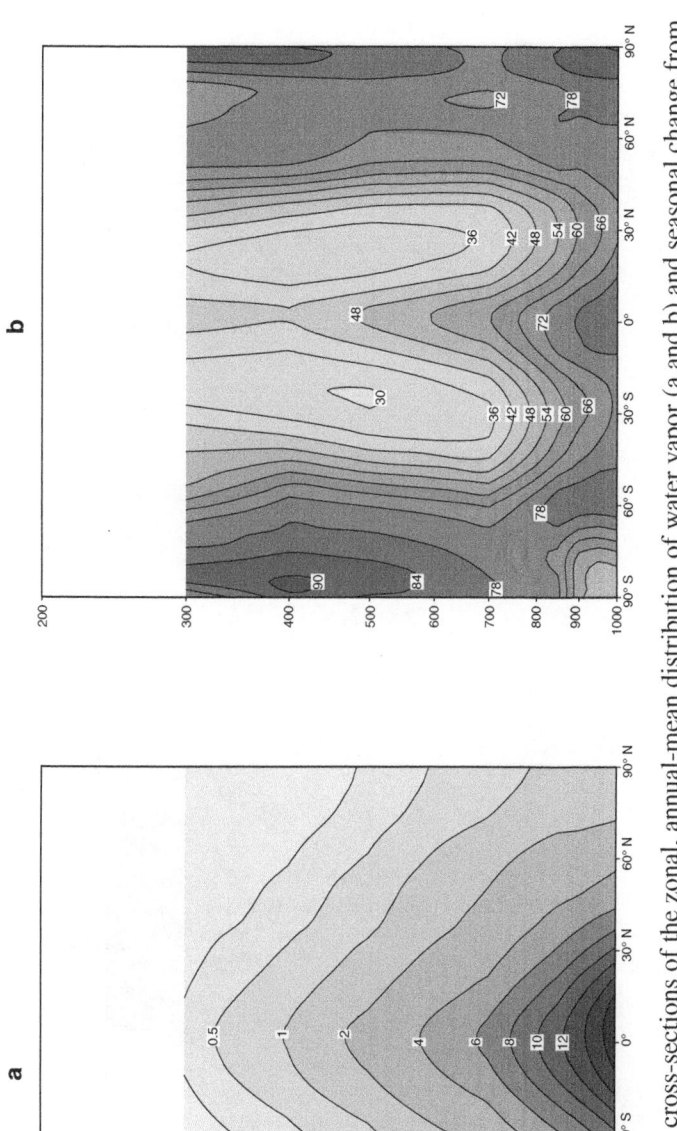

Figure 10.1. Vertical cross-sections of the zonal, annual-mean distribution of water vapor (a and b) and seasonal change from June, July, August to December, January, February (b and d) from the global radiosonde network (Piexoto and Oort, 1996). The water-vapor concentration is expressed in terms of both the mass mixing ratio, g, g water vapor per kg dry air, (a and c) and the relative humidity, %, (c and d).

287

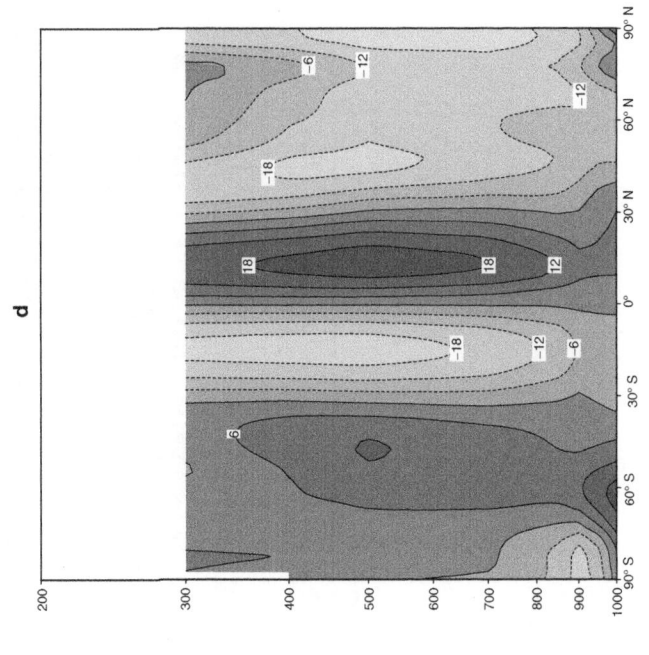

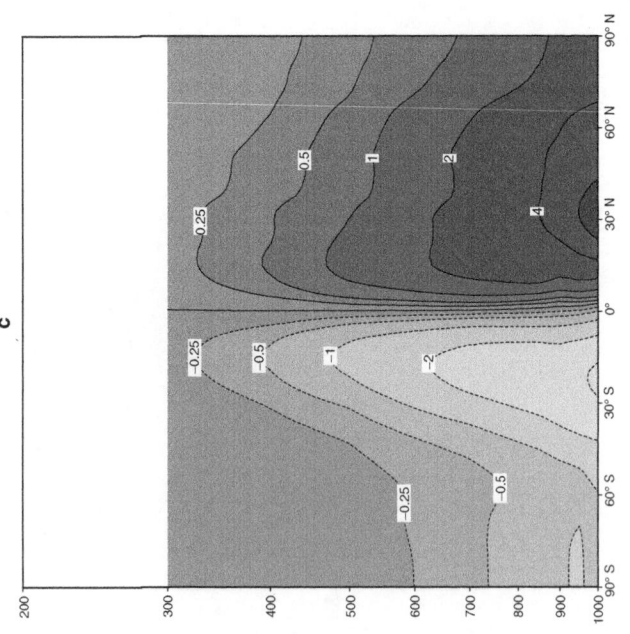

Figure 10.1. (*cont.*)

The relative humidity, on the other hand, contains no explicit dependence upon temperature and its distribution in the free troposphere is largely governed by the largescale atmospheric circulation. This connection between the largescale dynamics and relative humidity is evident by the contrast in relative humidity between the rising and sinking branches of the Hadley circulation. Similar dependences are reflected in the seasonal variations of these fields.

The radiosonde network has several unique advantages over other systems. Most notable is the ability of radiosondes to measure the vertical distribution of water-vapor concentrations with a high degree of vertical resolution (100 meters or less). Radiosondes also measure water vapor under both clear and cloudy weather conditions, in contrast to many remotely-sensed climatologies which are restricted to water-vapor measurements from cloud-free regions only. Perhaps the greatest asset of the radiosonde network is its longevity. Radiosonde observations provide the only continuous, long-term archive of water-vapor variations prior to the start of the operational satellite records in 1979. Consequently, much of our understanding of inter-decadal variations and long-term trends are, for better or worse, based on radiosonde records. Even attempts to estimate trends from satellite instruments have relied on radiosonde measurements to inter-calibrate successive satellite instruments (Schroeder and McGuirk, 1998), although the merits of this approach have been questioned (Ross and Gaffen, 1998). Careful analysis of the radiosonde measurements starting in the early 1970s have identified significant upward trends during the last several decades over North America (Ross and Elliott, 1996), China (Zhai and Eskridge, 1997) and the western tropical Pacific (Gutzler, 1996). An attempt to examine global moisture trends from radiosonde measurements (Oort, 1993) also suggests upward trends in the lower troposphere, particularly in the tropics. However, as discussed below, the radiosonde observing system has a number of shortcomings related to instrumentation changes and regional sampling restrictions which complicate the interpretation of their long-term trends.

There are two primary limitations of radiosonde water-vapor measurements. First, the accuracy of many humidity sensors is compromised above ~500 hPa due to the difficulty of measuring the low water-vapor concentrations at these altitudes (Elliott and Gaffen, 1991). In particular, older sensor designs typically respond more slowly to the rapid decrease in moisture concentrations with height, resulting in a spurious moist bias at high altitudes. The magnitude of this effect varies depending upon the sensor type and, as a result, different sensors report different absolute water-vapor values. Indeed, the accuracy of upper-air water-vapor measurements has been shown to vary significantly from one country to the next, often reflecting the type of instrument used. For example, Figure 10.2 displays the difference in upper-tropospheric relative humidity (UTH) measured at each radiosonde station with the corresponding retrieval obtained from the satellite-based HIRS instrument

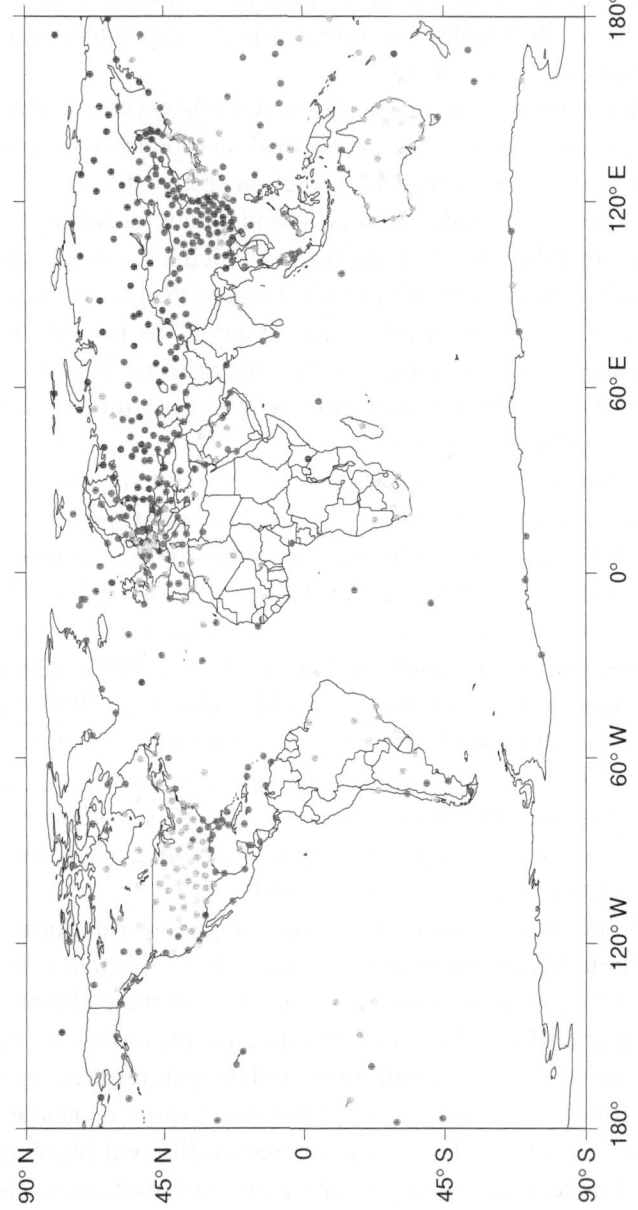
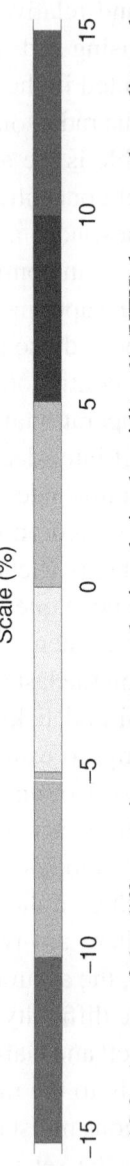

Scale (%)

Figure 10.2. A map of the difference in upper-tropospheric relative humidity, % (UTH) between the radiosonde observations and HIRS satellite retrievals for June–August, 1989 (adapted from Soden and Lanzante, 1996).

(see Section 10.3.2). Differences between the satellite- and radiosonde-reported UTH range from ± 15% in absolute values. However, upon closer inspection, the discrepancy in the UTH field exhibits a distinct geopolitical dependence. Over much of the former Soviet Union, China, and eastern Europe the radiosondes report a systematically moister upper troposphere relative to the satellite observations, whereas over most of the remaining stations, particularly western Europe and North America, the radiosonde measurements are systematically drier than the satellite observations. Of greater concern for the climate-monitoring problem is that the radiosonde instrumentation has changed over time at most locations. As countries upgrade their radiosonde instrumentation, the increased use of more accurate and faster-responding humidity sensors is believed to have introduced a spurious drying trend, particularly in the upper troposphere (Ross and Gaffen, 1998). The absence of reliable station-history information to document the timing of such changes has hindered attempts to distinguish spurious changes in moisture associated with instrumentation transitions from those which may be attributable to legitimate climate change.

Perhaps an even greater shortcoming of the radiosonde network is their relatively poor spatial coverage. While over 800 radiosonde stations are located around the globe, their geographical distribution is highly uneven (see Figure 10.2). Relatively good coverage is obtained over northern-hemisphere land regions. However, vast areas of the open oceans, particularly in the tropics and southern hemisphere, remain unobserved. Attempts have been made to fill these voids using objective analysis techniques which interpolate the station data onto a uniformly distributed grid (e.g., Oort, 1983). However, the highly heterogeneous nature of water vapor limits the utility of such analyses, as even the most sophisticated interpolation techniques are unable to compensate for such large data voids. For example, comparisons with satellite observations demonstrate that the current radiosonde network is unable to capture even the most basic features of El Niño southern oscillation (ENSO)-driven variations in water vapor over the tropical oceans (Soden and Lanzante, 1996). Thus, sampling errors become a large uncertainty when interpreting global or even regional water-vapor variations.

10.3 Satellite remote sensing

Due to their ability to provide global coverage and accurate measurements in both the troposphere and stratosphere, satellite observations provide a key supplement to the radiosonde network. Dating back to the late 1960s, numerous instruments have been developed for measuring water vapor from space-borne platforms (Table 10.1). Indeed satellite remote sensing of water vapor has been performed in nearly every portion of the electromagnetic spectrum – solar, thermal-infrared (IR), microwave,

Table 10.1. *Current remote-sensing capabilities*

Method	Instrument	Dates	Spectral band
Solar occultation	SAGE II	1984–present	0.94 μm
	HALOE	1991–present	0.94 μm
Thermal infrared	HIRS	1979–present	6.3 μm
	GOES	1982–present	6.3 μm
	ERBE	1984–89	5–50 μm
Microwave	SMMR	1979–84	22 GHz
	SSMI	1987–present	22 GHz
	SSMT2	1991–present	183 GHz
	AMSU A/B	1995–present	22, 183 GHz
	MLS	1991–present	200 GHz

and radiowave – attesting to the ubiquitous nature of water-vapor–radiative interactions. The following sections summarize our current remote-sensing capabilities in these spectral regions.

10.3.1 Solar measurements

One method of estimating the concentration of water vapor, referred to as solar occultation, measures the extinction of sunlight in a near-infrared (IR) (0.94 μm) absorption band of water vapor. This approach is also known as "limb sounding," since it measures the near-horizon or "limb" radiation. By viewing sideways through the atmosphere, occultation instruments look directly at the sun and compare the amount of the solar radiation transmitted through the atmosphere with unattenuated radiation measured above the atmosphere. The transmission of sunlight is computed as the instrument scans vertically through the atmosphere and, given knowledge of the extinction coefficients and concentrations of other radiatively active species, can be used to determine the corresponding water-vapor path lengths as a function of height.

The Limb Infrared Monitor Sounder (LIMS) was the first instrument to provide global measurements of water vapor based on this technique (Fischer *et al.*, 1981). More recently, the Stratospheric Aerosol and Gas Experiment II (SAGE II) and HALogen Occultation Experiment (HALOE) have provided much longer climatologies (see Table 10.1). To illustrate the observing capabilities of solar-occultation measurements, Figure 10.3a depicts a height–latitude cross-section of SAGE II water-vapor measurements for April, 1988. Note the contrast between tropospheric and stratospheric water-vapor concentrations as well as the presence of a distinct water-vapor minimum, or hygropause, near the tropical tropopause. The

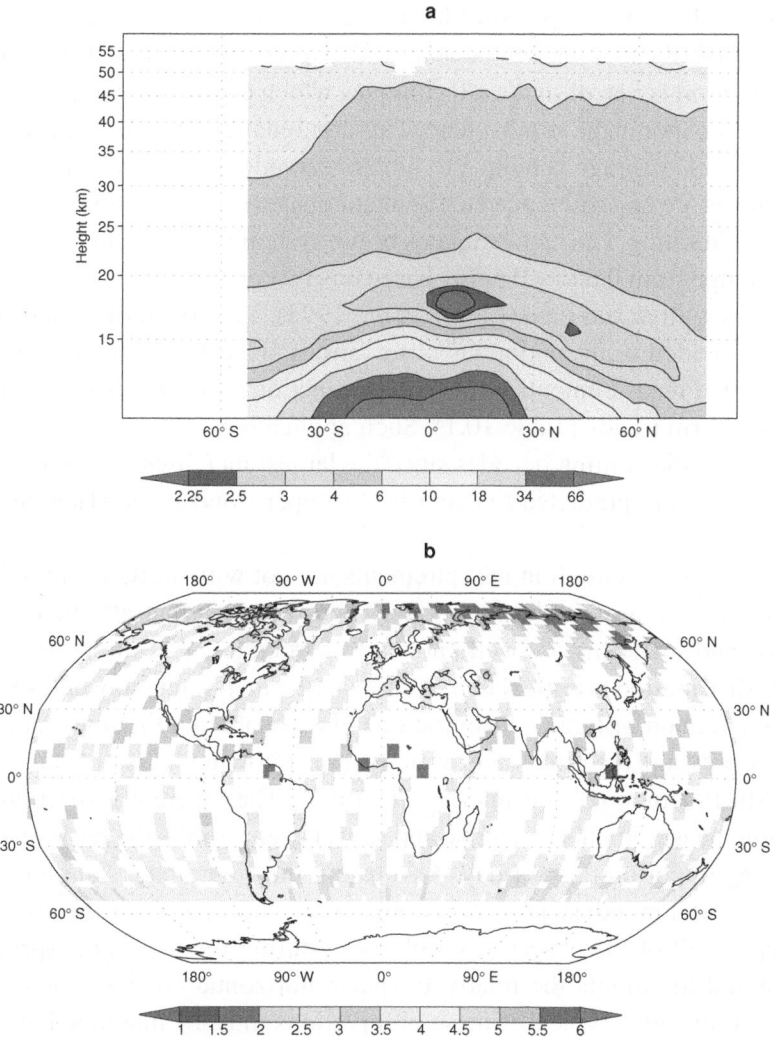

Figure 10.3. The vertical cross-section (a) and horizontal distribution at 20 km (b) of the water-vapor mixing ratio (in μg kg^{-1}) as observed by SAGE II for the month of April, 1988.

This figure is available for download in colour from
www.cambridge.org/9780521791328

ability to resolve the vertical structure of water vapor in the upper troposphere and lower stratosphere makes these measurements particularly useful for understanding the response of upper-tropospheric water vapor to climate change (Rind *et al.*, 1991; Del Genio *et al.*, 1994; Lin *et al.*, 1998), stratospheric transport (Mote *et al.*, 1995), and troposphere–stratosphere exchange (Rind *et al.*, 1993; Wang *et al.*, 1996).

The solar occultation technique has several unique advantages over other remote-sensing methods. Solar occultation measurements offer relatively high vertical resolution of the water-vapor profiles ($\sim$1 km) and, like most satellite observations, near-global coverage can be achieved. Moreover, as evident in Figure 10.3, solar occultation is one of the few techniques which can provide accurate observations of water vapor in the stratosphere. This combination of high vertical resolution and near-global coverage is unique to limb observations. Finally, since occultation measurements are first referenced to the unattenuated solar radiation, they are inherently self-calibrating. Current estimates of the systematic errors in the water-vapor retrievals range from 0.6 to 1.0 ppmv based on error analyses (Chu *et al.*, 1993) and comparisons with *in situ* data (Larsen *et al.*, 1993). Moreover, the SAGE II instrument has provided near-continuous observations since late 1984, offering stable, well-calibrated measurements of the solar extinction which are ideal for the detection of long-term trends (Table 10.1). Such applications will become increasingly important over the coming decades since the largest increases in water vapor from global warming are predicted to occur in the upper troposphere (Held and Soden, 2000).

However, solar occultation measurements are not without their limitations. Relating the extinction of sunlight to water vapor requires that all other sources of absorption and scattering are well known. In particular, an accurate description of the ozone distribution is critical to the retrieval of water-vapor concentrations. Similarly, the injection of large concentrations of aerosols into the lower stratosphere effectively prevents the retrieval of water vapor following large volcanic eruptions such as Mt. Pinatubo (McCormick *et al.*, 1995). The inherent restriction of solar occultation measurements to sunrise and sunset events limits the retrievals to about 30 sampling opportunities per day. As a result, the geographic sampling obtained for any particular month is quite low compared to other space-borne sensors. Figure 10.3b depicts all of the observations obtained during the month of April, 1988 at $\sim$ 20 km and highlights the relatively sparse horizontal coverage obtained from these measurements. Typically, these observations must be integrated over several months to obtain near-global coverage and therefore are most suited for describing zonal-mean cross-sections of water vapor rather than regional variations. The limb viewing geometry also results in a large horizontal footprint ($\sim$ 250 km). Since clouds (or aerosols) interfere with the retrieval of water vapor, occultation measurements are effectively limited to cloud-free regions of the upper troposphere and stratosphere. Indeed, typically fewer than half of the SAGE II water-vapor profiles extend below 15 km and those which do are heavily biased towards cloud-free conditions. This clear-sky sampling limitation introduces a significant bias in the climatology (Zhang, 1995). Consequently, solar occultation measurements are most suited for studies of water vapor in the stratosphere and upper-most portion of the troposphere, but are of limited value for describing the climatological distribution of water vapor at lower levels. As we shall see in the next two sections, the poor horizontal coverage and clear-sky sampling bias which hinders solar occultation measurements are well compensated for by thermal-IR and microwave observations. However, these advantages come at the cost of high vertical resolution.

10.3.2 Thermal-infrared measurements

Water vapor absorbs and emits thermal radiation at wavelengths throughout the infrared spectrum. Of particular importance for remote sensing is the 6.3 μm absorption band whose use dates back to the early 1960s in which the first space-borne measurements of water vapor were made from the Television InfraRed Observation Satellite (TIROS) II satellite (Bandeen *et al.*, 1961) and TIROS IV (Raschke and Bandeen, 1967). In this portion of the spectrum water vapor is the only significant absorber. Near the center of the band the absorption is so strong that the underlying surface is completely obscured and the upwelling radiance originates exclusively from thermal emission by water vapor in the upper half of the troposphere. As one moves away from the center, the attenuation weakens and the layers of peak emission shift to lower levels of the atmosphere. An example of the satellite-measured radiance at 6.7 μm, which lies near the center of this band, is provided in Figure 10.4a. Larger concentrations of water vapor increase the atmospheric absorption and, since temperatures decrease with height in the troposphere, reduce the radiance emitted to space. Therefore larger radiances (dark regions in Figure 10.4a) imply a drier atmosphere, whereas the smaller radiances (bright regions) imply a moister atmosphere. The initial applications of water-vapor imagery were dominated by such qualitative interpretations, largely unaware of early sensitivity studies (Möller, 1961; Raschke and Bandeen, 1967) which demonstrated that radiance measurements in the 6.3 μm band primarily respond to variations in relative humidity, rather than specific humidity as discussed in Chapter 9. This dependence stems from the interplay between temperature-dependent Planck emission and specific-humidity-dependent absorption which combine to make the radiance most sensitive to variations in relative humidity. For example, if the atmospheric temperature is increased while holding the specific humidity constant there will be a corresponding increase in the upwelling radiance which is associated not with a change in specific humidity, but rather with a reduction in relative humidity. Similarly, if the temperature is increased while holding the relative humidity constant, the increased Planck emission is largely offset by the increase in specific humidity which reduces the atmospheric transmissivity. Thus, under a constant relative-humidity constraint, large changes in specific humidity have relatively little impact on the 6.3 μm radiance. Consider Figure 10.4b, which shows the distribution of upper-tropospheric relative humidity retrieved from the corresponding radiance data in

a

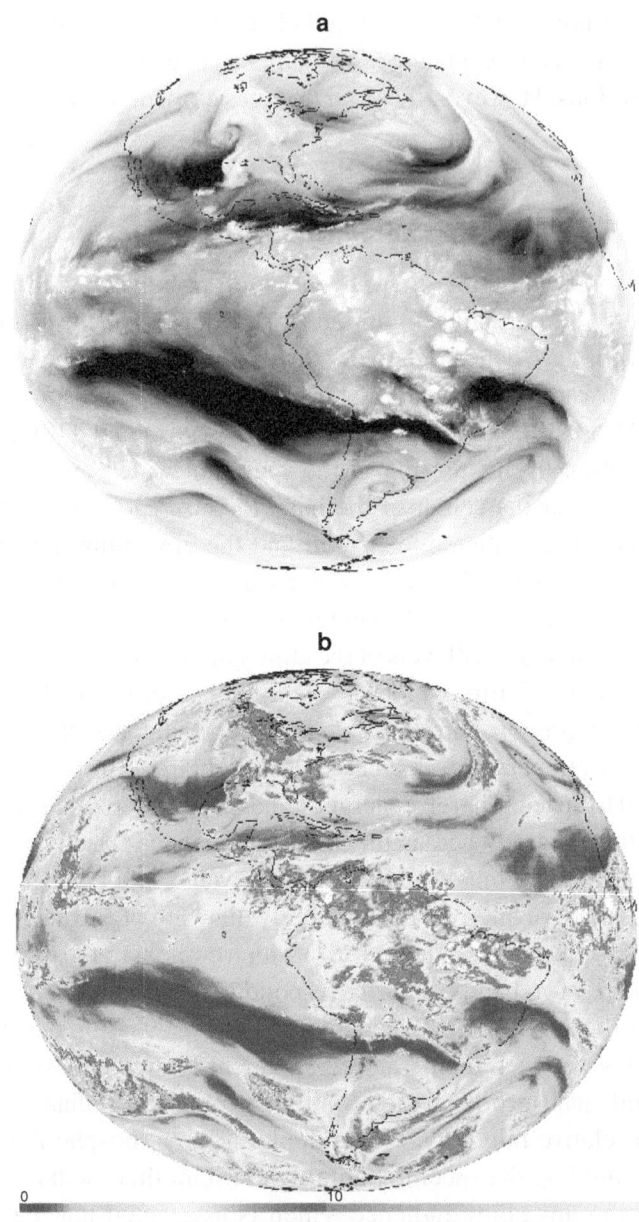

b

0   10

Upper-tropospheric relative humidity (%)   High-cloud cover

Figure 10.4. The 6.7 μm radiance field (a) and corresponding retrieval of upper-tropospheric relative humidity (b) from GOES-8 observations on April 27, 1999. This figure is available for download in colour from www.cambridge.org/9780521791328

Figure 10.4a. Note the close similarity between the radiance and relative-humidity fields, their large horizontal variations, and their close association with various features of the atmospheric circulation (e.g., tropical convective towers, extra-tropical eddies, etc.).

The first global observations of water vapor were made using thermal-IR measurements from the TIROS II–IV series of satellites (Bandeen *et al.*, 1961) and, shortly thereafter, from the Nimbus series of satellites (Schenk and Salomonson, 1972). Starting in 1979, the NOAA series of operational satellites have provided continuous, twice-daily, global measurements of the upwelling radiation using HIRS. This combination of high space–time coverage and lengthy duration is unmatched by any other water-vapor observing system. The HIRS instrument contains three water-vapor channels centered at 6.7, 7.3, and 8.3 μm which are sensitive to water-vapor concentrations in the upper, middle, and lower troposphere, respectively. As an example of the monitoring capabilities of this instrument, Figure 10.5 illustrates maps of the relative humidity (left) and layer-integrated precipitable water (right), as derived by Soden and Bretherton (1996) and Susskind *et al.* (1997), for three layers centered in the upper, middle, and lower troposphere. Due to their longevity and space–time coverage, HIRS measurements are used extensively for deriving water-vapor climatologies (Smith *et al.*, 1979; Wu *et al.*, 1993; Stephens *et al.*, 1996; Susskind *et al.*, 1997), describing its seasonal and inter-annual variations (e.g., Bates *et al.*, 1996; Soden and Bretherton, 1996), monitoring its long-term trends (e.g., Geer *et al.*, 1999; Schroeder and McGuirk, 1998), and evaluating GCM simulations (e.g., Salathé and Chesters, 1995; Chen *et al.*, 1996).

Similar radiance measurements are available on geostationary satellites such as GOES (an example of which is shown in Figure 10.4). An international network of such satellites now provides global coverage from geostationary orbit and compliments the polar-orbiting HIRS instruments by offering synchronous observations of water vapor with very high time resolution (30 min to 1 h). Such capabilities are particularly important for documenting the diurnal cycling of water vapor as well as tracking the evolution of water-vapor fields from a Lagrangian perspective (e.g., Soden, 1998).

The main drawback of thermal-IR measurements is their poor vertical resolution which is partly attributable to the relatively poor spectral resolution of current instruments. For example, the weighting functions of the HIRS instrument, which define the layers of atmosphere to which the radiance measurements are sensitive, are shown in Figure 10.6 for each of the three water-vapor channels. The broad depth of these curves and their mutual overlap with each other prevents inverting such radiances into a unique profile of moisture. Consequently, retrievals of water-vapor profiles rely on a-priori information to constrain the inversion (Rogers, 1976; Engelen and Stephens, 1999). Another limitation is that, due to the strong

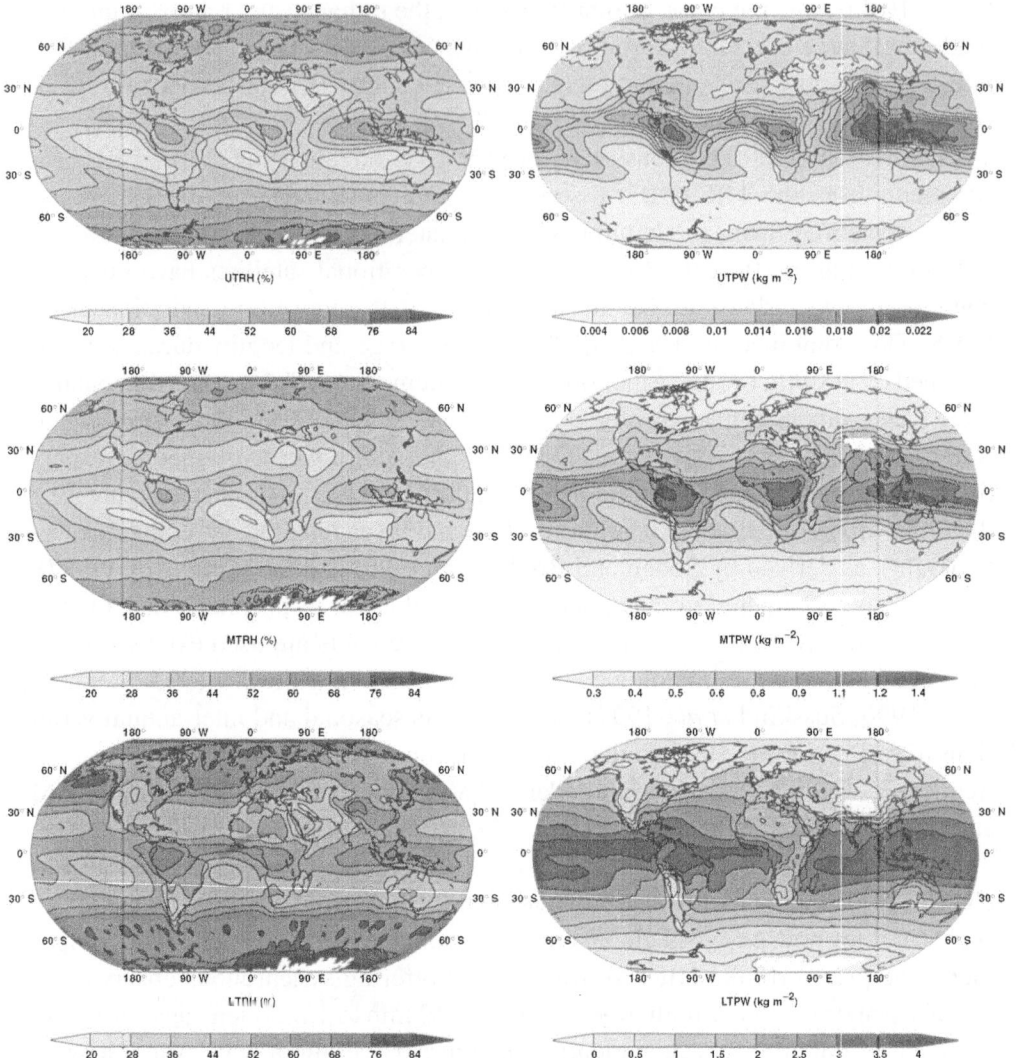

Figure 10.5. The distribution of layer-integrated relative humidity (left) and precipitable water (right) for the upper (top), middle (middle), and lower (bottom)
troposphere from HIRS observation at 6.7, 7.3, and 8.3 μm (Soden and Bretherton,
1996; Susskind *et al.*, 1997).

attenuation of IR radiation by clouds, most information on water vapor is lost when
clouds obscure the region of interest. However, the impact of this problem is less-
ened somewhat by the high horizontal resolution of the radiance measurements,
which vary from ∼ 4 km to ∼ 20 km, enabling the retrieval of water vapor around
the cloud boundaries. Therefore, the bias due to restricting observations to cloud-
free conditions is not nearly as severe as it is for solar occultation measurements,

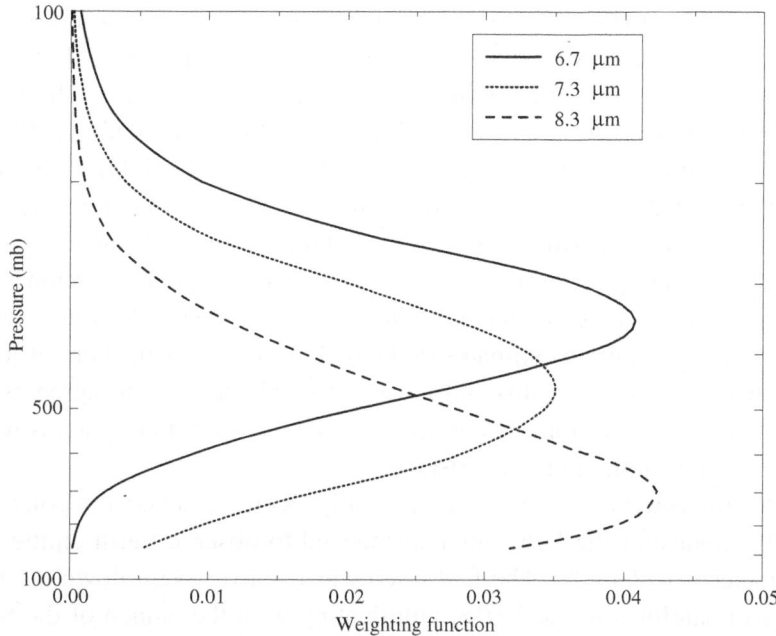

Figure 10.6. Weighting functions for the three HIRS water-vapor channels (6.7, 7.3, and 8.3 μm) for a typical tropical profile. The curves are normalized such that the sum of weights over pressure is equal to 1.

for which the footprint is an order of magnitude longer. In fact, careful inter-comparisons with radiosonde observations suggest that the cloud-free sampling restriction introduces biases of less than 10% in the mean climatology (Gaffen *et al.*, 1991; Soden and Lanzante, 1996). However, as demonstrated in the next section, satellite observations in the microwave spectrum are much less sensitive to clouds, permitting the retrieval of water vapor even under completely overcast conditions.

10.3.3 Microwave measurements

Although the amount of energy emitted in the microwave spectrum is quite small, in-teractions between microwave radiation and the atmosphere–surface system enable the retrieval of many important geophysical parameters. In particular, a water-vapor resonance line at 22.235 GHz has been used since the 1970s for the retrieval of the column-integrated water-vapor mass, hereafter referred to as the total precip-itable water (TPW). Although a variety of methods exist for retrieving TPW from radiance observations at 22 GHz, they all exploit the contrast in the microwave emission from water vapor with the very low emission from the highly reflective ocean surface. Due to the low emissivity of the ocean surface at these wavelengths, the emission of microwave radiation to space actually increases with water-vapor mass. Figure 10.7a shows the brightness temperature at 22 GHz as observed from the descending passes of the Advanced Microwave Sounding Unit (AMSU)-A instrument on January 27, 2000. The narrow bands of warmer brightness temperature (T_b) over the ocean result from plumes of high water-vapor content, whereas the much lower oceanic emission over the high-latitude oceans indicate lower moisture content. By comparing the emission from a weak water-vapor absorption band near 23 GHz with that in adjacent channels which are unattenuated by water vapor, one can obtain highly accurate estimates of TPW. Indeed, errors in such retrievals are typically within 10% ($\sim$1 mm) and have historically proven to agree as closely with radiosonde observations as possible for two measurement systems with such different spatial sampling characteristics.

To date, microwave instruments have only been deployed on polar-orbiting spacecraft because of the large antennas needed to observe Earth-emitted energy at micrometer wavelengths. The first microwave sensors were flown on the Nimbus series of satellites in the 1970s, culminating with the launch of the Scanning Multichannel Microwave Radiometer (SMMR) on Nimbus-7 in 1979 which provided the first widely used climatology of TPW derived from microwave measurements (Prabhakara *et al.*, 1985). More recently, the Special Sensor Microwave Imager (SSMI) (launched in June, 1987) and the AMSU (launched in May, 1998) both carry similar channels and have provided continuous observations of TPW to the present. This rapidly growing archive of microwave observations, with their high accuracy and excellent space–time coverage, have been widely used for characterizing the distribution and variations of TPW (Wentz and Schabel, 2000), and understanding its coupling to surface temperature, atmospheric dynamics, and the hydrologic and energy budgets of the atmosphere (Stephens, 1990; Raval and Ramanathan, 1989; Soden, 2000). With continued maintenance, one would also expect the microwave TPW retrievals to play an important role in the detection and attribution of model-predicted changes in water vapor associated with global warming.

While the microwave observations at 22 GHz provided accurate information on the distribution of TPW, they were limited to ocean surfaces due to the larger and more variable emissivity of land and ice (see Figure 10.7). More importantly, they offered no information on the vertical distribution of water vapor due to the weakness of the 22 GHz line. These limitations were overcome with the launch of the Special Sensor Microwave Temperature (SSMT)2 in 1993 and AMSU-B in 1998, both of which measure the radiation in different portions of the stronger absorption line of water vapor located at 183 GHz. By combining multiple channels which sense from the center to the wings of this line, radiance observations at 183

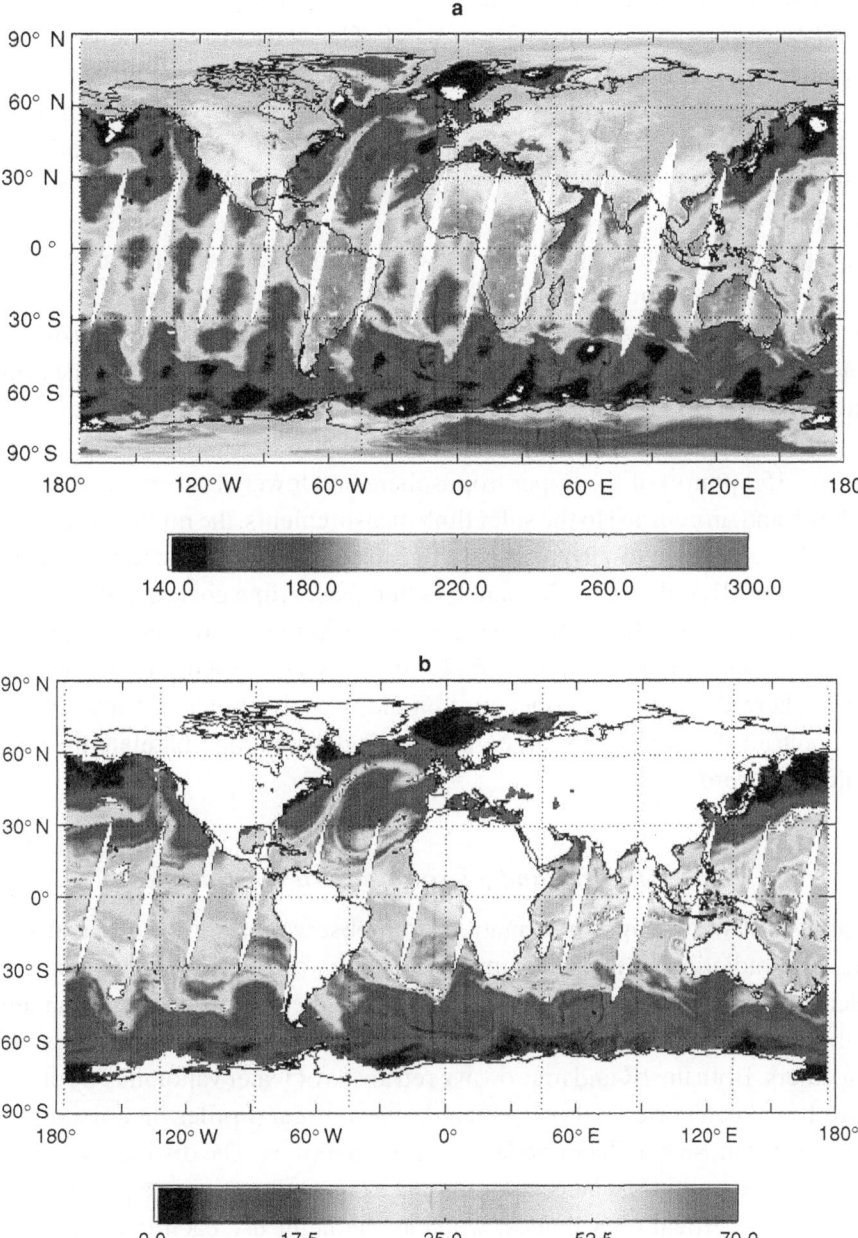

Figure 10.7. Observations of the 22 GHz brightness temperature, in K, (a) and retrieved distribution of total precipitable water (b) from AMSU observations of January 28, 2000.

This figure is available for download in colour from www.cambridge.org/9780521791328

GHz provide information on the vertical distribution of water vapor much like the thermal-IR observations in the 6.3 μm band. The vertical resolution of the 183 GHz channels is very similar to that of the IR measurements (Figures 10.5 and 10.6), and thus only offer limited information on the vertical distribution of water vapor. However unlike the IR measurements, the 183 GHz observations are largely insensitive to cirrus cloud, enabling observations of water vapor under both clear and cloudy conditions.

The Microwave Limb Sounder (MLS) offers yet another capability for retrieving water vapor from radiance measurements in the microwave spectrum (Read *et al.*, 1995). Operating since September 1991, the MLS is a limb-viewing microwave radiometer which measures the emission from water vapor at ∼ 205 GHz. The primary advantages of MLS is that it offers somewhat higher vertical resolution (∼ 3 km) compared to nadir-viewing instruments, it is sensitive to low water-vapor concentrations (∼ 150 ppmv) of the upper troposphere and lower stratosphere (Pumphrey *et al.*, 2000) and, in contrast to the solar limb measurements, the microwave emission is relatively insensitive to cirrus clouds. Because it is not restricted to solar viewing events, the MLS also provides much better space–time coverage than SAGE II observations, although the spatial resolution (∼ 100 km) is still much coarser than nadir-viewing instruments such as SSMT2 and HIRS (∼ 20 km). However, at high latitudes where the moisture concentration along a given pressure level is highly variable, retrievals under such large viewing angles and long pathlengths become difficult to interpret.

10.3.4 Infrared and microwave retrieval strategies

The retrieval of water-vapor information from observations of the attenuated electromagnetic radiation is one of many applications of the field of inverse theory. Detailed discussion of the inversion problem and methods for its solution are beyond the scope of this chapter, but introductions may be found in Rogers (1976) among others. Both the IR and microwave retrievals of water vapor may be classified into two broad categories – those which derive vertical profiles on a fixed vertical coordinate and those which derive layer-mean quantities. The distinction is that the former combine multiple water-vapor radiance channels to derive a profile of water vapor on fixed vertical coordinates, whereas the latter use each radiance channel independently to derive a layer-integrated water-vapor quantity corresponding to the depth of that particular channel's weighting function (e.g., Figure 10.6). Both retrieval strategies require accurate forward models for relating water-vapor concentrations to the atmospheric transmittance and both depend, to varying degrees, on a-priori information to constrain the retrieval. In general, the dependence on a-priori information is greater for profile inversions in which the retrieved quantities are derived with higher vertical resolution than the channel response functions. Typically these approaches perform a direct linear inversion (e.g., Eyre, 1987) or non-linear iterative perturbations (e.g., Engelen and Stephens, 1999) of an initial "first-guess" water-vapor profile to bring the radiance calculations corresponding to that profile into closer agreement with the observed values. This strategy was originally motivated by the desire to convert the radiances into "model-like" quantities, such as moisture profiles, in order to facilitate their assimilation into, and comparison with, numerical models. The dependence of such retrievals on auxiliary information often introduces substantial discrepancies between retrievals even when based upon the same radiance information. For example, the dryness of the sub-tropical upper troposphere can differ by 50% or more between various estimates (Spencer and Braswell, 1997). However this uncertainty is not a result of uncertainty in the radiance measurement, but rather in the auxiliary information used to constrain the retrieval (Engelen and Stephens, 1998). More recently, attention has shifted towards using the radiances directly for both data assimilation (e.g., McNally and Vesperini, 1996; Derber and Wu, 1998) and model evaluation (e.g., Soden and Bretherton, 1996; Salathé and Chesters, 1995; Chen *et al.*, 1996). As noted below, the direct use of radiances insures a more consistent treatment between the observations and model simulations.

The second strategy, based on single-channel retrievals, dates back to the earliest attempts to measure water vapor from space (Raschke and Bandeen, 1967) in which look-up tables, derived from a-priori radiative transfer calculations, were used to infer a layer-mean relative humidity from the observed radiance at 6.7 μm. More recently, Soden and Bretherton (1993) derived an analytical expression, based on simplified radiative theory, for relating the observed radiance to the layer-mean relative humidity. The general form of this relationship is $\ln(\frac{RH}{\cos\theta}) \propto T_b$, where RH is the retrieved layer-mean relative humidity, T_b is the observed brightness temperature, and θ is the satellite-viewing zenith angle. Despite its simplicity, this analytical relationship has been shown to be accurate to within 1 K (or $\sim$ 10% relative uncertainty) for both the thermal-IR (6.3 μm) and microwave (183 GHz) spectral channels. The advantage of this approach is that it enables a simple, yet accurate method for interpreting the observed radiances in terms of a more familiar measure of water vapor and has resulted in wide-spread application, particularly for retrieving upper-tropospheric humidity (e.g., Bates *et al.*, 1996; Stephens *et al.*, 1996; Schmetz and Turpeinen, 1996; Slingo and Webb, 1997; Spencer and Braswell, 1997; Engelen and Stephens, 1998; Moody *et al.*, 1999). This approach was used to translate the radiance data from Figure 10.4a into the upper-tropospheric relative-humidity distribution in Figure 10.4b. Provided that the radiative properties of this spectral region are well known and a representative set of training profiles is available, the retrieval of layer-mean quantities, such as UTH, is

Table 10.2. *Earth Observing System remote-sensing capabilities*

Method	Instrument	Platform	Spectral band
Solar occultation	SAGE III	Meteor 3	0.94 μm
Thermal infrared	MODIS	Terra/Aqua	6.3 μm
	AIRS	Aqua	6.3 μm
	HIRDLS	Aura	6.3 μm
	CERES	Terra/Aqua	5–50 μm
Microwave	AMSR	Aqua	22 GHz
	HSB	Aqua	183 GHz
	MLS	Aura	183 GHz

possible to within approximately 10% relative uncertainty (Soden and Bretherton, 1993).

10.4 Outlook for the future

A multitude of systems currently exist for observing water vapor, each with their own strengths and weaknesses. Taken independently, each system has serious limitations. However when combined, the complementary nature of the various remotely-sensed and radiosonde observations offer a capable and robust network for monitoring the distribution and variations of atmospheric water vapor. Despite their many shortcomings, radiosondes still provide an important network of *in situ* measurements for evaluating and inter-calibrating satellite retrievals. The ability to perform satellite–radiosonde comparisons before and after transitions from successive satellites and between different instruments is critical to developing the stable, long-term climatologies necessary for the detection and attribution of water-vapor trends. To facilitate this task, future contributions of the radiosonde observing system will primarily involve developing a network of accurate "reference" radiosondes to serve as ground truth to validate and calibrate water-vapor data from remotely-sensed platforms.

10.4.1 Emerging satellite systems

The deployment of the Earth Observing System (EOS), beginning in late 1999, provided several improvements in our ability to monitor variations in water vapor from space. The EOS platform consists of a wide variety of instruments (Table 10.2). In most cases these instruments primarily provide an extension of previous observing capabilities, albeit with greater accuracy and higher resolution, rather than a true advance in measurements. For example, an improved SAGE instrument,

SAGE III, is scheduled for flight as part of the EOS mission. It will employ multi-wavelength narrow-band absorption measurements with enhanced dynamic range, thereby enabling more accurate and better calibrated water-vapor measurements. Most importantly, it will insure the extension of the SAGE II water-vapor record with hopefully enough overlap to permit reliable inter-calibration and continued monitoring of long-term trends in upper-tropospheric and stratospheric water vapor. Similarly, the MODIS, Advanced Microwave Scanning Radiometer (AMSR), and Humidity Sounder for Brazil (HSB) instruments provide more accurate and higher-resolution versions of the HIRS, SSMI, and SSMT2, respectively.

One exception to the above is the Atmospheric InfraRed Sounder (AIRS) which will fundamentally advance our remote-sensing capabilities. In contrast to current IR sensors which are limited to a few, coarse-resolution water-vapor channels, AIRS is a high-spectral-resolution spectrometer with nearly 2400 channels spanning both the visible (0.4 – 1 mm) and infrared (3.7 – 15 mm). The increased spectral resolution and greater number of IR sounding channels – nearly two orders of magnitude beyond current operational sensors – represent a substantial increase in the ability to describe the vertical distribution of moisture. However, the vertical resolution of the observations does not increase in proportion to the enhanced spectral resolution or number of channels. Even at these high spectral resolutions ($\lambda/\Delta\lambda \sim 1200$), the half-power depth of the weighting functions are still relatively coarse (2–3 km). Consequently, as the number of channels increase, the amount of independent information per channel decreases, further "sickening" an already ill-conditioned retrieval problem. While the increase in vertical resolution of the measurements is modest, the increased number of channels does provide additional constraints on the shape of the moisture profile necessary to satisfy the observed spectrum of radiances, which does translate into a substantial increase in the information content.

Similarly the National Polar-Orbiting Operational Environmental Satellite System (NPOESS), scheduled for launch near the end of this decade, will provide more accurate and higher-resolution versions of current operational sensors while also insuring the continuation of high-spectral-resolution observations initiated by AIRS. Perhaps the greatest asset of the EOS and NPOESS missions is that many of these instruments will be flown together, enabling one to utilize radiance information in multiple spectral regions simultaneously to compare the distribution of water vapor and its radiative effects. Indeed, the combination of high-spectral-resolution information in the IR, with narrow-band microwave channels and broad-band solar and thermal radiometers, will provide unprecedented opportunities for observing the distribution and radiative effects of atmospheric water vapor, as well as describing its coupling to other components of the hydrological cycle.

10.4.2 *Global positioning system*

One of the more recent developments in remote sensing is the use of radiowave measurements from the Global Positioning System (GPS). The GPS consists of a network of 24 satellites that transmit L-band radio signals (at 19 and 24 cm wavelengths) for the primary purpose of navigation. In addition, GPS signals are delayed and refracted by the atmosphere as they propagate from GPS satellites to either Earth-based or orbiting satellite receivers. In particular, water vapor retards the propagation of the radiowaves through the atmosphere thereby introducing a detectable delay in the radio signal, which is nearly proportional to the integrated pathlength of water-vapor mass. Compared to radiosonde observations, surface-based GPS observations offer increased spatial and temporal resolution, which offers promise in a range of operational and research applications (Businger *et al.*, 1996). While surface-based GPS does not provide global coverage, a GPS receiver in low Earth orbit (LEO) can provide refractivity profiles from approximately 500 GPS radio occultations per day distributed over the globe (Businger *et al.*, 1996). The refraction of the radio signals transmitted by one of the GPS satellites as they pass through Earth's atmosphere is measured by the LEO GPS receiver. Given accurate knowledge of the temperature and pressure from other sources, these refractivity profiles can, in principle, be converted into profiles of water vapor with a vertical resolution of ∼ 1 km. The GPS results may also offer information on the isotopic distribution of water vapor, which is valuable for understanding the transport of moisture into the upper troposphere and lower stratosphere. Such radio occultation techniques have been widely used for the study of planetary atmospheres (Marouf *et al.*, 1986) and, with the recent launch of a GPS receiver on Microlab 1, the application of radio occultation is now beginning to be explored for Earth's atmosphere.

10.4.3 *Combined active and passive sounding*

While the current suite of observing systems is adequate for many purposes, gaps in our knowledge of the distribution of water vapor still remain. In particular, the ability to observe water vapor with high vertical resolution (∼ 1 km or less), good space- (< 10 km) and time-resolution (< 1 day), and global coverage does not currently exist. Our understanding of the vertical distribution of water vapor over the open oceans, particularly in the lower troposphere, is quite limited. For example, the moisture structure within the oceanic boundary layer and its transition between the boundary layer and free atmosphere remains largely unknown. Such information is crucial to determining the evaporative fluxes of heat and moisture from the surface, the net radiative flux at the surface, and the formation and dissipation of boundary layer clouds (which are known to be problematic in many general circulation models (GCMs)). Similarly, the mechanisms which transport moisture between the upper troposphere and lower stratosphere are not well understood. A solution to the current observational gaps may lie with the development of a combined active and passive observing system (Browell *et al.*, 1999). The proposed instrument combines a DIfferential Absorption Lidar (DIAL) with high-resolution IR measurements from a Fourier Transform Spectrometer (FTS). The lidar system is capable of measuring accurately ($\pm$10%) water-vapor profiles with a vertical resolution of a few hundred meters – something that is unachievable from passive sensors. While current space-borne lidar designs would be lacking in horizontal resolution and be restricted to nadir views (thus limiting its horizontal sampling), combining it with the FTS yields a highly complementary remote-sensing duet, capable of both high space–time resolution and high vertical resolution. For example, information on the vertical structure of moisture from the lidar would provide ideal independent first-guess information for a structure-preserving retrieval algorithm for inverting the high-horizontal-resolution IR radiances. The resulting data could, in principle, offer the best of both sensors – detailed vertical profiles with mesoscale horizontal resolution.

10.4.4 Final remarks

Water vapor represents a key variable for climate studies. The importance of water vapor is reflected in the vast number of observing systems designed to measure its distribution over a wide range of space and time scales. Such observations are a central ingredient in advancing our understanding of the role of water vapor in coupling the atmospheric hydrological cycle and shaping Earth's climate. However, observations are only one component of a balanced research strategy and their greatest benefit comes when they are analyzed in concert with model simulations and theoretical studies. Indeed, satellite and radiosonde observations of water vapor have been invaluable in testing the ability of GCMs to reproduce the observed variations in moisture. To date, the empirical and model–data comparisons largely support the ability of models to simulate changes in the moisture distribution (see Chapter 9). However, such tests are limited to observations of natural climate variability and thus provide information on the mechanisms that maintain the current distribution of water vapor, rather than a direct confirmation of the predictions of increased moisture that are to accompany global warming. This limitation will remain until observations are compiled with sufficient accuracy and longevity to detect global trends in water vapor. If the growth rates of water vapor currently predicted by GCMs are correct (2–6% per decade, depending upon altitude), observational confirmation of anthropogenically enhanced water-vapor concentrations should be feasible within the next 20 years. However, this requires that radiance measurements from existing and future satellite programs are inter-calibrated, and that sufficient redundancy is achieved, either through overlapping satellites or through ground-based and *in situ* reference networks, to establish the stability of these radiance records. It is the development and verification of stable, long-term radiance records that represents the greatest challenge to the future of water-vapor observations.

References

Bandeen, W. R., R. A. Hanel, J. Licht, R. Stampfl, and W. Stroud (1961). Infrared and reflected solar radiation measurements from the TIROS II satellite. *J. Geophys. Res.* **66**, 3169–85.

Bates, J. J., X. Wu, and D. L. Jackson (1996). Interannual variability of upper troposphere water vapor band brightness temperature. *J. Clim.* **9**, 427–38.

Bevis, M., S. Businger, T. A. Herring *et al.* (1992). GPS meteorology: remote sensing of atmospheric water vapor using the Global Positioning System. *J. Geophys. Res.* **97**, 15787–801.

Browell, E. V. (1999). Airborne lidar measurements of water vapor, aerosol, and cloud distribution for climate studies. In *Proceedings of Water Vapor in the Climate System*, Jekyll Island, Georgia, October 25–28, American Geophysical Union, pp. 31–2.

Chen, C. T., E. Roeckner, and B. J. Soden (1996). A comparison of satellite observations and model simulations of column-integrated moisture and upper-tropospheric humidity. *J. Clim.* **9**, 1561–85.

Chu, W. P., E. W. Chiou, J. C. Larsen *et al.* (1993). Algorithms and sensitivity analyses for SAGE II water vapor retrieval. *J. Geophys. Res.* **98**, 4857–66.

Del Genio, A. D., W. Kovari, Jr., and M. S. Yao (1994). Climatic implications of the seasonal variation of upper troposphere water vapor. *Geophys. Res. Lett.* **21**, 2701–4.

Derber, J. C. and W. Wu (1998). The use of TOVS cloud-cleared radiances in the NCEP SSI analysis system. *Mon. Wea. Rev.* **126**, 2287–99.

Elliott, W. P. and D. J. Gaffen (1991). On the utility of radiosonde humidity archives for climate studies. *Bull. Amer. Meteor. Soc.* **72**, 1507–20.

Engelen, R. J. and G. L. Stephens (1998). Comparison between TOVS/HIRS and SSM/T-2 derived upper-tropospheric humidity. *Bull. Amer. Meteor. Soc.* **79**, 2748–51.
  (1999). Characterization of water vapor retrievals from TOVS/HIRS and SSMT2 measurements. *Quart. J. Roy. Meteor. Soc.* **50**, 5–33.

Eyre, J. R. (1987). Inversion of cloudy satellite sounding radiances by nonlinear optimal estimation. *Quart. J. Roy. Meteor. Soc.* **115**, 1001–26.

Fischer, H., J. Gille, and J. Russell (1981). Water vapor in the stratosphere: preliminary results of the LIMS experiment aboard Nimbus-7. *Adv. Space Res.* **1**, 279–81.

Gaffen, D. J. and W. P. Elliott (1993). Column water vapor content in clear and cloudy skies. *J. Clim.* **6**, 2278–87.

Gaffen, D. J., T. P. Barnett, and W. P. Elliott (1991). Space and time scales of global tropospheric moisture. *J. Clim.* **10**, 989–1008.

Geer, A. J., J. E. Harries, and H. Brindley (1999). Spatial patterns of climate variability in upper-tropospheric water vapor radiances from satellite data and climate model simulations. *J. Clim.* **12**, 1940–55.

Gutzler, D. S. (1996). Low frequency ocean–atmosphere variability across the tropical western Pacific. *J. Atmos. Sci.* **53**, 2773–85.

Held, I. M. and B. J. Soden (2000). Water vapor feedback and global warming. *Ann. Rev. Energy Env.* **25**, 441–75.

IPCC (1990). *Climate Change: the IPCC Scientific Assessment*, Cambridge, Cambridge University Press.

Larsen, J. C., E. W. Chiou, W. P. Chu *et al.* (1993). A comparison of SAGE II tropospheric water vapor measurements to radiosonde observations. *J. Geophys. Res.* **98**, 4897–917.

Lin, W. Y., M. Zhang, and M. Geller (1998). Diabatic subsidence in the subtropical upper troposphere derived from SAGE II measurements. *Geophys. Res. Lett.* **25**, 4181–4.

Marouf, E. A., G. Tyler, and P. Rosen (1986). Profiling Saturn's rings by radio occultation. *Icarus.* **68**, 120–66.

McCormick, M. P., L. W. Thomason, and C. R. Trepte (1995). Atmospheric effects of the Mt. Pinatubo eruption. *Nature* **373**, 399–404.

McNally, A. P. and M. Vesperini (1996). Variational analysis of humidity information from TOVS radiances. *Quart. J. Roy. Meteor. Soc.* **122**, 1521–44.

Möller, F. (1961). Atmospheric water vapor measurements at 6–7 microns from a satellite. *Plan. Space Sci.* **5**, 202–6.

Moody, J. L., A. J. Wimmers, and D. J. Clay (1999). Remotely sensed specific humidity: development of a derived product from GOES imager channel 3. *Geophys. Res. Lett.* **26**, 59–62.

Mote, P. W., K. H. Rosenlof, J. R. Holton, R. S. Harwood, and J. Waters (1995). Seasonal variations of water vapor in the tropical lower stratosphere. *Geophys. Res. Lett.* **22**, 1093–6.

National Research Council (1999). *The GEWEX Global Water Vapor Project (GVaP) – US opportunities*, Washington, DC, National Academy Press.

Oort, A. H. (1983). *Global Atmospheric Circulation Statistics, 1958–1973*, Rockville, MD, NOAA, US Department of Commerce, NOAA Professional Paper 14.

(1993). Observed humidity trends in the atmosphere. In *Proceedings of the Seventeenth Annual Climate Diagnostics Workshop*, Norman, OK, American Meteorological Society, pp. 24–30.

Peixoto, J. P. and A. H. Oort (1983). The atmospheric branch of the hydrological cycle and climate. In *Variation of the Global Water Budget*, London, Reidel, pp. 5–65.

(1996). The climatology of relative humidity in the atmosphere. *J. Clim.* **9**, 3443–63.

Prabhakara, C., D. Short, and B. Vollmer (1985). El Niño and atmospheric water vapor: observations from Nimbus 7 SMMR. *J. Clim. Appl. Meteor.* **24**, 1311–24.

Pumphrey, H. C., H. L. Clark, and R. S. Harwood (2000). Lower stratopsheric water vapour measured by UARS MLS. *Geophys. Res. Lett.* **27**, 1691–4.

Raschke, E. and W. R. Bandeen (1967). A quasi-global analysis of tropospheric water vapor content from TIROS IV radiation data. *J. Appl. Meteor.* **6**, 468–81.

Rasmusson, E. M. (1967). Atmospheric water vapor transport and the water balance of North America. Part 1. Characteristics of the water vapor flux field. *Mon. Wea. Rev.* **95**, 403–26.

Raval, A. and V. Ramanathan (1989). Observational determination of the greenhouse effect. *Nature* **342**, 758–62.

Read, W. G., W. G. Waters, D. A. Flower, L. Frouidevaux, and R. J. Jarnot (1995). Upper tropospheric water vapor from UARS MLS. *Bull. Amer. Meteor. Soc.* **76**, 2381–9.

Rind, D., E. W. Chiou, W. Chu *et al.* (1991). Positive water vapour feedback in climate models confirmed by satellite data. *Nature* **349**, 500–3.

(1993). Overview of the SAGE II water vapor observation characteristics. *J. Geophys. Res.* **98**, 4835–56.

Rogers, C. D. (1976). Retrieval of atmospheric temperature and composition from remote measurements of thermal radiation. *Rev. Geophys. Space Phys.* **14**, 609–43.

Ross, R. J. and W. P. Elliott (1996). Tropospheric water vapor climatology and trends over North America. *J. Clim.* **9**, 3561–74.

Ross, R. J. and D. J. Gaffen (1998). Comment on widespread tropical atmospheric drying from 1979–1995. *Geophys. Res. Lett.* **25**, 4357–8.

Salathé, E. P. and D. Chesters (1995). Variability of moisture in the upper troposphere as inferred from TOVS satellite observations and the ECMWF model analyses in 1989. *J. Clim.* **8**, 120–32.

Schenk, W. E. and V. V. Salomonson (1972). A multispectral determination of sea surface temperature using Nimbus 2 Data. *J. Phys. Oceanogr.* **2**, 157–67.

Schmetz, J. and O. M. Turpeinen (1988). Estimation of the upper tropospheric relative humidity field from METEOSAT water vapor image data. *J. Appl. Meteor.* **27**, 889–99.

Schroeder, S. R. and J. P. McGuirk (1998). Widespread tropical atmospheric drying from 1979–1995. *Geophys. Res. Lett.* **25**, 1301–4.

Slingo, A. and M. J. Webb (1997). The spectral signature of global warming. *Quart. J. Roy. Meteor. Soc.* **123**, 293–307.

Smith, W. L., H. B. Howell, and H. M. Woolf (1979). Use of interferometric radiance measurements for sounding the atmosphere. *J. Atmos. Sci.* **36**, 5660–75.

Soden, B. J. (1998). Tracking upper tropospheric water vapor radiances: a satellite perspective. *J. Geophys. Res.* **103**, 17069–81.

Soden, B. J. and F. P. Bretherton (1993). Upper tropospheric relative humidity from the GOES 6.7 μm channel: method and climatology for July 1987. *J. Geophys. Res.* **98**, 16669–88.

(1996). Interpretation of TOVS water vapor radiances in terms of layer-average relative humidities: method and climatology for the upper, middle, and lower troposphere. *J. Geophys. Res.* **101**, 9333–43.

Soden B. J. and J. R. Lanzante (1996). An assessment of satellite and radiosonde climatologies of upper tropospheric water vapor. *J. Clim.* **9**, 1235–50.

(2000). The sensitivity of the tropical hydrological cycle to ENSO. *J. Clim.* **13**, 538–49.

Spencer, R. W. and W. D. Braswell (1997). How dry is the tropical free troposphere? Implications for global warming theory. *Bull. Amer. Meteor. Soc.* **78**, 1097–106.

Stephens, G. L. (1990). On the relationship between water vapor over oceans and sea surface temperature. *J. Clim.* **3**, 634–45.

Stephens, G. L., D. L. Jackson, and I. Wittmeyer (1996). Global observations of upper tropospheric water vapor derived from TOVS. *J. Clim.* **9**, 305–26.

Susskind, J., P. Peraino, L. Rokke, L. Iredell, and A. Mehta (1997). Characteristics of the TOVS Pathfinder A dataset. *Bull. Amer. Meteor. Soc.* **78**, 1449–72.

Wang, P., P. Minnis, P. McCormick, G. Kent, K. Skeens (1996). A 6-year climatology of cloud occurrence frequency from Stratospheric Aerosol and Gas Experiment II. *J. Geophys. Res.* **101**, 29407–29.

Wentz, F. J. and M. Schabel (2000). Precise climate monitoring using complementary satellite data sets. *Nature* **403**, 414–15.

Wu, X., J. J. Bates, and S. J. S. Khalsa (1993). A climatology of the water vapor band brightness temperature from NOAA operational satellites. *J. Clim.* **6**, 1282–300.

Zhai, P. and R. E. Eskridge (1997). Atmospheric water vapor over China. *J. Clim.* **10**, 2643–52.

Zhang, M. H. (1995). Assessment of the SAGE sampling strategy in the derivation of tropospheric water vapor distribution in a general circulation model. *Geophys. Res. Lett.* **22**, 1353–6.

11

New frontiers in remote sensing of aerosols and their radiative forcing of climate

YORAM J. KAUFMAN

NASA Goddard Space Flight Center, Greenbelt, MD

LORRAINE A. REMER

NASA Goddard Space Flight Center, Greenbelt, MD

DIDIER TANRÉ

University of Lille 1 - Sciences and Technology, Villeneuve D'Ascq Cedex, France

11.1 Introduction

As shown in the previous chapter, remote sensing offers a tremendous opportunity in observing Earth's climate system. This chapter further explores the unique capabilities of satellite observations. Remote sensing of aerosol and aerosol radiative forcing of climate is being revolutionized as we write these lines. We are moving from a qualitative global description of the aerosol presence, available now for several years (Jankowiak and Tanré, 1992; Husar *et al.*, 1997; Herman *et al.*, 1997a), to quantitative measurements of laboratory precision (Chu *et al.*, 1998; Tanré *et al.*, 2001). These measurements will not only tell us the aerosol loading or optical thickness (Tanré *et al.*, 1997), but also the aerosol intrinsic optical properties and radiative forcing of climate (Kaufman *et al.*, 2001). The revolution is accomplished by designing and using new satellite sensors (Kaufman *et al.*, 1997b), by re-analyzing images from the old operational sensors using new science understanding (Nakajima and Higurashi, 1998), and by forming a federated network of instruments for ground-based remote sensing of aerosols and their properties (Holben *et al.*, 1998). This maturity of the aerosol remote-sensing science resembles and follows the development in remote sensing of the Earth radiation budget, which pioneered the use of high-precision space-borne and ground-based radiation measurements (Cess *et al.*, 1995; Ramanathan *et al.*, 1989; Wielicki *et al.*, 1996).

Here we summarize two examples that extend remote sensing from the traditional role of qualitative monitoring of the aerosol distribution. We show that a remote-sensing system that combines spectral measurements from satellites (e.g., the Moderate Resolution Imaging Spectroradiometer, MODIS, on the Earth Observing

Frontiers of Climate Modeling, eds. J. T. Kiehl and V. Ramanathan.
Published by Cambridge University Press. © Cambridge University Press 2006.

System "Terra" and "Aqua" missions) with spectral and angular measurements of the Sun and sky brightness from the ground by the Aerosol Robotic Network (AERONET) is the most accurate system to derive the aerosol. The two examples are the following.

(1) Spectral single-scattering albedo, a measure of the aerosol absorption that defines if the aerosol cools or heats (or both) the climate system.
(2) Spectral radiative forcing at the top and bottom of the atmosphere.

11.2 Background

During the past two decades we have realized that aerosol science is critically important to understand radiative forcing of climate, and to predict future developments of our climate system (e.g., Twomey *et al.*, 1984; Coakley and Cess, 1985; Hansen and Lacis, 1990; Charlson *et al.*, 1992; IPCC, 1995). Modeling the aerosol radiative forcing of climate requires knowledge of both the spatial distribution, temporal evolution *and* the optical properties of atmospheric aerosol. The common approach to address this issue is via modeling the distribution of aerosol sources, transport and processing in the atmosphere and in clouds, estimating the optical properties of the aerosol, and calculating the regional and global radiative impact (Penner *et al.*, 1992; Kiehl and Briegleb, 1993; Boucher and Lohmann, 1995; Liousse *et al.*, 1996; Tegen *et al.*, 1996; Hobbs *et al.*, 1997). Traditionally, satellite data are required in these studies only to supply the global distribution of aerosols that can be used to validate or initiate the models. Information on aerosol optical properties and vertical distribution is obtained from laboratory and one-time, localized field experiments. The global coverage and temporal frequency of satellite and ground-based remote-sensing instruments have been underutilized in terms of providing the essential information on aerosol properties.

The reason for omitting remote-sensing information on aerosol properties is the quality of the traditional sensors. Five years ago, before this revolution in remote sensing began, satellites were not designed for remote sensing of aerosols. Most of their solar channels were not calibrated, the channels were not optimized for aerosol retrievals; for example AVHRR (Advanced Very High Resolution Radiometer) has wide spectral channels in the solar spectrum, which include water-vapor absorption that varies the signal independently of the presence of aerosol. Their spectral separation is small making it difficult to assess the aerosol particles' size. Aerosol optical thickness (a measure of its loading in the atmospheric column) was derived only over the oceans, with low accuracy as a by-product of AVHRR measurements (Husar *et al.*, 1997). In a similar way the METEOSAT single solar channel was used on a regional scale (Jankowiak and Tanré, 1992) to derive dust loading over the Atlantic Ocean. However these satellite qualitative images already revealed very important first global pictures of the distribution and sources of aerosols (see Figure 11.1). We saw that dust from the Sahara is a dominant aerosol feature. We learned that Australia is unable to produce large dust amounts despite the large size of its deserts. We saw the seasonal variation in the amount and latitude of smoke emitted from Africa and pollution from China. As we shall show in this chapter, a combination of precise satellite measurements together with ground-based world-wide radiation measurements, can provide, in addition to the distribution of aerosol loading, also the main aerosol properties and even direct measurements of the effect of aerosols on the spectral radiative fluxes. Aerosol particles have a short lifetime in the troposphere (a few days), and therefore have a highly variable spatial and temporal signal. To estimate the regional and global effects of aerosols there is a need for continuous global monitoring of aerosols and their radiative impact (Kaufman *et al.*, 1997a). As a result new satellite instruments have been designed with new capabilities to measure precisely the global distribution of aerosols and their radiative properties. New capabilities include polarization on POLDER (POLarization and Directionality of the Earth's Reflectances; Deschamps *et al.*, 1994; Herman *et al.*, 1997b; Deuze *et al.*, 1999), multiview observations on Along Track Scanning Radiometer (ATSR; Veefkind *et al.*, 1998), and Multi-angle Imaging Spectro Radiometer (MISR; Martonchik and Diner, 1992; Diner *et al.*, 1998), and multi-spectral observations by MODIS (Kaufman *et al.*, 1997a; Tanré *et al.*, 1997) and Global Imager (GLI; Nakajima *et al.*, 1998). A combination of angular, spectral, and polarization capabilities over each pixel was suggested on EOSP (Earth Observing Scanning Polarimeter; Mishchenko and Travis, 1997). Lidars in space provide the much needed aerosol vertical distribution. Some of these instruments (POLDER and ATSR) have been launched in the last five years and others are planned for launch in the next few years.

In parallel to the design of new satellite sensors for precise measurement of aerosol loading and properties, better inversion of data from existing instruments has also been developed. The AVHRR's two spectral channels are used to derive a measure of the aerosol size and the optical thickness (Nakajima and Higurashi, 1998; Higurashi and Nakajima, 1999). The Total Ozone Mapping Spectrometer (TOMS) UV measurements (Herman *et al.*, 1997a; Hsu *et al.*, 1996) are used to derive the aerosol index (see Figure 11.1) over land and ocean. New analysis of satellite data produced some interesting research results. Analysis of aerosol and clouds from AVHRR showed the impact of aerosol on cloud microphysics and reflectance (Coakley *et al.*, 1987; Kaufman and Fraser, 1997). Combining this technique with new space-borne measurements of precipitation from TRMM resulted in measurements of the impact of smoke aerosols from large fires on precipitation in the tropics (Rosenfeld, 1999). Comparison between aerosol frequency from the

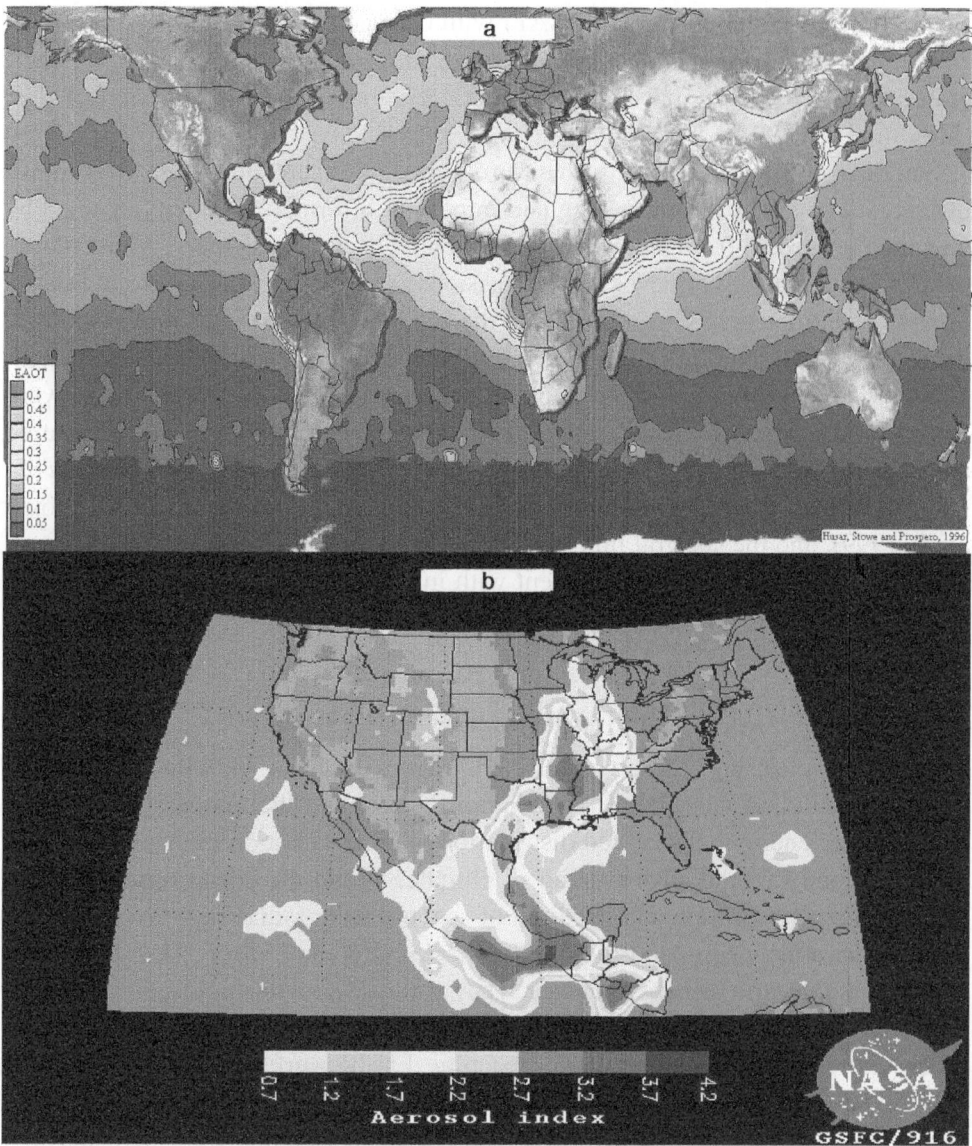

Figure 11.1. Examples of derivation of global aerosol distributions from satellite sensors that were not designed for aerosol work. (a) Global AVHRR aerosol data for one season (Husar *et al.*, 1997) showing dust and smoke from Africa, pollution and dust in the Arabian sea, pollution off the east coast of the USA and east coast of China; and (b) regional TOMS data (Herman *et al.*, 1997) showing the spread of smoke from biomass burning in Mexico across North America.

Meteorological Satellite (METEOSAT) with climate models and temperature as-similations showed the impact of dust on heating of the boundary layer (Alpert *et al.*, 1998) and the impact of atmospheric circulation on the emission of dust (Moulin *et al.*, 1997).

Satellites by themselves only measure part of the aerosol picture by observing aerosol from above. Interesting and important information can be obtained by viewing the aerosol from below against the black backdrop of space. In parallel to the design and launch of new satellite missions, a global network of ground-based Sun/sky radiometers has been formed in the last five years to view the aerosol from below. The AERONET (Holben *et al.*, 1998) is a federated system designed to communicate in real time the spectral optical thickness from 100–200 locations every day, every 15 minutes. The sky angular and spectral information is also measured and used to retrieve the aerosol size distribution (Tanré et al, 2001; Nakajima *et al.*, 1989; Kaufman *et al.*, 1994), refractive index, single-scattering albedo (Dubovik *et al.*, 1999), and the spectral flux reaching the surface. The network is used to validate the satellite data and supplement with information on the aerosol properties and diurnal cycle.

11.3 Remote sensing of aerosol absorption

Aerosol particles, e.g., desert dust, smoke from biomass burning, and urban–industrial pollution can affect the radiation budget and the temperature field by changing the energy balance and distribution of solar radiation in the atmosphere. To understand this radiative forcing of climate, we need to determine the effect of aerosols on absorption and partition of spectral solar radiation between Earth's surface, the atmosphere, and the fraction backscattered to space. Aerosols play a dual role in the earth's energy balance. They increase both the atmospheric absorption of solar radiation and the backscattering of sunlight to space. Both of these processes reduce the solar radiation reaching Earth's surface. However, it is the delicate balance between aerosol backscattering of radiation to space and aerosol absorption of solar radiation in the atmosphere that determines if the aerosols cool or warm the planetary system. This balance is best described by the aerosol single-scattering albedo, ω_0, the ratio between scattering and scattering + absorption (Hansen and Lacis, 1990) and depends upon the surface reflectance. The larger the ω_0 the less absorption by the aerosol. A value of $\omega_0 = 1.0$ indicates the aerosols only scatter and do not absorb. In cloud-free conditions over the oceans, a single-scattering albedo $\omega_0 > 0.84$ will cool the Earth system, while $\omega_0 < 0.84$ will heat it. However, over the bright land and due to the effect of aerosols on cloud formation this balance may be reached for a single-scattering albedo of 0.91 (Hansen *et al.*, 1997). Therefore accurate determination of the single-scattering albedo is critical to model the impact of aerosol on climate and to predict climate change. Presently there appear to be no accurate methods to determine the aerosol single-scattering albedo and its spatial and temporal variability with the required precision. *In situ* measurements of aerosol absorption are uncertain, in particular for dust, because of size-dependent sampling efficiency (Huebert *et al.*, 1990), complex sample preparation, and deduction of the absorption from the sample reflectivity that depends on several weak assumptions (Sokolik *et al.*, 1992). Such measurements tend to exaggerate the dust absorption (Heintzenberg *et al.*, 1997). Remote-sensing techniques measure the properties of the undisturbed aerosols in the entire atmospheric column, relevant to aerosol direct radiative forcing of climate.

Satellites have a unique vantage point to measure precisely the aerosol single-scattering albedo (Kaufman, 1987). We shall demonstrate the technique for the favorable conditions of a heavy outbreak of Saharan dust as it crosses the coastline into the Atlantic from the bright desert of Senegal. What is the basis for the accurate measurement of aerosol absorption from satellites? Figure 11.2 demonstrates the principle for the satellite measurements. For dark surface (ocean), dust increases the apparent reflectance of the Earth + atmosphere system, as observed from space, by backscattering solar radiation to space. Typical variations in ω_o (e.g., $\Delta\omega_o = \pm 0.1$ Figure 11.2) have relatively only a minor influence on the radiation field over the ocean. As a result, the dark ocean is an excellent background against which to measure the aerosol optical thickness or loading, in the presence of uncertainty in the single-scattering albedo. Over brighter land (e.g., reflectance of 0.25) the effect of absorption is much stronger. Like a balance scale used in a laboratory for precise weighing, the single-scattering albedo determines if dust will increase (higher ω_o) or decrease (lower ω_o) the apparent reflectance of the surface–atmosphere system from the clear to dusty day. Higher absorption will lower the radiance while higher scattering will increase the radiance. The apparent reflectance is the total radiance reflected to space divided by the solar insolation. The sensitivity of the apparent reflectance to ω_o is even higher for the brighter desert surfaces (e.g., reflectance of 0.4).

In mathematical form, the dependence of a change in the apparent reflectance of the Earth–atmosphere system, $\Delta\rho^*$, on the aerosol optical thickness, τ_a, is given in Equation (11.1), where ρ is the surface reflectance and the optical thickness, τ_a, is

$$\Delta\rho^* = \rho^*(\tau_a, \rho) - \rho^*(0, \rho) \sim +\omega_o\tau_a P/(4\mu\mu_o)$$
$$- \rho\tau_a[1 - \omega_o(1 + \beta)](1/\mu + 1/\mu_o) \qquad (11.1)$$

defined so that the direct vertical transmission through the aerosol layer is $T = \exp(-\tau_a)$; P is the scattering phase function that is dependent primarily on particle size and β is the backscattering coefficient (an integral on the phase function

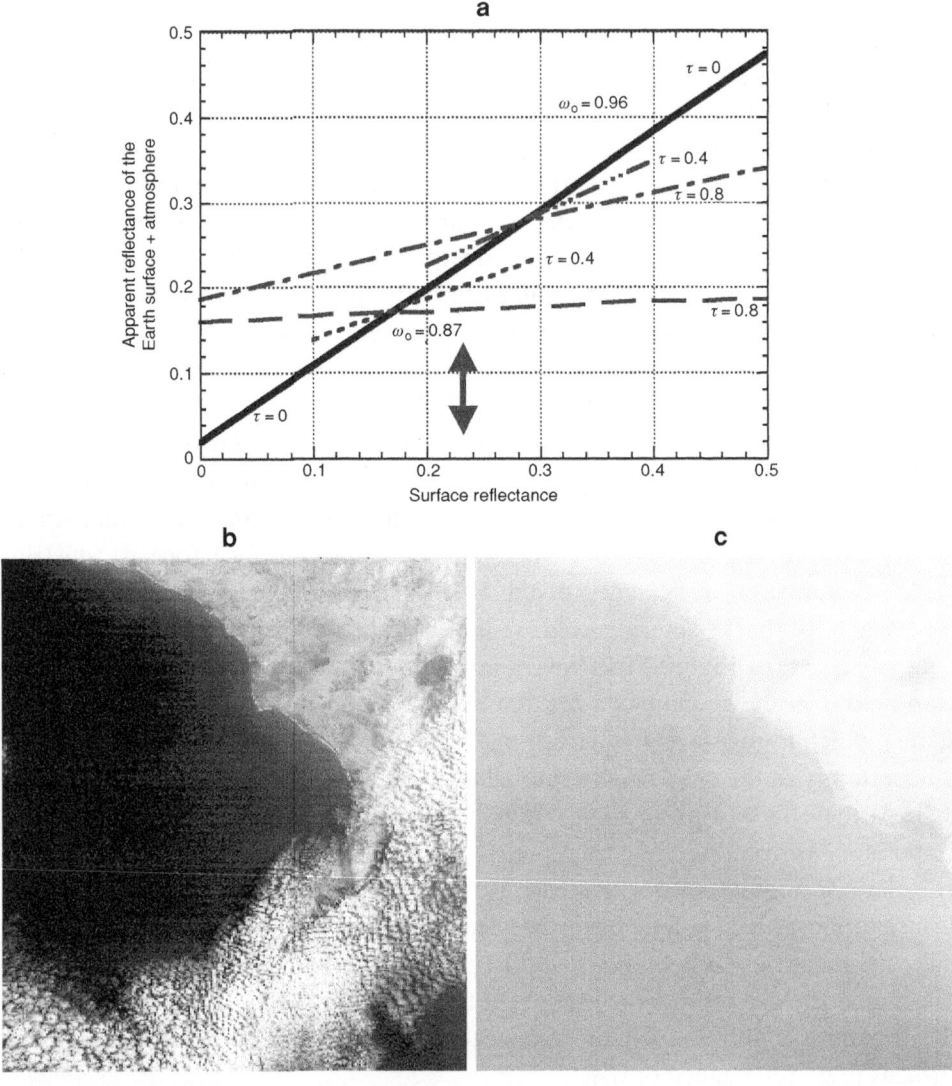

Figure 11.2. (a) The apparent reflectance of the Earth surface as observed from space and influenced by the atmosphere at 0.66 μm. Solid line: no dust ($\tau = 0$) only molecular scattering; broken lines: dust with low absorption, $\omega_0 = 0.96$, and high absorption, $\omega_0 = 0.87$, respectively. Optical thickness, τ, of 0.4 and 0.8 is indicated. The apparent reflectance serves like a laboratory scale to measure the balance between scattering and absorption, determining precisely the aerosol single-scattering albedo. The concept was explored by Fraser and Kaufman (1985), and Kaufman (1987). Bottom: Landsat Thematic Mapper (TM) images over the coast of Senegal. (b) May 3, 1987 – a relatively clear day (optical thickness of 0.8 at 0.64 μm), (c) April 17, 1987 – a severe dust storm (optical thickness of 2.4). The images were produced using the same gray levels. Dust increases the image brightness over the land and ocean indicating low absorption.

following Wiscombe and Grams, 1976). Parameters μ and μ_o are the cosines of the view- and solar-zenith-angles respectively. Equation 11.1 is applicable for the single-scattering approximation, which is used here for illustration purposes only. The actual calculations shown later involve rigorous radiative transfer using Mie theory for spherical particles. The delicate balance is achieved on the right-hand side of (11.1). The first term is positive and is due to the sunlight interacting only with the atmosphere. The second term is negative; it is due to sunlight reflected by the surface and is proportional to the aerosol absorption and backscattering. The satellite technique is based on this delicate balance between these two terms, and therefore is very sensitive to small changes in the single-scattering albedo. A balance is reached for $\Delta\rho^* = 0$ as shown in (11.2), when $\Delta\rho^*$ is independent of the aerosol concentration or optical thickness.

$$\Delta\rho^* = 0 \rightarrow \omega_o = \rho(\mu + \mu_o)/[P/4 + (1 + \beta)(\mu + \mu_o)] \qquad (11.2)$$

The two Landsat images used in the analysis are also shown in Figure 11.2. As expected the heavy dust on April 17, 1987 (Figure 11.2c) increases the apparent ocean brightness, but it also increases the brightness of the near-by desert. This increase in the desert brightness cannot be explained by estimates of dust absorption found in the literature. These estimates are based on *in situ* measurements that suggest strong dust absorption across the visible and near-infrared (IR) parts of the solar spectrum (Carlson and Benjamin, 1980; WMO, 1983). The desert brightening does agree with lower dust absorption derived from radiation measurements (Fouquart *et al.*, 1986; Otterman *et al.*, 1982), and may resolve some disagreements between *in situ* measurements and ERBE flux measurements at the top of the atmosphere reported by Ackerman and Chung (1992).

To derive the aerosol single-scattering albedo from the change in the spectral brightness observed by Landsat (Figure 11.2) we need to determine the aerosol scattering properties and therefore the size distribution. Inversion of AERONET sky angular-radiance data measured from the ground at three locations in and around the Sahara (Cap Verde of the African Atlantic coast, Banizombu in Africa, Sede Boker in the Israeli desert) are used to define the effective dust-particle size (Tanré *et al.*, 2001). The results are also compared with dust measurements downwind from China at San Nicholas Island off the coast of California. The dust effective radius is between 1.5 and 2.5 μm. For this range of the effective radius we derive the dust single-scattering albedo shown in Figure 11.3 using radiative-transfer calculations. First the radiance in the clear day is used to derive the surface reflectance for the optical thickness measured in that day. Then radiance at the top of the atmosphere is calculated for the dust assumed-size distribution and refractive index, for several

New frontiers in remote sensing

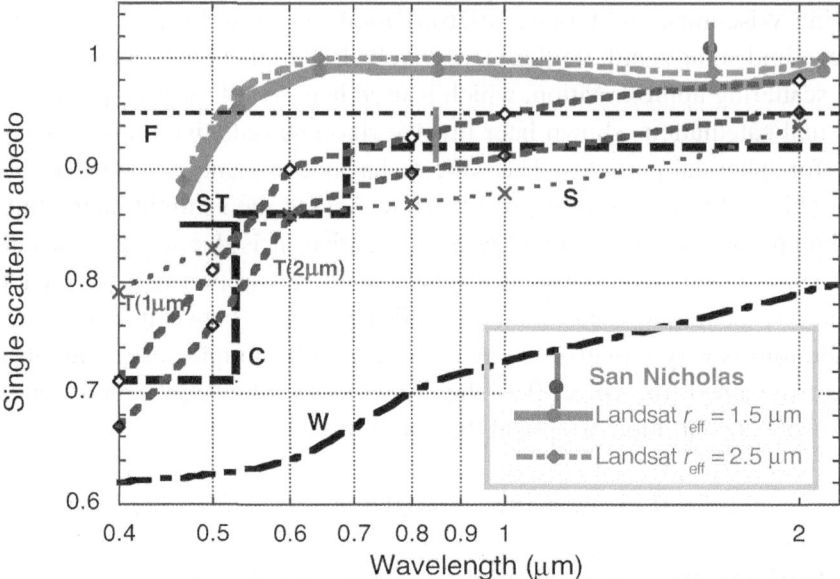

Figure 11.3. Dust spectral single-scattering albedo, ω_0, derived for the effective radius of 1.5 and 2.5 μm (top two lines), that fit the change in the Landsat spectral brightness ($\omega_0 = 1$: non-absorbing and $\omega_0 = 0$: fully absorbing). Data from San Nicholas Island off the coast of California are also shown, representing Asian dust after long transport. Results are compared to ω_0 values from the literature: F, Fouquart *et al.* (1986); W, WMO (1983); T(1μm) and T(2μm), Tegen *et al.* (1996) for 1 μm and 2 μm effective radius respectively; C, Carlson and Benjamin (1980); ST, Sokolik and Toon (1996); S, Sokolik *et al.* (1993).

values of the aerosol single-scattering albedo and for the surface reflectance determined in the clear day. The single-scattering albedo is derived from this relationship and the value of the radiance in the dusty day measured by the satellite. The real part of the refractive index was assumed to decrease with increasing wavelength from 1.53 in the visible down to 1.46 in the mid IR. The method is described by Kaufman *et al.* (2001). The remote-sensing values of ω_0 are much higher than the values found in the literature, mainly in the red part of the spectrum (around 0.6 μm). They are not affected significantly by the uncertainty in the dust effective radius. Application of the satellite technique to Asian dust arriving at the California coast shows that the properties are similar to that of Saharan dust at 1.65 μm but the dust is probably accompanied by an accumulation mode with black carbon that absorbs solar radiation at submicron wavelengths.

Dust non-sphericity can affect the dust scattering properties, change the balance of scattering to absorption, and distort the determination of the single-scattering albedo. We were lucky that the Landsat images are obtained for a solar zenith angle of 32° corresponding to scattering angle of 148° where dust non-sphericity is expected to have a minimal effect on the phase function (Nakajima *et al.*, 1989; Mishchenko *et al.*, 1997). Closure of the dust effect on the radiation field over the ocean also reveals that the non-sphericity effect should be undetected (Kaufman *et al.*, 2001).

This technique, demonstrated here with Landsat data, can be applied to data from MODIS and other next-generation sensors. Landsat revisits a site only once every 16 days and collects 10% of the available global data; MODIS will provide additional solar channels and nearly daily global coverage. With this remote-sensing tool we will be able to determine the single-scattering albedo for various types of aerosols and better clarify the warming or cooling nature of the aerosols in different regions.

11.4 Remote sensing of aerosol fluxes and radiative forcing of climate

The aerosol effect on radiative forcing of climate manifests itself in absorption of solar radiation, and in redistribution of the direct solar radiation by backscattering solar radiation to space and increasing the diffuse solar radiation reaching the surface. The net effect is reduction of total solar radiation reaching the surface. However aerosol effect on the radiation reflected to space depends on the reflectivity of Earth's surface, presence of aerosol layers over clouds, and the aerosol single-scattering albedo (Fraser and Kaufman 1995; Hansen and Ruedy, 1997). Climate models can include these aerosol radiative processes by using measured or calculated aerosol loading and optical properties. However, the same measured radiation field, which is used to retrieve the aerosol loading, can be used directly to derive the aerosol impact on radiative fluxes, and this can be used directly in climate models, or used to verify their treatment of the aerosol effect. In this section we demonstrate this application using spectral radiation measured at the top of the atmosphere and spectral radiation measured from ground-based radiometers. The advantage of this approach is that the derived fluxes depend only marginally on the assumed aerosol properties (Kaufman *et al.*, 1997a). This advantage can be explained using single-scattering approximation for the radiance at the top of the atmosphere for zero surface reflectance, (11.3):

$$\rho_{top}^*(\tau_a, \rho) = \omega_o \tau_a P/(4\mu\mu_o) \tag{11.3}$$

The corresponding normalized fluxes are given in (11.4):

$$F_{top}(\tau_a, \rho)/F_o = \omega_o \tau_a \beta/\mu_o. \tag{11.4}$$

Therefore we obtain

$$F_{top}(\tau_a, 0)/\rho_{top}^*(\tau_a, 0) = \mu\beta/P. \tag{11.5}$$

This ratio of flux to radiance is independent of the aerosol loading or single-scattering albedo. The backscattering function β is an integral on part of the angular range of P. Therefore averaging the ratio in Equation (11.4) over the view direction will further decrease the dependence on the aerosol properties.

11.4.1 Remote sensing of aerosol fluxes and radiative forcing at the top of atmosphere

Figure 11.4 demonstrates the application of the concept to data from a high-flying spectral imager. In this case we use data from the AVIRIS instrument flown on the ER-2 aircraft at 20 km altitude in the Sulfates, Clouds and Radiation Brazil (SCAR-B) experiment, to simulate the use of satellite data, and estimate the effect of aerosol on backscattering solar radiation to space. The method is based on two principles. (1) Smoke is visible in the short solar wavelengths and transparent in the mid IR. (2) Surface reflectance in the visible part of the spectrum is highly correlated to the reflectance in the mid IR (Kaufman *et al.*, 1997a). Therefore the images in the long wavelengths (1.65 and 2.1 μm) are used, in smoke-free locations, to estimate the surface–molecular scattering contribution to the flux, L, in the 0.4–0.7 μm range, Equation (11.6), and to estimate the surface reflectance at 0.66 μm, (11.7):

$$\pi L^s_{0.4-0.7 \ \mu m} = 19 + 291 \cdot \rho_{2.13} - 44 \cdot \rho_{1.65} (\text{W m}^{-2}), \tag{11.6}$$

$$\rho_{0.66} = \rho_{2.13}/2. \tag{11.7}$$

The apparent reflectance at the top of the atmosphere, $\rho^*_{0.66}$ is then used to estimate the aerosol optical thickness, Equation (11.8):

$$\rho^*_{0.66} - \rho_{0.66} \rightarrow \tau \tag{11.8}$$

The aerosol effect on reflected sunlight $\pi \Delta L_{0.4-0.7 \ \mu m}$ is derived from the difference between the integral on the measured radiance $L_{0.4-0.7 \ \mu m}$ and the estimate surface–molecular scattering contribution, Equation (11.9):

$$\pi \Delta L_{0.4-0.7 \ \mu m} = \pi L_{0.4-0.7 \ \mu m} - \pi L^s_{0.4-0.7 \ \mu m}. \tag{11.9}$$

The results are plotted in Figure 11.4 for several locations in Brazil. The data are fitted with an aerosol model that is used to convert the measured radiance difference $\pi \Delta L_{0.4-0.7 \ \mu m}$ to the instantaneous flux for unit optical thickness: $\Delta F_{0.4-0.7 \ \mu m} = 70 \ \text{W m}^{-2}$. Since this flux is of reflected solar radiation, the corresponding radiative forcing is negative: $-70 \ \text{W m}^{-2}$. The forcing for 0.7–2.1 μm is insignificantly different from zero, due to a balance between an increase in the backscattering to space for $\lambda < 0.8$ μm and decrease for $\lambda > 0.8$ μm (see Figure 11.4d). For a wide range of solar zenith angles we found forcing of -50 to $-70 \ \text{W m}^{-2}$, $28° \leqslant \theta_o \leqslant 67°$, and averaged on 24 hours: $\Delta F_{24hr} = -25 \ \text{W m}^{-2}$.

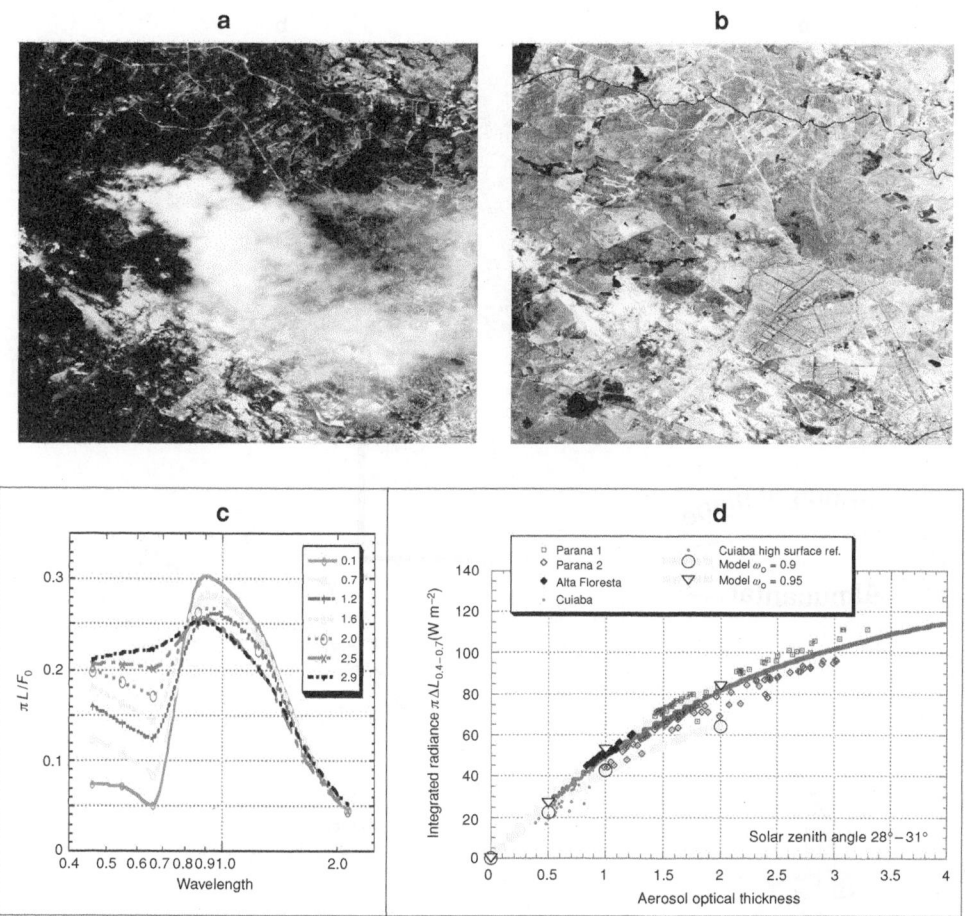

Figure 11.4. The AVIRIS spectral images collected by the ER-2 during the SCAR-B experiment in Brazil in 1995, showing fire and smoke in the region around Cuiaba, Brazil. (a) Red (0.66), green (0.55), blue (0.47 μm) color composite. The smoke is transparent in (b) Near-IR color composite: red (1.65), green (1.2) and blue (2.1 μm) image produced for wavelengths > 1 μm, therefore this spectral range is used to observe the surface features, and to derive the smoke optical thickness at short wavelengths (Chu *et al.*, 1998). (c) The spectral radiance for a range of smoke optical thicknesses derived from this image. (d) Radiance integrated on the spectral range 0.4-0.7 μm of AVIRIS is used to plot the effect of the smoke on the upward radiance expressed in flux units, $\pi \Delta L$, as a function of the optical thickness. Results are shown for several locations, for solar zenith angles 28–31°. The black line is the best fit to the data. Using two models that fit the data (for $\omega_o = 0.90$ and 0.95), the nadir radiance is converted into flux. The resulting instantaneous flux is 70 W m^{-2} and daily average aerosol radiative forcing of -25 W m^{-2}.

This figure is available for download in colour from
www.cambridge.org/9780521791328

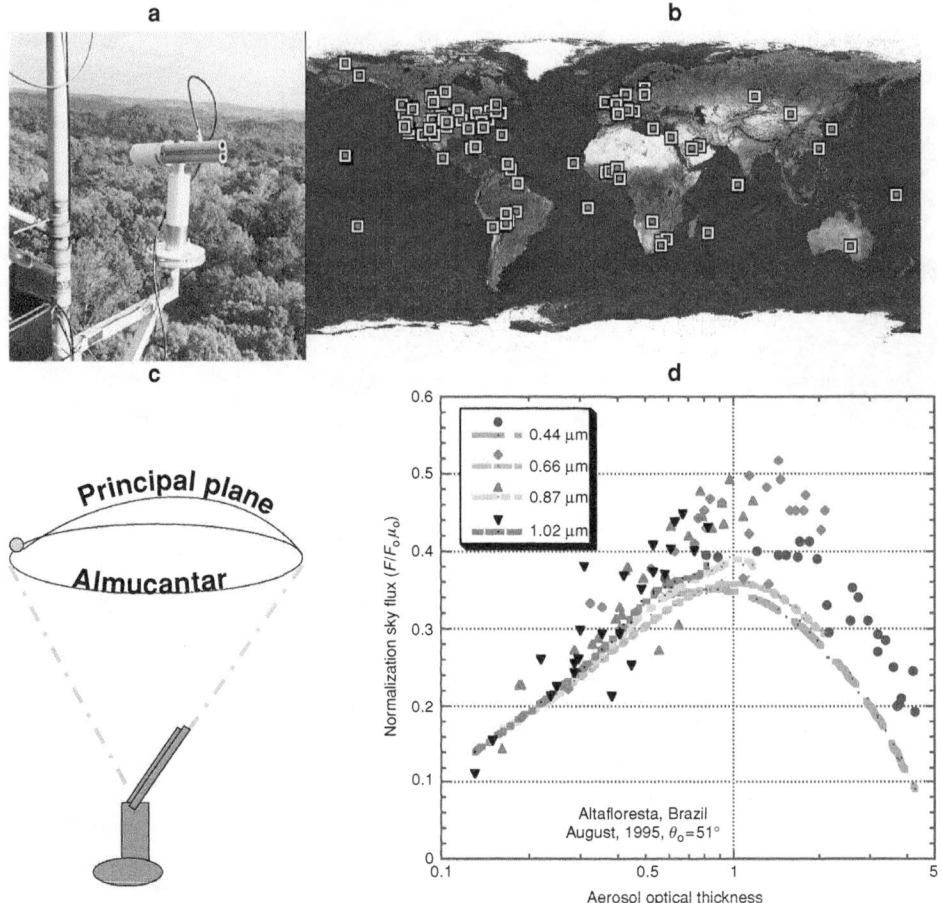

Figure 11.5. (a) A member of the AERONET federated global network of Sun/sky radiometers, and (b) a map of the global distribution of these instruments. These autonomous instruments measure the direct transmission of sunlight to the surface and the sky brightness in the almucantar plane and the principal plane passing through the Sun (c). The latter is used here to derive the solar diffuse spectral flux reaching the Earth surface at Alta Floresta, Brazil, as a function of the smoke optical thickness (d). The sky flux F is normalized by the solar extra-terrestrial flux F_o and the cosine of the solar zenith angle μ_o. Data are for the 1995 SCAR-B experiment. The symbols stand for fluxes derived from the AERONET data and the lines for the smoke model of Remer *et al.*, 1998.

11.4.2 *Remote sensing of aerosol fluxes and radiative forcing at the surface*

Principal plane radiances measured by the AERONET federated global network of Sun/sky radiometers are used to derive the effect of aerosol on diffuse solar radiation reaching Earth's surface. A photograph of a unit of the AERONET network and a map of the global distribution of AERONET are shown in Figure 11.5a,b. These autonomous instruments measure the direct transmission of sunlight to the surface and the sky brightness in the almucantar plane and the principal plane passing through the Sun (see Figure 11.5c). Though the principal plane is only one line in the sky, it contains all the scattering angles present in the sky. Since all the physical and chemical parameters that determine the optical properties of the aerosol and the sky radiance are only a function of the scattering angle, we should be able to reconstruct accurately the sky brightness from the principal-plane measurements. The advantage of using fluxes derived from just the radiances in the principal plane is that it is easier to avoid scattered clouds and derive the fluxes in their presence. The remote-sensing technique is illustrated using the single-scattering approximation for a non-reflective surface. The diffuse radiance at the surface is given in Equation (11.10), where F_o is the extra-terrestrial solar spectral irradiance,

$$L_{sky}(\tau, \theta, \theta_o, \phi) = F_o \omega_o P(\Theta) \tau / 4 \cos(\theta) \tag{11.10}$$

τ the optical thickness, and θ is the scattering angle related to the view zenith angle θ, the solar zenith angle θ_o, and the azimuth between them ϕ, by (11.11):

$$\cos(\Theta) = \cos(\theta_o) \cos(\theta) + \sin(\theta_o) \sin(\theta) \cos(\phi) \tag{11.11}$$

The diffuse flux reaching the surface is the integral of the sky radiance over the downward hemisphere, (11.12):

$$F_{sky} = \int\int L_{sky}(\theta, \theta_o, \phi) \sin(\theta) d\theta / d\phi \tag{11.12}$$

The steps of the method to derive the downward spectral flux are the following:

- Calculate the sky illumination $L_a(\tau, \rho, \theta, \phi)$ for an arbitrary aerosol type and the optical thickness, τ_a, derived from the solar measurements.
- Reconstruct the true sky radiance, $L_s(\tau, \rho, \theta, \phi)$, by scaling L_a with the radiance measured in the principal plane, $L_{pp-m}(\Theta)$ for the same scattering angle, Equation (11.13):

$$L_s(\tau, \rho, \theta, \phi) = L_a(\tau, \rho, \theta, \phi)[L_{pp-m}(\Theta)/L_{pp-a}(\tau, \Theta)],$$
$$\Theta = f(\theta, \theta_o, \phi), \tag{11.13}$$

where $L_{pp-a}(\Theta)$ is the value of L_a in the principal plane. The illustration with the single-scattering approximation allowed the derivation of this analytical expression. Substituting L_a and L_{pp} from (11.10), we obtain (11.14) and (11.15).

$$L_s = (F_o/4)[\omega_{oa} P_a(\theta, \phi) \tau_a / \cos(\theta_s)][\omega_{om} P_m(\Theta) \tau_m / \cos(\theta_{pp})]/$$
$$[\omega_{oa} P_a(\Theta) \tau_a / \cos(\theta_{pp})] \tag{11.14}$$
$$= F_o \omega_{om} P_m(\theta, \phi) \tau_m / 4 \cos(\theta_s) \tag{11.15}$$

Here ω_{om}, τ_m, and $P_m(\theta, \phi)$ are the exact, though unknown quantities of the sky; θ_s is the view zenith angle of the part of the sky for which the radiance L_s is calculated and θ_{pp} is the angle for the part of the principal plane used to calculate it. Therefore, in the single-scattering approximation the reconstruction of the sky brightness removed any memory of the arbitrarily assumed aerosol properties. Multiple scattering can be expected to retain some of the memory for these properties.

- Integrate the radiance $L_s(\tau, \rho, \theta, \phi)$ to get the diffuse sunlight flux reaching the surface.

The effect of multiple scattering on the accuracy is studied for optical thickness of 1.0. The errors in the study are due to uncertainty in aerosol size of an order of magnitude, in surface reflectance of 0.1, in single-scattering albedo of 0.13, and in the optical thickness of 0.05. Despite these large errors, the root mean square of the error is 2.5%, smaller than calibration errors.

To apply the method to measurements, a cloud-screening technique was developed based on the smoothness of the angular distribution of the sky data. Figure 11.5d shows results of the measurements in Alta Floresta, Brazil and compares them with the smoke model of Remer *et al.* (1998). Sky radiance increases with the smoke optical thickness, reaching a maximum, then decreases due to the overwhelming effects of smoke absorption and backscattering to space. The diffuse solar fluxes derived from the AERONET radiometer fit the model for optical thickness less than 0.6 (for which the model was derived) and are higher than the model for higher optical thicknesses. The AERONET data collected in Cuiaba, Brazil were compared with the model and with shadow band fluxes. The results show that the model correlates very well with the measurements ($r = 0.98$) with a slight non-linearity for high optical thicknesses Equation (11.6):

$$F_{model}/F_o\mu_o = 0.006 + 1.02(F_{AERONET}/F_o\mu_o) - 0.32(F_{AERONET}/F_o\mu_o)^2$$
(11.16)

Comparison with shadow band measurements show the relation given in Equation (11.17):

$$F_{shadow}/F_o\mu_o = -0.03 + 1.05(F_{AERONET}/F_o\mu_o) - 0.04(F_{AERONET}/F_o\mu_o)^2,$$
$$r = 0.90. \quad (11.17)$$

The high correlation of the AERONET fluxes with the model shows the strength of the method. It is also very well correlated with the shadow band, but the shadow band is shifted from the AERONET measurements and the model by $\Delta F/F_o\mu_o = -0.03$.

The attenuation of total solar radiation (diffuse + direct), or the fraction of sunlight reaching Earth's surface is shown as a function of the optical thickness in Figure 11.6 for the four AERONET spectral bands, five locations, and for solar zenith angle of $50 \pm 5°$. Note that dust in Cape Verde attenuates the solar radiation

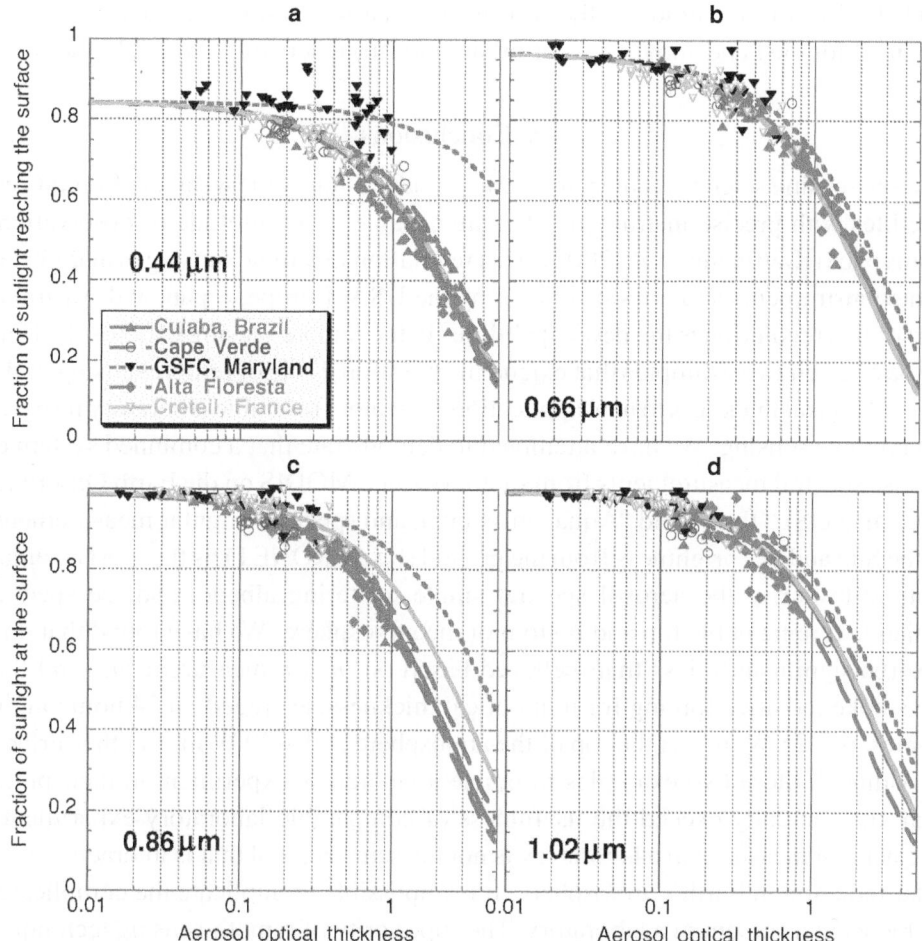

Figure 11.6. Attenuation of solar radiation, or fraction of sunlight reaching Earth's surface. Results are given for five locations as a function of the aerosol optical thickness for the four AERONET spectral bands. The solar zenith is $50 \pm 5°$. (a) 0.44; (b) 0.66; (c) 0.86; and (d) 1.02 μm.

less than smoke in Brazil (Cuiaba and Alta Floresta). Lowest attenuation is by urban–industrial pollution, e.g., GSFC, Maryland, USA and Creteil, France. The numerical fits to the data are: $f = \exp(-a - b\tau)$, where a is the Rayleigh scattering effect, $b \sim 0.37$ for smoke, 0.32 for dust and surprising only 0.19 for Goddard Space Flight Center (GSFC) urban pollution, which requires further studies.

Aerosol forcing at the surface is defined here as the difference in the total solar flux reaching the surface for an atmosphere with and without aerosol. Integrating on the solar spectrum, for a tropical atmospheric model, we get smoke forcing of $\Delta F_{24hr}/\Delta \tau = -80 \pm 5 \text{ W m}^{-2}$, assuming equal distribution of solar zenith angles across the day, and assuming that the spectral average values of b, derived for the

0.44–1.02 μm range hold for the rest of the solar spectrum. Similar results were obtained for dust and some cases of urban–industrial aerosol (Creteil, France).

11.5 Conclusions

Remote sensing science is entering a new era, substituting old "weather observation" satellites with precise instruments that are capable of a multitude of observations from space (Kaufmann *et al.*, 2002). This revolution in remote sensing is manifested in new instruments and missions flown by the USA, Europe, Japan, and Australia. Part of this revolution is the new capabilities in the remote sensing of aerosols. Here we have given two examples that extend the traditional role of remote sensing of the aerosol optical thickness to new applications that more resemble laboratory research than remote sensing. We have attempted to demonstrate that a combined system of precise spectral measurements from satellites (e.g., MODIS on the Earth Observing System (EOS) "Terra" and "Aqua" missions), and spectral–angular measurements of the Sun and sky brightness from the ground (by AERONET), is the most accurate method to derive the aerosol spectral single-scattering albedo, and the spectral radiative forcing at the top and bottom of the atmosphere. We have found that dust absorbs significantly less than believed based on *in situ* measurements, and that the smoke radiative forcing for unit optical thickness, averaged on 24 hours in the tropics, is -25 W m^{-2} at the top of the atmosphere and -80 W m^{-2} at the surface.

Although these two examples resemble a laboratory experiment in their precision, they actually exceed the usefulness of comparable laboratory experiments. The aerosol measured in a laboratory is not the same aerosol that is interacting with solar radiation in Earth's atmosphere. It is impossible to duplicate the complicated ambient environment in a laboratory. The superiority of remote-sensing techniques is that they measure the total column of ambient aerosol, and provide the important characteristics of the aerosols that actually affect our climate.

The remote-sensing revolution continues with the launch of EOS-Terra in the USA, the Environmental Satellite (ENVISAT) in Europe, and the Advanced Earth Observing Satellite, ADEOS-II, in Japan. The next step in the revolution will be the introduction of remote sensing from lidars that will add the third dimension to aerosol remote sensing. These new capabilities are expected to put remote sensing in the forefront of aerosol and climate studies, by practically generating "space global laboratories" for precise and diverse aerosol observations from space and from the ground. Together with field experiments, chemical analysis, chemical-transport models, and climate models, we anticipate, in the next decade, to be able to resolve some of the outstanding questions regarding the role of aerosols in climate, in atmospheric chemistry, and its influence on human health and life on this planet.

References

Ackerman, S. A. and H. Chung (1992). Radiative effects of airborne dust on regional energy budgets at the top of the atmosphere. *J. Appl. Meteor.* **31**, 223–33.

Alpert, P., Y. J. Kaufman, Y. Shay-El *et al.* (1998). Quantification of dust-forced heating of the lower troposphere. *Nature* **395**, 367–70.

Boucher, O. and U. Lohmann (1995). The sulfate–CCN–cloud albedo effect, a sensitivity study with two general circulation models. *Tellus* **47B**, 281–300.

Carlson, T. N. and S.G. Benjamin (1980). Radiative heating rates of Saharan dust. *J. Atmos. Sci.* **37**, 193–213.

Cess R. D., M. H. Zhang, P. Minnis *et al.* (1995). Absorption of solar-radiation by clouds – observations versus models. *Science* **267**, 496–9.

Charlson, R. J., S. E. Schwartz, J. M. Hales *et al.* (1992). Climate forcing by anthropogenic aerosols. *Science* **255**, 423–30.

Chu, A., Y. J. Kaufman, L. A. Remer, and B. N. Holben (1998). Remote sensing of smoke from MODIS airborne simulator during SCAR-B experiment. *J. Geophys. Res.* **103**, 31979–88.

Coakley J. A. and R. D. Cess (1985). Response of the NCAR Community Climate Model to the radiative forcing by the naturally-occurring tropospheric aerosol. *J. Atmos. Sci.* **42**, 1677–92.

Coakley, J. A., Jr., R. L. Bernstein, and P. A. Durkee (1987). Effect of ship stack effluents on cloud reflectance. *Science* **237**, 953–6.

Deschamps, P. Y., F. M. Bréon, M. Leroy *et al.* (1994). The POLDER mission: instrument characteristics and scientific objectives. *IEEE Trans. Geosci. Rem. Sens.* **32**, 598–615.

Deuzé, J. L., M. Herman, P. Goloub, D. Tanré, and A. Marchand (1999). Characterization of aerosols over the ocean from POLDER. *Geophys. Res. Lett.* **26**, 1421–4.

Diner, D. J., J. C. Beckert, T. H. Reilly *et al.* (1998). Multi-angle Imaging SpectroRadiometer (MISR) – instrument description and experiment overview. *IEEE Trans. Geosci. Rem. Sens.* **36**, 1339.

Dubovik, O., A. Smirnov, B. N. Holben *et al.* (1999). Accuracy assessment of aerosol optical properties retrieval from AERONET sun and sky radiance measurements. *J. Geophys. Res.* **105**, 9791–806.

Fouquart, Y., B. Bonnel, J. C. Brogniez *et al.* (1986). Observations of Saharan aerosols: results of ECLATS field experiment. II: broadband radiative characteristics of the aerosols and vertical radiative flux divergence. *J. Clim. Appl. Meteor.* **25**, 28–37.

Fraser, R. S. and Y. J. Kaufman (1985). The relative importance of aerosol scattering and absorption in remote sensing. *IEEE J. Geosc. Rem. Sens.* **23**, 625–33.

Hansen J. E. and A. A. Lacis (1990). Sun and dust versus greenhouse gases: an assessment of their relative roles in global climate change. *Nature* **346**, 713–19.

Hansen, J., M. Sato, and R. Ruedy (1997). Radiative forcing and climate response. *J. Geophys. Res.* **102**, 6831–64.

Heintzenberg, J., S. E. Schwartz, J. M. Hales *et al.* (1997). Measurements and modeling of aerosol single scattering albedo: progress, problems and prospects. *Beitr. Phys. Atmos.* **70**, 249–63.

Herman, J. R., P. K. Barthia, O. Torres *et al.* (1997a). Global distribution of UV absorbing aerosol from Nimbus 7/TOMS data. *J. Geophys. Res.* **102**, 16911–22.

Herman, M., J. L. Deuzé, C. Devaux *et al.* (1997b). Remote sensing of aerosol over land surfaces including polarization measurements and application to POLDER measurements. *J. Geophys. Res.* **102**, 17039–50.

Higurashi, A. and T. Nakajima (1999). Development of a two channel aerosol retrieval algorithm on global scale using NOAA AVHRR. *J. Atmos. Sci.* **56**, 924–41.

Hobbs, P. V., J. S. Reid, R. A. Kotchenruther, R. J. Ferek, and R. Weiss (1997). Direct radiative forcing by smoke from biomass burning. *Science* **275**, 1776–8.

Holben, B. N., T. F. Eck, I. Slutsker *et al.* (1998). AERONET-A federated instrument network and data archive for aerosol characterization. *Rem. Sens. Environ.* **66**, 1–16.

Hsu, N. C., J. R. Herman, P. K. Bhartia *et al.* (1996). Detection of biomass burning smoke from TOMS measurements. *Geophys. Res. Lett.* **23**, 745–8.

Huebert, B. J., G. Lee, and W. L. Warren (1990). Airborne aerosol inlet passing efficiency measurement. *J. Geophys. Res.* **95**, 16369–81.

Husar, R. B., J. Prospero, and L. L. Stowe (1997). Characterization of tropospheric aerosols over the oceans with the NOAA advanced very high resolution radiometer optical thickness operational product. *J. Geophys. Res.* **102**, 16,889–910.

IPCC (1995). *Climate Change 1995*, eds. J. T. Houghton, L.G. Meira Filho, B. A. Callandar, N. Haris, A. Kattenberg, and K. Maskell, Cambridge, Cambridge University Press.

Jankowiak, I. and D. Tanré (1992). Climatology of Saharan dust events observed from Meteosat imagery over Atlantic Ocean. Method and preliminary results. *J. Clim.* **5**, 646–56.

Kaufman, Y. J. (1987). Satellite sensing of aerosol absorption. *J. Geophys. Res.* **92**, 4307–17.

Kaufman, Y. J. and R. S. Fraser (1997). The effects of smoke particles on clouds and climate forcing. *Science* **277**, 1636–9.

Kaufman, Y. J., A. Gitelson, A. Karnieli *et al.* (1994). Size distribution and scattering phase function of aerosol particles retrieved from sky brightness measurements. *JGR-Atmos.* **99**, 10341–56.

Kaufman, Y. J., D. Tanré, L. Remer *et al.* (1997a). Remote sensing of tropospheric aerosol from EOS-MODIS over the land using dark targets and dynamic aerosol models. *J. Geophys. Res.* **102**, 17051–67.

Kaufman, Y. J., D. Tanré and O. Boucher (2002). A satellite view of aerosols in the climate system, *Nature* **419**, 215–23.

Kaufman, Y. J., D. Tanré, H. R. Gordon *et al.* (1997b). Passive remote sensing of tropospheric aerosol and atmospheric correction. *J. Geophys. Res.* **102**, 16815–30.

Kaufman, Y. J., D. Tanré, A. Karnieli, and L. A. Remer (2001). Satellite and ground-based radiometers reveal much lower dust absorption of sunlight than used in climate models. *Geophys. Res. Lett.* **28**, 1479–83.

Kiehl, J. T. and B. P. Briegleb (1993). The relative roles of sulfate aerosols and greenhouse gases in climate forcing. *Science* **260**, 311–14.

Liousse, C., J. E. Penner, C. Chuang *et al.* (1996). A global 3-D model of carbonaceous aerosols. *J. Geophys. Res.* **101**, 19411–32.

Martonchik, J. V. and D. J. Diner (1992). Retrieval of aerosol and land surface optical properties from multi-angle satellite imagery. *IEEE Trans. Geosci. Rem. Sens.* **30**, 223–30.

Mishchenko, M. I. and L. D. Travis (1997). Satellite retrieval of aerosol properties over the ocean using polarization as well as intensity of reflected sunlight. *J. Geophys. Res.* **102**, 16989–17013.

Mishchenko, M. I., L. D. Travis, R. A. Kahn, and R. A. West (1997). Modeling phase function for dustlike tropospheric aerosol using a shape mixture of randomly oriented polydisperse spheroids. *J. Geophys. Res.* **102**, 16831–48.

Moulin C., C. E. Lambert, F. Dulac, and U. Dayan (1997). Control of atmospheric export of dust from North Africa by the North Atlantic oscillation. *Nature* **387**, 691–4.

Nakajima, T. and A. Higurashi (1998). A use of two channel radiances for an aerosol characterization from space. *Geophys. Res. Lett.* **25**, 3815–18.

Nakajima, T., M. Tanaka, M. Yamano *et al.* (1989). Aerosol optical characteristics in the yellow sand events observed in May, 1982 at Nagasaki. Part 2. Models. *J. Meteor. Soc. Jpn.* **67**, 279–91.

Nakajima T. Y., T. Nakajima, M. Nakajima *et al.* (1998). Optimization of the Advanced Earth Observing Satellite II Global Imager channels by use of radiative transfer calculations. *Appl. Optics* **37**, 3149–63.

Otterman, J., R. S. Fraser, and O. P. Bahethi (1982). Characterization of tropospheric desert aerosols at solar wavelengths by multispectral radiometry from Landsat. *J. Geophys. Res.* **87**, 1270–8.

Penner, J. E., R. E. Dickinson, and C. A. O'Neill (1992). Effects of aerosol from biomass burning on the global radiation budget. *Science* **256**, 1432–3.

Ramanathan, V., R. D. Cess, E. F. Harrison *et al.* (1989). Cloud-radiative forcing and climate: results from the Earth Radiation Budget Experiment. *Science* **243**, 57–63.

Remer, L. A., Y. J. Kaufman, B. N. Holben, A. M. Thompson, and D. McNamara (1998). Tropical biomass burning smoke aerosol size distribution model. *J. Geophys. Res.* **103**, 31879–92.

Rosenfeld, D. (1999). TRMM observed first direct evidence of smoke from forest fires inhibiting rainfall. *Geophys. Res. Lett.* **26**, 3105–8.

Sokolik, I., A. Andronove, and T. C. Johnson (1992). Complex refractive index of atmospheric dust aerosols. *Atmos. Environ.* **27A**, 2495–502.

Sokolik, I. N. and O. B. Toon (1996). Direct radiative forcing by anthropogenic airborne mineral aerosol. *Nature* **381**, 681–3.

Tanré, D., Y. J. Kaufman, M. Herman, and S. Mattoo (1997). Remote sensing of aerosol over oceans from EOS-MODIS. *J. Geophys. Res.* **102**, 16971–88.

Tanré, D., C. Devaux, M. Herman, R. Santer, and J. Y. Gac (1988). Radiative properties of desert aerosols by optical ground based measurements at solar wavelengths. *J. Geophys. Res.* **93**, 14223–31.

Tanré, D., Y. J. Kaufman, B. N. Holben *et al.* (2001). Climatology of dust aerosol size distribution and optical properties derived from remotely sensed data in the solar spectrum. *J. Geophys. Res.* **106**, 18205–17.

Tanré, D., L. R. Remer, Y. J. Kaufman *et al.* (1999). Retrieval of aerosol optical thickness and size distribution over ocean from the MODIS airborne simulator during Tarfox. *J. Geophys. Res.* **104**, 2261–78.

Tegen, I., A. A. Lacis, and I. Fung (1996). The influence on climate forcing of mineral aerosols from disturbed soils. *Nature* **380**, 419–22.

Twomey, S. A., M. Piepgrass, and T. L. Wolfe (1984). An assessment of the impact of pollution on the global albedo. *Tellus* **36B**, 356–66.

Veefkind, J. P., G. de Leeuw, and P. A. Durkee (1998). Retrieval of aerosol optical depth over land using two-angle view satellite radiometry during TARFOX. *Geophys. Res. Lett.* **25**, 3135–8.

Wielicki, B. A., B. R. Barkstrom, E. F. Harrison *et al.* (1996). Clouds and the Earth's radiant energy system (CERES): an Earth observing system experiment. *Bull. Amer. Meteor. Soc.* **77**, 853–68.

Wiscombe, W. J. and G. W. Grams (1976). The backscatter fraction in two stream approximations. *J. Atmos. Sci.* **33**, 2440–51.

WMO (1983). Radiation commission of IAMAP meeting of experts on aerosol and their climatic effects, WCP55, Williamsburg, VA, March 28–30.

12

Cloud–climate feedback: lessons learned from two El Niño events

MINGHUA ZHANG

Institute for Terrestrial and Planetary Atmosphere, SUNY Stony Brook, Stony Brook, NY

12.1 Introduction

As shown in Chapters 5 and 8, clouds as one of the moist fluid dynamical phenomena play a subtle role in Earth's climate. Clouds act as a greenhouse ingredient to warm Earth; they also reflect solar radiation to space to cool Earth. The net radiative effect of these two competing processes depends on the amount, height, type, and the optical properties of clouds. All these characteristics vary in a climate change. Clouds thus exert a feedback to any forced climate change (see Chapter 8). This cloud–climate feedback problem was first studied by Schneider (1972) and Cess (1975). Later, it was found by Cess *et al.* (1989) that a difference in cloud–climate feedback contributes to a three-fold difference in the sensitivity of a large group of general circulation models (GCMs). At the same time, Mitchell *et al.* (1989) showed that different treatments of clouds in the GCMs can lead to either amplification or damping of the global-warming scenario in response to the increasing level of carbon dioxide in the atmosphere. These studies stimulated much subsequent research on this topic and they helped to initiate national programs such as the Atmospheric Radiation Measurement (ARM) Program of the Department of Energy.

Cloud-radiative forcing (CRF), first introduced by Charlock and Ramanathan (1985), can be conveniently used to quantify the radiative impact of clouds. It is defined as the influence of clouds on the input of radiant energy to the Earth–atmosphere system at the top of the atmosphere (TOA). If we use F to denote the outgoing longwave radiation at the TOA, and Q to denote the net downward shortwave radiation at the TOA, and if we use subscript c to denote clear-sky quantities, the longwave (LW) CRF is defined as $F_c - F$. It is typically positive, describing the trapping of longwave by clouds (Ramanathan *et al.*, 1989). The

Frontiers of Climate Modeling, eds. J. T. Kiehl and V. Ramanathan.
Published by Cambridge University Press. © Cambridge University Press 2006.

shortwave (SW) CRF is defined as $Q - Q_c$. It is typically negative, describing the reduction of solar energy due to cloud reflections. The net CRF, or the total CRF, is simply the sum of the two.

The feedback of clouds on the climate system can then be defined as the variation of CRFs in a climate change. In a global-warming scenario, if the change of CRF is positive, variation of clouds acts to trap more radiant energy in the climate system and thus it amplifies the climate warming, a positive feedback. If the change is negative, clouds act to mediate the climate change by disposing more radiant energy to space.

Cess and Potter (1988) designed an elegant method of understanding cloud feedback mechanisms in GCMs. They imposed a change in the sea-surface temperature (SST) in the models to diagnose the response of CRF at the TOA. This chapter follows Cess and Potter (1988) with the attempt to study the process of cloud-forcing changes by using observational data. The Earth Radiation Budget Experiment (ERBE) measurements, and the Clouds and the Earth's Radiant Energy System (CERES) measurements, which provided the timely needed CRF data, span the two El Niño events of 1987 and 1998. These two El Niño events are used as case studies to learn how cloud forcing responds to SST changes. Our interest is in the energy budget of the Earth–atmosphere system rather than in the local responses. Zhang *et al.* (1996) have reported the relationship between variations of cloud forcing and SSTs in the 1987 El Niño event using five years of ERBE data from January, 1985 to December, 1989. This chapter extends the Zhang *et al.* (1996) analysis to include CERES measurements. It also differs from Zhang *et al.* (1996) in that only ERBS measurements are used to minimize differences caused by the different satellite sampling methods.

The purpose of the chapter is not to answer whether cloud feedback is positive or negative in a global climate change. Instead, it is to show how cloud feedback is affected by the compensating geographical distribution of cloud changes in a large domain. It is shown from the analysis of the two El Niño events that the warmer tropics are associated with a reduced longwave greenhouse effect of clouds, and less reflection of solar radiation to space, which is in sharp contrast to the SST relationship in the eastern tropical Pacific or in the western tropical warm pool. Furthermore, we will show that different mechanisms may have caused this relationship between SST and cloud forcing for the two El Niño events.

12.2 Data and procedure

12.2.1 Cloud forcing

The ERBE S-4G product, released from the NASA Langley Research Center, consists of measurements from three satellites: ERBS from November, 1984 to

December, 1989, NOAA 9 from February, 1985 to January, 1987, and NOAA 10 from November, 1986 to May, 1989. The Earth Radiation Budget Satellite (ERBS) is a precessing orbit satellite that samples a complete local diurnal cycle in about 36 days, while NOAA 9 and NOAA 10 are approximately Sun-synchronous satellites that sample fixed local hours (around 14:30 and 07:30 local mean solar time, respectively). The gridded monthly CRF data in the S-4G product was derived from the instantaneous radiance measurements through several steps. One of them is the modeling of the diurnal cycle. Because of the difference in the diurnal sampling among the three satellites, potential differences among ERBS and NOAA-9, NOAA-10 can be interpreted as El Niño signatures. We therefore restrict the analysis of ERBE data to ERBS measurements. The data include fluxes under both all-sky and clear-sky conditions at a $2.5° \times 2.5°$ resolution. The implementation of ERBE, including the technique to measure all-sky and clear-sky fluxes separately, can be found in Barkstrom and Smith (1986), Ramanathan *et al.* (1989), and Harrison *et al.* (1990).

The CERES ES-4G dataset covers the eight-month period from January, 1998 to August, 1998. An overview of the CERES instrument and data algorithm can be found in Wielicki *et al.* (1995). The CERES instrument was on board the TRMM satellite, which also uses a precessing orbit that samples the whole diurnal cycle in about a month. The CERES ERBE-like product was processed using the same procedure as that used for the ERBE products. Validation of the CERES measurements against ERBE is not straightforward, because the CERES period covers the strong 1997–98 El Niño event and thus the underlying meteorological conditions are different. Wong *et al.* (2001) used a radiation model to calculate clear-sky LW flux by using the operational analysis and found consistencies between the calculated fluxes and the ERBE and CERES measurements. The difference is reported to be less than 1 W m^{-2} in the whole tropics. The comparison for cloudy skies cannot be easily made in of lack of cloud microphysical and optical information. Potential differences are possible, which are caused by different instruments and samplings from ERBE and CERES. There are no known causes, however. In the present study, we will use measurements from these two experiments to analyze the response of cloud forcing to the two El Niño events, bearing in mind these sources of potential uncertainties.

There are grid points where CRF data are missing in the ERBE and CERES products. These missing grids are potentially important in contributing to the area averages of the cloud forcings, because missing CRF can occur in heavily overcast conditions where clear-sky scenes cannot be defined. Cloud-forcing data at these missing grid points are estimated by first filling the missing clear-sky fluxes with values at the nearest grid boxes, and then by calculating the CRFs at these grid boxes.

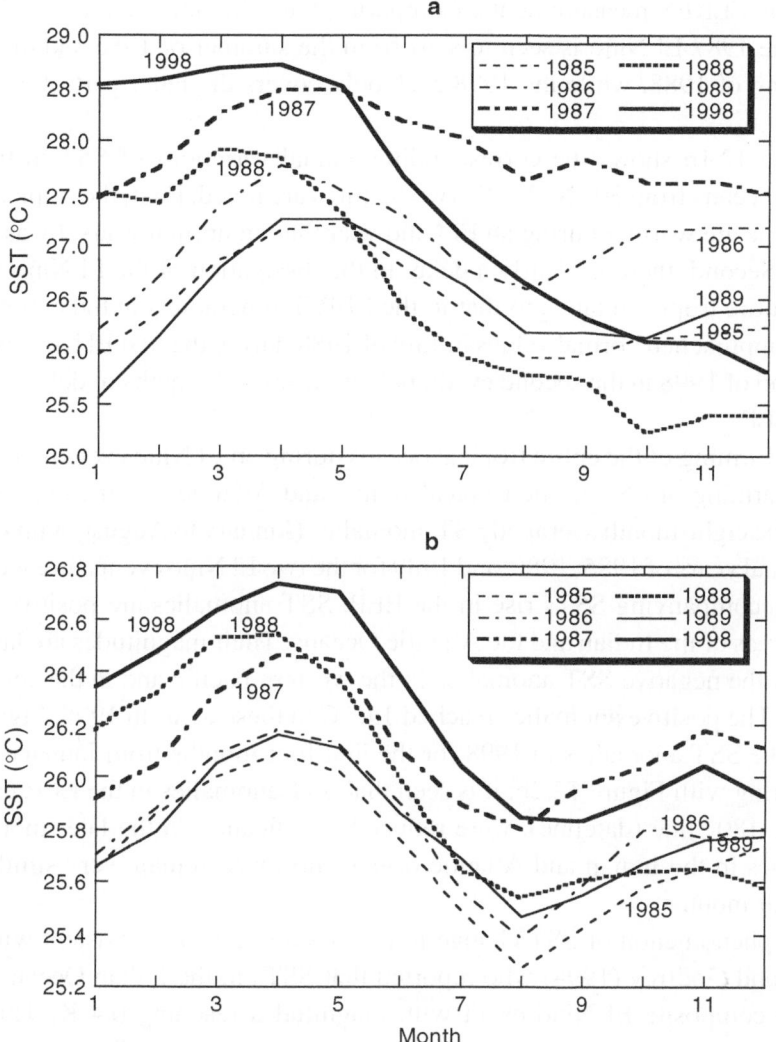

Figure 12.1. Sea-surface temperatures averaged over (a) the EEP and (b) the entire tropics.

12.2.2 Sea-surface temperature

The SST data are taken from the monthly National Meteorological Center analysis which is derived from ship, buoy, and satellite measurements. Detailed description of the SST data can be found in Reynolds (1988), Reynolds and Smith (1994; 1995) and Smith and Reynolds (1998).

Figure 12.1a shows the annual variation of SST averaged over the equatorial eastern Pacific (EEP, 10° N–10° S, 180° W–90° W) for the years when there are

ERBE or CERES measurements. Comparing with the normal years of 1985 and 1989, the 1987 El Niño is seen to start from the summer of 1986 and dissipate in the spring of 1988, while the 1998 SST only covers the latter part of an El Niño event.

Figure 12.1b shows the corresponding annual variations of SST in the entire tropical oceans from 30° N–30° S. Two features are noted. First, the tropical oceans as a whole are warmer during an El Niño year than in normal years, by as much as 0.7 °C. Second, there is clearly a delay in the dissipation of the El Niño warming of the entire tropics relative to that in the EEP. Temperatures in the entire tropical oceans approached normal values in July of 1988 during the first El Niño event, and at the end of 1998 in the second event, both with several months of delay compared to the EEP.

The warming of the entire tropical oceans during an El Niño event is contributed to by warming of SST in the tropical Indian and Atlantic oceans. Figure 12.2a,b shows the eight-month averaged SST anomalies (January to August) with respect to the normal years of 1985, 1986, and 1989 for the two El Niño events. In each El Niño event, accompanying SSTs rise in the EEP: SST anomalies are positive over the wider areas of the Indian and the Atlantic Oceans. Their magnitudes are larger than those of the negative SST anomalies in the western pacific and in the sub-tropical Pacific. The positive anomalies reached 1.0 °C in these areas in 1998. Figure 12.2c shows the SST anomalies in 1998 for the first four months from January to April. Comparing with Figure 12.2b, it is seen that SST anomalies in the EEP, especially near the 180° line (dateline), were reduced significantly in the later months. The anomalies in the Indian and Atlantic oceans, however, remain very similar in the later four months.

This phenomenon of SST change in the broad oceans is consistent with Meehl (1987) and Godfrey (1994) who reported that SSTs in the Indian Ocean warm up in their composite El Niño event with magnitudes reaching 0.4 K. This is also consistent with Villwock and Latif (1994) who reported that SST variations in the central and eastern tropical Pacific are typically followed by SST anomalies of the same sign in the Indian Ocean. The reason for this warming is not clear. Hirst and Godfrey (1993) suggested that the open passage at the southern tip of Indonesia allows warm ocean currents cross the Indian Ocean and raise the SSTs there during an El Niño event. To the contrary, Villwock and Latif (1994) argued that the surface wind forced by SST anomalies in the equatorial Pacific is the primary cause of the SST warming in the Indian Ocean. Sea-surface-temperature anomaly may even be a result of cloud variations.

Although the SST-anomaly patterns in these two El Niño events are qualitatively similar, the difference in their magnitudes corresponds to potentially important differences in the SST forcing to the atmospheric circulation and clouds. Figure 12.3

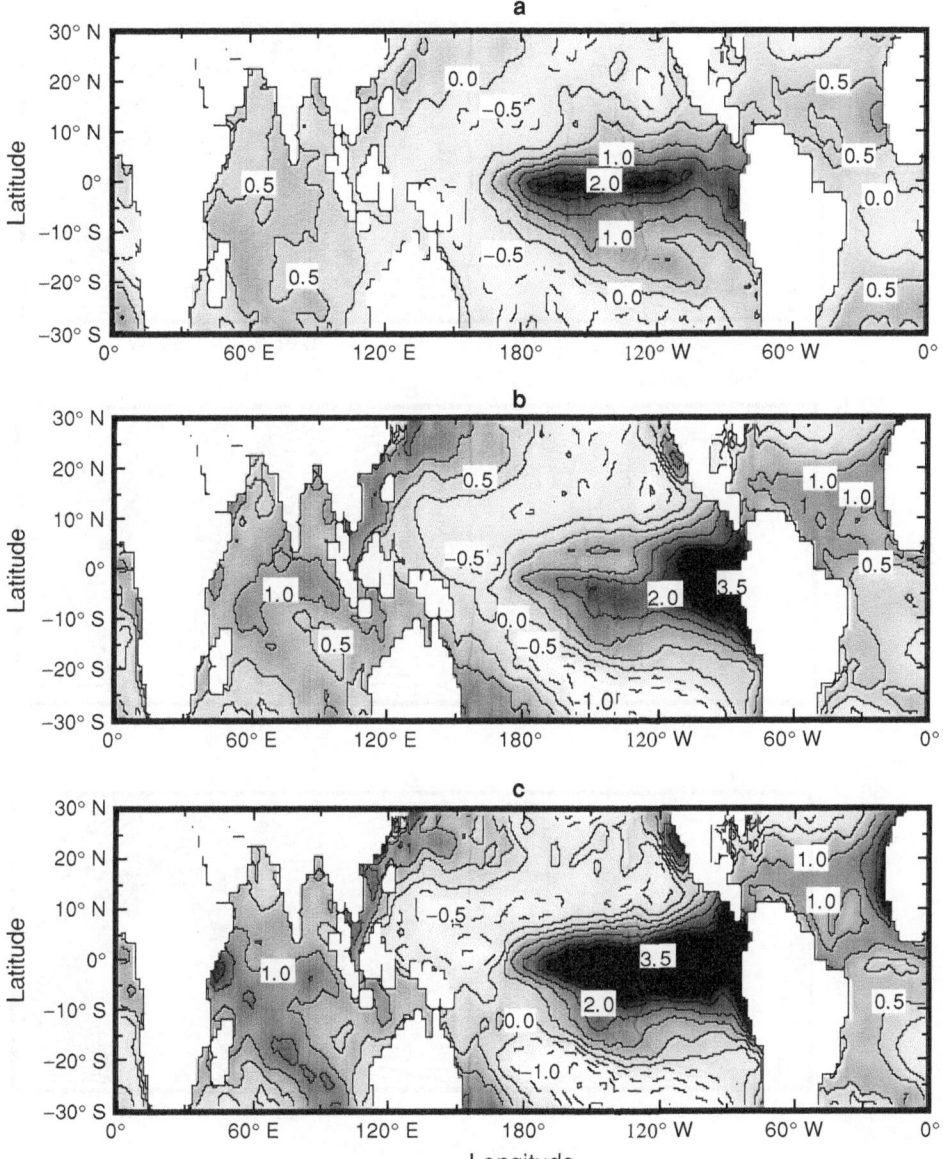

Figure 12.2. Sea-surface temperature anomalies (°C) relative to the normal years of 1985, 1986, and 1989: (a) January to August, 1987; (b) January to August, 1998; (c) January to April, 1998.

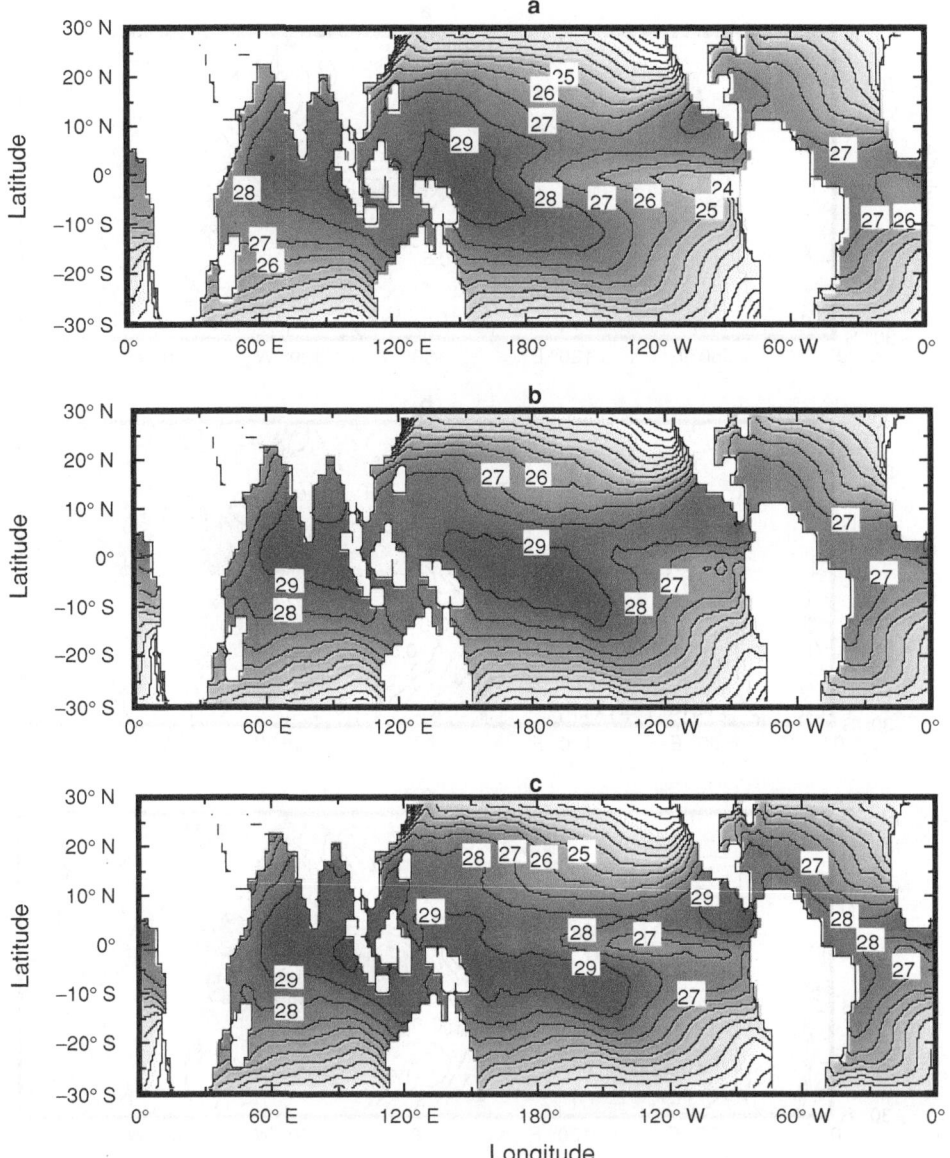

Figure 12.3. Distribution of SST (°C) averaged from January to August. (a) Mean of 1985, 1986, and 1989; (b) 1987; (c) 1998.

shows the SST distributions in the normal years and in the two El Niño years averaged from January to August. In 1987, the negative SST anomaly in the western Pacific and the positive SST anomaly in the central to eastern Pacific actually reflect a shift of the warm pool toward the dateline. In 1998, the zonal SST gradient along the tropical Pacific equator, i.e., the temperature difference between the warm pool and cold tongue, is 2 °C versus 5 °C in a normal year. This decrease in the zonal SST gradient has significant consequences on the atmospheric Walker circulation and subsequently the associated cloud processes, as will be discussed later.

12.3 Results

We first show the relationships of cloud forcing and SST in the EEP. Figure 12.4 shows the LW CRF and SW CRF against the SST in the EEP. Each solid circle in the figure represents an eight-month average for a given year. Consistent with several early studies, warmer SSTs are found to be associated with larger LW CRF, and smaller SW CRF, and thus more clouds. The 1998 data point from using the eight-month CERES measurements shows a remarkable consistency. This indirectly suggests that the CERES data are comparable with the ERBE data.

There is a large cancellation between the LW and SW components (Kiehl, 1994). Zhang *et al.* (1996) showed that with the 2.5° × 2.5° grid data in the EEP, the sign of the net cloud forcing variation with SST follows that of the shortwave. This change in cloud forcing, however, became statistically insignificant as the size of the area increases up to the whole tropical Pacific. In view of the small magnitude of the natural climate change in these years and the data length, we use the LW and SW components as a vehicle to understand the cloud process.

Figure 12.5a shows the corresponding relationship between the area-averaged LW CRF and the SST in the entire tropics. In contrast to the SST–CRF relationship for the EEP, the warmer oceans are associated with smaller LW CRF. This is true for both the 1987 El Niño event, and the 1998 event. It is seen that the tropical oceans as a whole were much warmer in 1998 than in normal years, and the LW CRF over the entire tropics is much smaller. Figure 12.5b shows the relationship between SW CRF and SST. Warm SSTs correspond to less reflection of radiation by clouds, even though the data points are more scattered than those in Figure 12.5a.

To gain some insight about the cause of the change of LW CRF with SST, Figure 12.6 shows the geographical distributions of the anomalies of the LW CRF for the two El Niño events. For both events the LW CRF increases in the central to eastern tropical Pacific where the SST anomaly is the maximum. Over many regions in the tropics, however, LW CRF is decreased. This is in sharp contrast to

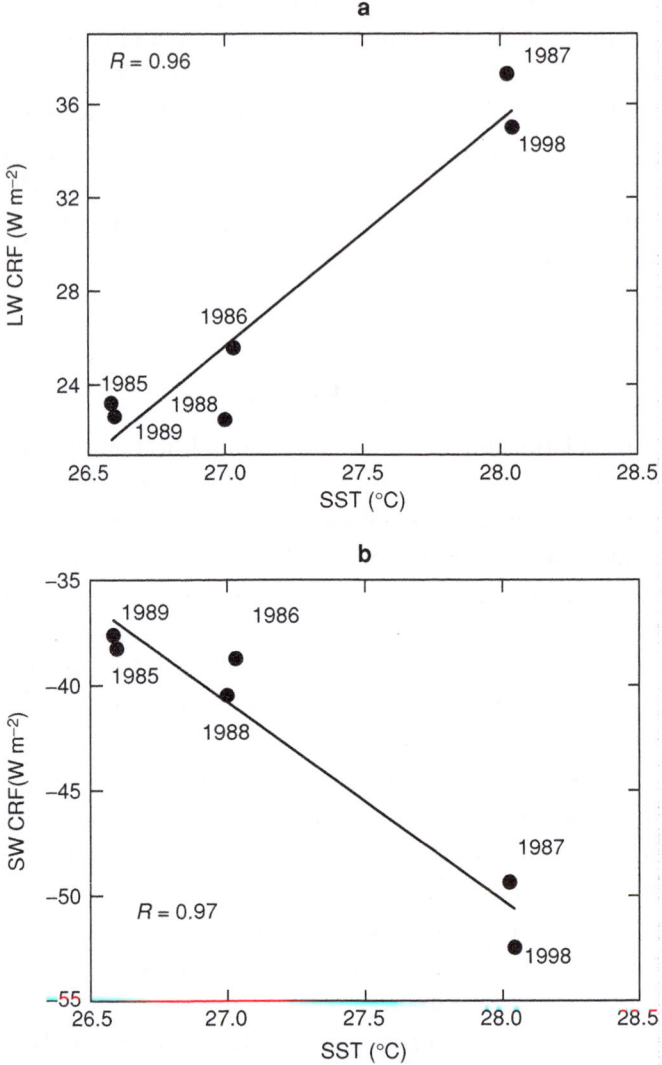

Figure 12.4. Relationship between CRFs and SST in the eastern equatorial Pacific: (a) LW CRF versus SST; (b) SW CRF versus SST.

the SST anomalies shown in Figure 12.2a,b, in which the anomalies of SST are positive over the majority of the tropics.

The mechanisms of the decrease of the LW CRF in these two El Niño events are, however, somewhat different. Figure 12.7 shows the basic distribution of the LW CRF. In a normal year, a well-defined convective center is located in the western tropical Pacific over the maritime continents, extending southeast along the South Pacific Convergence Zone (SPCZ) and eastward along the Intertropical

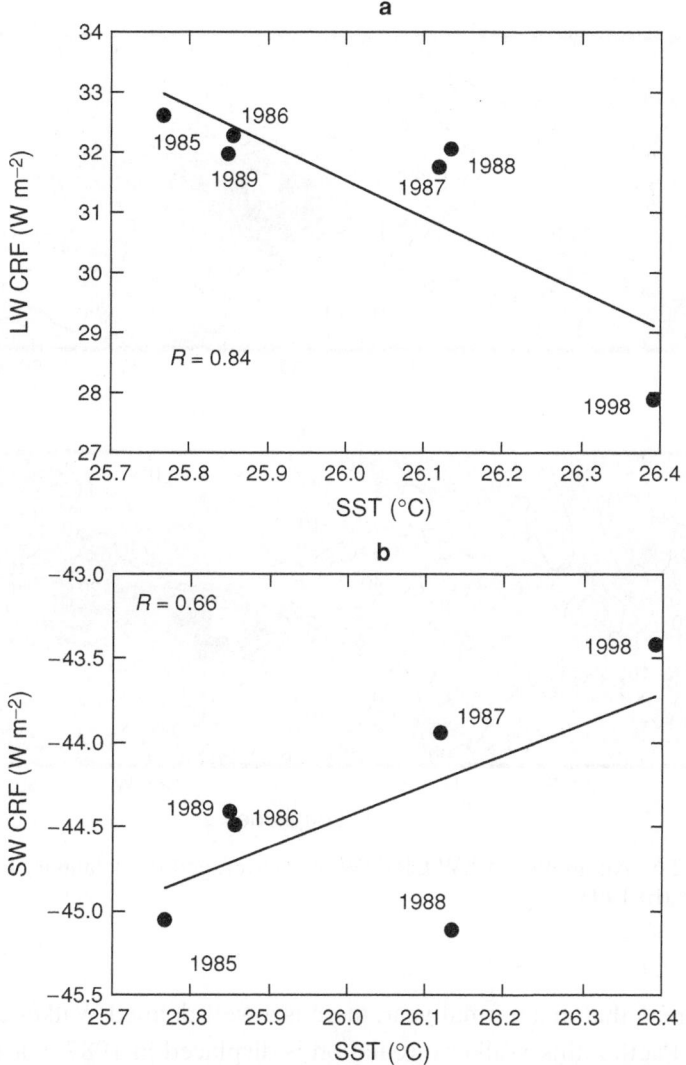

Figure 12.5. Relationship between CRFs and SST in the entire tropics: (a) LW CRF versus SST; (b) SW CRF versus SST.

Convergence Zone (ITCZ) in the eastern Pacific north of the equator. In the 1987 El Niño event, the convective center is displaced toward the dateline, in response to the shift of the 29 °C line of warm pool shown in Figure 12.3. The overall structure of the center and the two extensions in the SPCZ and ITCZ are still similar to those in a normal year. In 1998, however, one can hardly define a convection center in the Pacific. A narrow line of maximum convection spans the whole tropical Pacific, apparently associated with the small zonal SST gradient shown in Figure 12.3c.

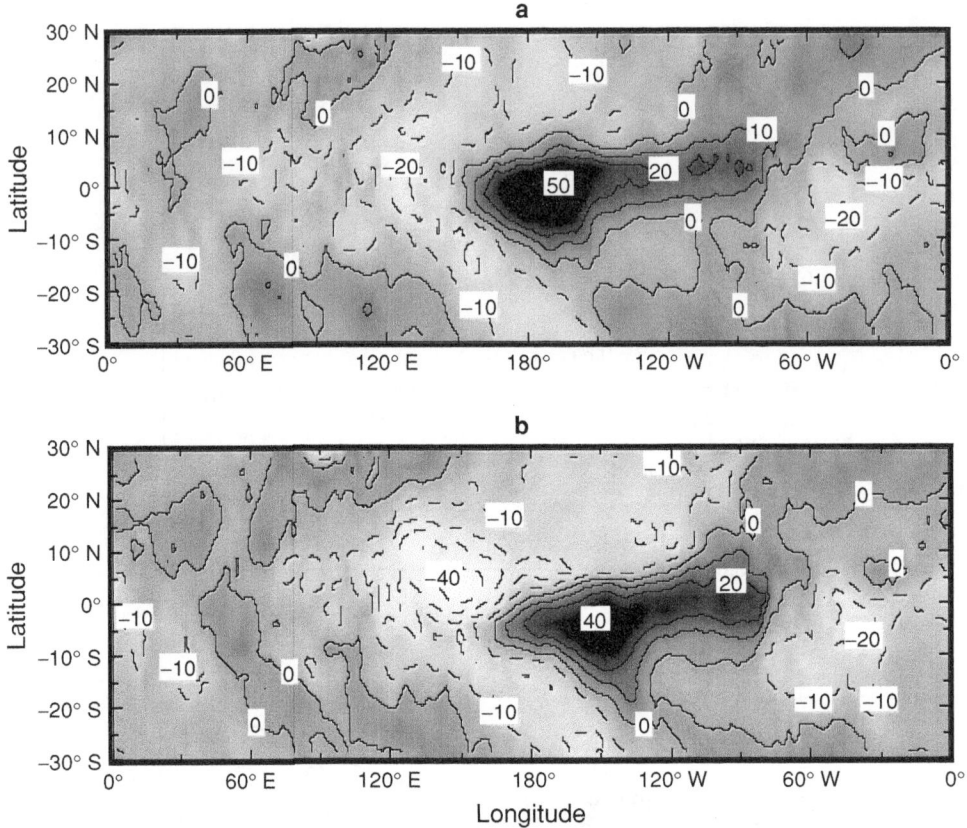

Figure 12.6. Anomalies of LW CRF (W m^{-2}) averaged from January to August:
(a) 1987; (b) 1998.

We can show that in a normal year, there is a well-defined Walker circulation in
the tropical Pacific; this Walker circulation is displaced in 1987, but it has almost
collapsed in 1998. Figure 12.8a shows a height–longitude cross-section of the zonal
wind averaged from 5° S to 5° N and from January to August in a normal year. The
wind data are from the NCEP/NCAR re-analysis. It is seen that east of 140° E in
the tropical Pacific, there are well defined easterlies in the lower troposphere and
westerlies in the upper troposphere, while to the west, there is a circulation cell
that is opposite in direction. The 140° E longitude line roughly corresponds to the
center of the upward branch of the Walker circulation and thus the maximum LW
CRF. In 1987, the structure of the Walker circulation is changed (Figure 12.8b).
Not only is the magnitude of the easterlies and the westerlies decreased; there is a
shift in the upward branch toward the dateline, as indicated by the zero zonal-wind
line in the lower troposphere. This is also consistent with the displaced center in

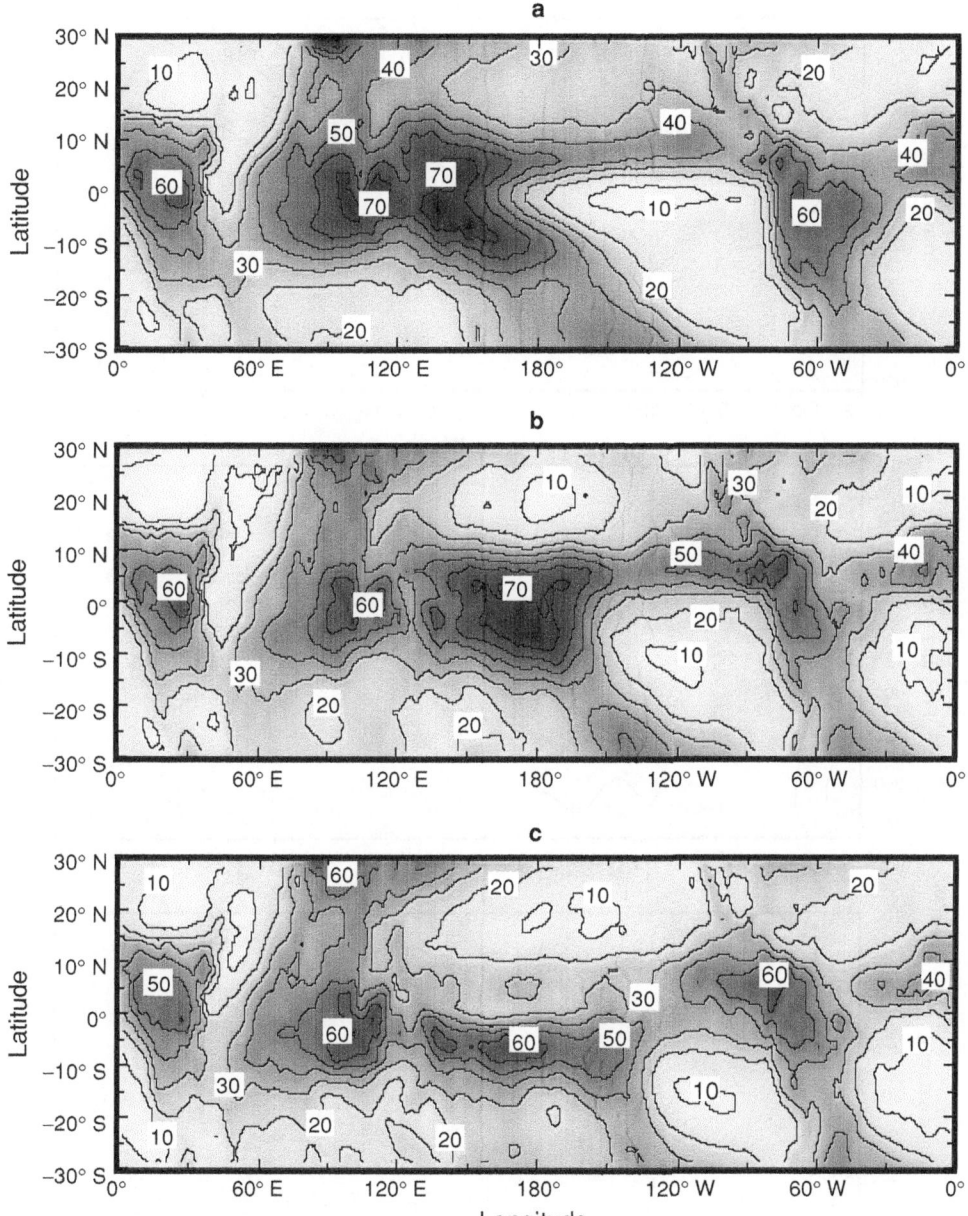

Figure 12.7. Distributions of LW CRF (W m^{-2}) averaged from January to August: (a) Mean of 1985, 1986, and 1989; (b) 1987; (c) 1998.

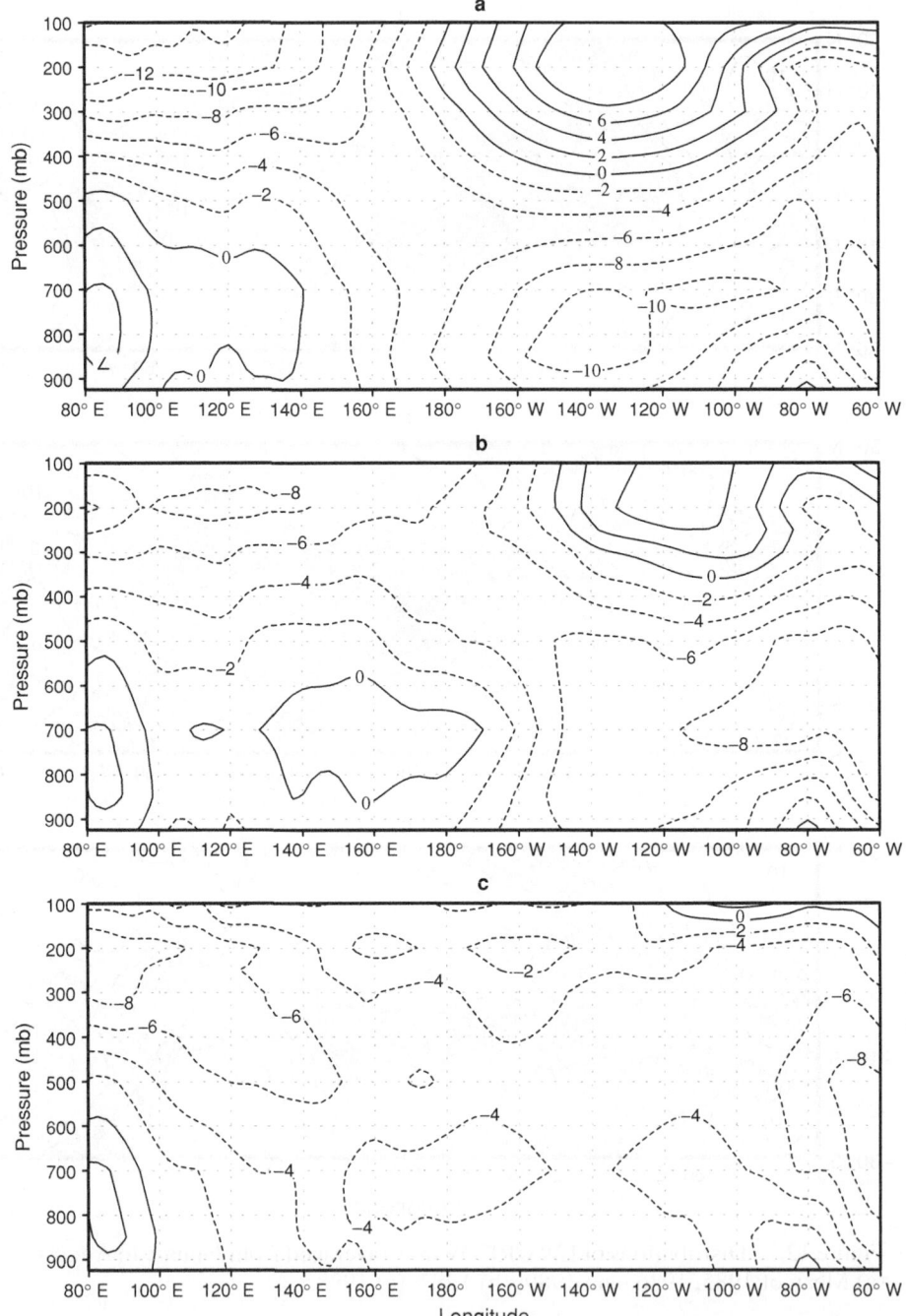

Figure 12.8. Pressure–longitude cross sections of zonal wind (m s^{-1}) averaged from 5° S to 5° N from January to August: (a) Mean of 1985, 1986, and 1989; (b) 1987; (c) 1998.

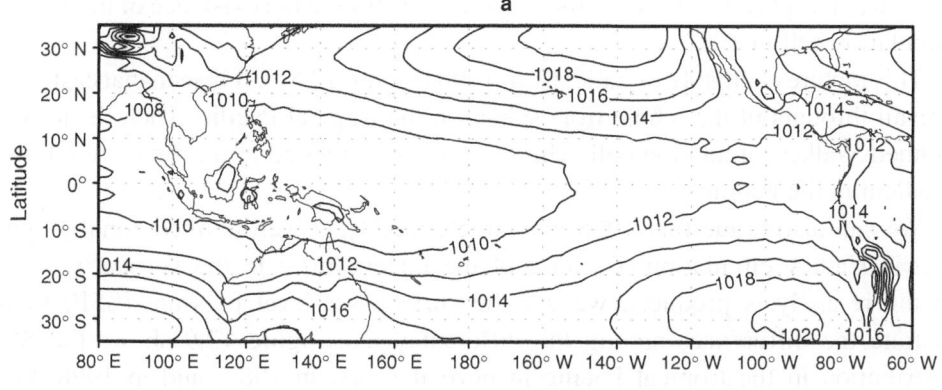

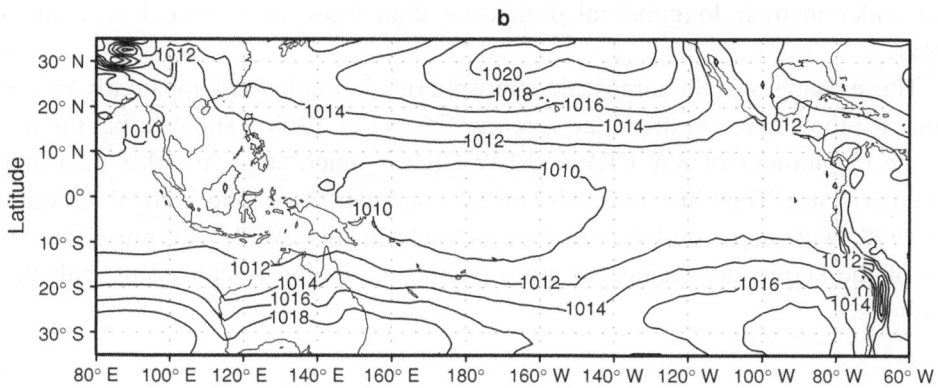

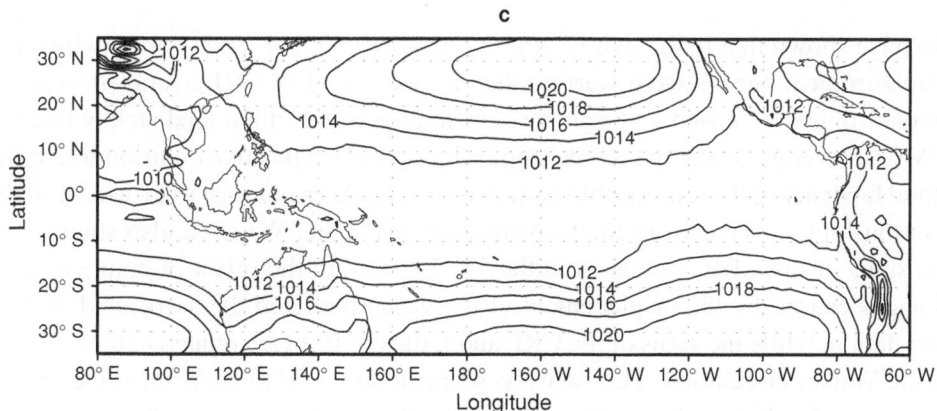

Figure 12.9. Distribution of sea-level pressure (mb). (a) Mean of 1985, 1986, and 1989; (b) 1987; (c) 1998.

LW CRF in Figure 12.6b. Nevertheless, one can still see the existence of the Walker circulation cell in 1987.

This picture changed dramatically in 1998 (Figure 12.8c). It is seen that easterlies prevail throughout the whole troposphere in the tropical Pacific. There is no well defined Walker circulation cell. This is apparently related to the small zonal SST gradient in the Pacific.

We can also argue that 1998 is characterized by a well defined zonally symmetric Hadley circulation. Because of the uncertainties in the vertical velocity in the re-analysis products, we use the surface sea-level pressure (SSP) to illustrate the differences among the different years. Figure 12.9 shows the SSP distribution in the tropical Pacific in normal years, in 1987, and in 1998. One can see that the sub-tropical highs in 1998 in both hemispheres are stronger and are wider in their longitudinal dimension than those in a normal year and in 1987.

These changes in the largescale atmospheric circulation are apparently responsible for the change in the cloud forcing. Cess *et al.* (2001) showed that the ratio of the magnitudes of SW CRF and LW CRF is much larger in 1998 than those in other years. They inferred a significant change of the cloud vertical structures in 1998, with more middle and lower clouds in this event. The present analysis of the significant change in atmospheric circulation is consistent with their findings.

12.4 Summary

We have shown that in the two El Niño events of 1987 and 1998, during which the whole tropical oceans were warmer than normal, the LW CRF averaged over the whole tropics decreased – a negative contribution to the cloud feedback, while the SW CRF averaged over the whole tropics increased – a positive contribution to the cloud feedback. This relationship is shown to be in sharp contrast to local SST–CRF relationships inferred from the tropical eastern Pacific. We have also shown that the signs of the CRF anomalies in the 1987 event are the result of compensating changes caused by the displacement of the warm pool and the associated Walker circulation; while the signs of the CRF anomalies in 1998 are related to the collapse of the Walker circulation. This analysis points to the need to understand the spatial structures of a climate change in order to study the cloud–climate feedback problem, since cloud feedbacks in a global-warming scenario with SST changed to a more spatially uniform distribution versus a scenario with larger spatial SST gradient are likely to be very different.

Acknowledgment

This research is supported by the National Science Foundation under grant ATM9701950 and by the Department of Energy under grant DEFG0298ER62570 to the State University of New York at Stony Brook.

References

Barkstrom, B. R. and G. L. Smith (1986). The Earth Radiation Budget Experiment: science and implementation. *Rev. Geophys.* **24**, 379–90.

Cess, R. D. (1975). Global climate change – investigation of atmospheric feedback mechanisms. *Tellus* **27**, 193–8.

Cess, R. D. and G. L. Potter (1988). Exploratory studies of cloud radiative forcing with a general circulation model. *Tellus* **39A**, 460–73.

Cess, R. D., G. L. Potter, J. P. Blanchet *et al.* (1989). Interpretation of cloud–climate feedback as produced by 14 atmospheric general circulation models. *Science* **245**, 513–16.

Cess, R. D., M. H. Zhang, B. A. Wielicki *et al.* (2001). The influence of the 1998 El Niño upon cloud radiative forcing over the Pacific warm pool. *J. Clim.* **14**, 2129–37.

Charlock, T. P. and V. Ramanathan (1985). The albedo field and cloud radiative forcing produced by a general circulation model with internally generated cloud optics. *J. Atmos. Sci.* **42**, 1408–29.

Godfrey, J. S. (1994). A literature review of SST anomalies in the Indian Ocean: their effects on the atmosphere and their causes. Proceedings of the International Conference on Monsoon Variability and Prediction, Trieste, Italy, May 1994, pp. 524–9.

Harrison, E. F., P. Minnis, B. R. Barkstrom *et al.* (1990). Seasonal variation of cloud radiative forcing derived from the Earth Radiation Budget Experiment. *J. Geophys. Res.* **95**, 687–18, 703.

Hirst, A. C. and J. S. Godfrey (1993). The role of the Indonesia Throughflow in a global ocean GCM. *J. Phys. Oceanogr.* **23**, 1057–86.

Kiehl, J. T. (1994). On the observed near cancelation between longwave and shortwave cloud forcing in tropical regions. *J. Clim.* **7**, 559–65.

Meehl, G. A. (1987). The annual cycle and interannual variability in the tropical Pacific and Indian Ocean regions. *Mon. Wea. Rev.* **115**, 1057–86.

Mitchell, J. F. B., C. A. Senior, and W. J. Ingram (1989). CO_2 and climate: a missing feedback? *Nature* **341**, 132–4.

Ramanathan, V., R. D. Cess, E. F. Harrison *et al.* (1989). Cloud-radiative forcing and climate: results from the Earth Radiation Budget Experiment. *Science* **243**, 57–63.

Reynolds, R. W. (1988). A real-time global sea-surface temperature analysis. *J. Clim.* **1**, 75–86.

Reynolds, R. W. and T. M. Smith (1994). Improved global sea–surface temperature analysis using optimum interpolation. *J. Clim.* **7**, 929–48.

(1995). A high-resolution global sea–surface temperature climatology. *J. Clim.* **8**, 1571–83.

Schneidar, S. H. (1972). Cloudiness as a global climate feedback mechanism: the effects on radiation balance and surface temperature of variations in cloudiness. *J. Atmos. Sci.* **29**, 1413–22.

Smith, T. M. and R. W. Reynolds (1998). A high resolution global sea-surface temperature climatology for the 1961-90 base period. *J. Clim.* **11**, 3320–3.

Villwock, A. and M. Latif (1994). Indian Ocean response to ENSO. Proceedings of the International Conference on Monsoon Variability and Prediction, Trieste, Italy, May 1994, pp. 530–7.

Wielicki, B. A., E. F. Harrison, R. D. Cess, M. D. King, and D. A. Randall (1995). Mission to Planet Earth: role of clouds and radiation in climate. *Bull. Amer. Meteor. Soc.* **76**, 2125–54.

Wong, T. D., D. F. Young, M. Haefflin, and S. Weckmann (2001). On the validation of the CERES/TRMM ERBE-like monthly mean clear-sky dataset and the effects of the 1998 ENSO event. *J. Clim.,* in press.

Zhang, M. H., S. C. Xie, and R. D. Cess (1996). Relationship between cloud-radiative-forcing and sea-surface temperature over the entire tropical oceans. *J. Clim.* **9**, 1374–84.

13

Runaway greenhouses and runaway glaciations: how stable is Earth's climate?

JAMES F. KASTING

Department of Geosciences, Pennsylvania State University, PA

13.1 Introduction

Is Earth's climate stable? At some level the answer is almost certainly "yes." The evidence for this is two-fold. First, the geologic record indicates that liquid water has been present on Earth's surface more or less continuously since about 4 Ga. ("Ga" stands for "giga-aeon," which means "billions of years ago.") We say "more or less" because, as discussed below, there appear to have been brief periods in Earth's history when the planet was almost entirely frozen. And, second, life appears to have been present since at least 3.5 Ga (Schopf, 1993) and perhaps 3.9 Ga, if carbon isotopes are admitted as indirect evidence (Mojzsis *et al.*, 1996). This latter requirement overlaps the first one to some extent because all organisms require liquid water during at least part of their life cycle. It is more stringent, however, in that liquid water can exist right up to the critical point (374 °C, 220 bar for pure water), whereas the upper temperature limit for life is $\sim$ 113 °C. (A common misconception is that liquid water requires temperatures below 100 °C, but this is only the boiling point at one atmosphere pressure. The ocean contains the equivalent of $\sim$ 270 bar of water vapor and so, like water in a pressure cooker, it would not boil until the temperature exceeded the critical temperature.)

Another way of evaluating Earth's climate stability is to compare Earth to its neighboring planets, Venus and Mars. Venus' climate is unbearably hot: the mean surface temperature is 730 K, or 457 °C. Mars' mean surface temperature is unbearably cold, about 218 K, or −55 °C. Neither of these planets' surfaces is suitable for life as we know it, although Mars could conceivably harbor life at some depth. Why is Venus too hot, Mars too cold, and Earth just right for life? Lynn Margulis has termed this "the Goldilocks question." Here, we examine the factors that stabilize

Frontiers of Climate Modeling, eds. J. T. Kiehl and V. Ramanathan.
Published by Cambridge University Press. © Cambridge University Press 2006.

Earth's climate and allow it to remain habitable, along with the destabilizing factors that can cause a planet's climate to move out of the life-supporting range.

13.2 Positive and negative feedback loops

Both the stabilization and the destabilization of climate can be discussed in terms of feedback loops (see Chapter 8). A feedback loop occurs when the components of a system are linked in such a way that information, or signals, can be transmitted around in a loop. Feedback loops can be either positive or negative. In a *positive feedback loop*, an initial perturbation to one component of the system is amplified as it propagates around the loop. A commonplace example is the feedback that occurs when a microphone and amplifier are hooked up to a loudspeaker and the speaker and microphone are placed too close together. Sound entering the microphone is amplified and emitted from the loudspeaker. The microphone picks up the amplified sound, amplifies it yet again, and the result is the high-pitched shriek that audiences find extremely annoying. In a *negative feedback loop*, just the opposite occurs: an initial perturbation is diminished as it propagates around the loop. Indeed, it will eventually disappear entirely if that is all there is to the system. An example of a negative feedback loop would be a room equipped with a heater and a thermostat. If the room gets too cold, the thermostat turns the heater on, and the room warms up. If the room gets too hot, the heater is turned off and the room cools back down. The room temperature is kept stable because the system responds to perturbations by adjusting the temperature in the opposite direction.

13.2.1 Positive feedbacks in the climate system

Earth's climate system contains two strong positive feedback loops, either of which can in principle lead to a climate catastrophe. The two loops are illustrated in Figure 13.1. In the diagram, the boxes represent system *components*, and the arrows between them represent *couplings* between them. An arrow with a normal (pointy) arrowhead represents a positive coupling, indicating that an increase (decrease) in component 1 causes a corresponding increase (decrease) in component 2. An arrow with a circular arrowhead indicates a negative coupling, meaning that an increase (decrease) in component one causes a decrease (increase) in component 2.

In the water vapor feedback loop (Figure 13.1a), the coupling is between surface temperature (T_s), atmospheric water vapor $f(H_2O)$, and the greenhouse effect (ΔT_g), see also the discussion in Chapter 9. The notation here follows Kump *et al.* (1999). Increasing T_s increases $f(H_2O)$ because of the Clausius–Clapeyron relation. Water vapor is a good greenhouse gas, so increasing H_2O increases T_s. Three positive couplings in the system produce a positive feedback loop.

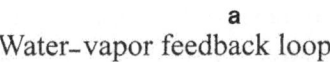

a
Water–vapor feedback loop

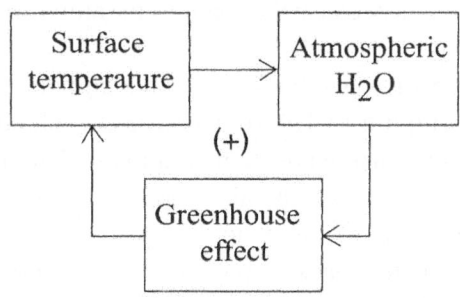

b
Snow and ice albedo feedback loop

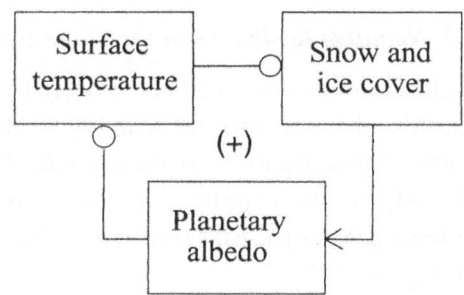

Figure 13.1. Positive feedback loops in the climate system. Arrows with pointy heads signify positive couplings; arrows with circular heads signify negative coupling.

It should be noted parenthetically that Lindzen (1990) has challenged the validity of this positive feedback loop. Lindzen suggested that the upper troposphere will dry out as surface temperature increases because convective updrafts will penetrate higher into the upper troposphere where the air is colder, causing more of the entrained water vapor to condense out before that air begins to subside. The relative humidity of the upper troposphere might therefore decrease as the climate warms. Although such a negative feedback is conceivable for small increases in T_s, it is easy to show that for large increases in T_s, the feedback must be positive (see Chapter 9). Ingersoll (1969) showed that if the H_2O mixing ratio near the ground is greater than about 10–20 percent by volume, the amount of latent heat stored in an air parcel is so great that the tropospheric lapse rate becomes very shallow, and water vapor remains a major constituent at all altitudes (including the stratosphere). Under these circumstances, any decrease in relative humidity would be dwarfed by the overall increase in the saturation vapor pressure of water at all altitudes. Thus, the water-vapor feedback is clearly positive for variations in T_s of the magnitude considered here. It is probably positive for small increases in T_s as well; it is just more difficult to prove this because the changes in saturation vapor pressure are not as pronounced.

The other strong positive feedback in Earth's climate system is the snow–ice albedo feedback loop (Figure 13.1b). An increase in T_s should decrease the amount of snow and ice cover by warming the higher latitudes. This decreases the planetary albedo which, in turn, increases T_s by allowing more of the incident sunlight to be absorbed. The two negative couplings in this system result in a positive feedback loop. More generally, any feedback loop with an even number of negative couplings must be positive; a feedback loop with an odd number of negative couplings will always be negative.

13.2.2 Negative feedbacks in the climate system

If the two positive feedback loops described above were the dominant features of the climate system, Earth's inhabitants would be in big trouble because the climate system would be unstable. As noted earlier, for the most part Earth's climate exhibits remarkable stability. In order for this to be true, the climate system must also contain some strong negative feedback loops. And so it does. The two most important of these are illustrated in Figure 13.2.

On short time scales, the important negative feedback is between surface temperature and the amount of IR radiation emitted to space (F_{IR}). As T_s increases, F_{IR} increases as well. If Earth had no atmosphere, F_{IR} would follow the Stefan–Boltzmann law: $F_{IR} = \sigma T_s^4$, where $\sigma = 5.67 \times 10^{-8}$ W m^{-2} K^{-4}. In the presence of an atmosphere containing greenhouse gases this relationship does not apply. Instead, the relationship between F_{IR} and T_s is approximately linear (North, 1975; Chapter 3; see also Figure 3 below). Outgoing infrared radiation cools the Earth. Therefore, as F_{IR} increases, T_s decreases, and this results in a negative feedback loop (Figure 13.2a). This feedback loop is so fundamental that it is often overlooked. However, even though we generally take it for granted, this negative feedback can break down at high values of T_s. The resulting *runaway greenhouse effect* is potentially catastrophic to habitable planets like Earth.

The other important negative feedback in the climate system operates on long time scales and helps to prevent global glaciation. This feedback involves the carbonate–silicate cycle and the controls on atmospheric CO_2. Its significance was first pointed out by Walker *et al.* (1981), but it has also been extensively discussed by R. Berner (Berner *et al.*, 1983; Berner, 1993, and refs. therein) and myself (e.g., Kasting *et al.*, 1988). On long (million-to-billion year) time scales, the concentration of atmospheric CO_2 is largely regulated by the carbonate–silicate cycle, also

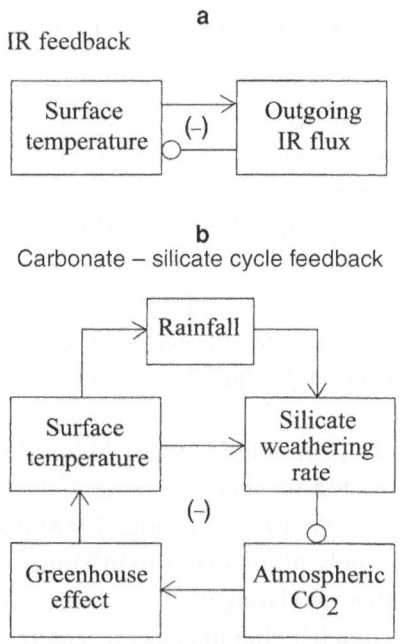

Figure 13.2. Negative feedback loops in the climate system.

know as the inorganic carbon cycle. Carbon dioxide from the atmosphere dissolves in rainwater, forming carbonic acid. This carbonic acid weathers silicate rocks on the continents, releasing Ca^{2+} and Mg^{2+} ions into solution, along with bicarbonate ions (HCO_3^-) and dissolved silica (SiO_2). These dissolved species are carried by streams and rivers to the ocean, where certain organisms, e.g., the planktonic foraminifera, use the Ca^{2+} and bicarbonate to make shells of calcium carbonate ($CaCO_3$). When the organisms die, some of them accumulate on the seafloor, forming $CaCO_3$ sediments. The seafloor spreads from the mid-ocean ridges as part of the cycle of plate tectonics, and at certain plate boundaries both the seafloor and the overlying carbonate sediments are carried down into subduction zones. The carbonate sediments undergo metamorphism, reforming silicate minerals and releasing gaseous CO_2, which is returned to the atmosphere by volcanism.

The negative feedback in this cycle involves the relationship between T_s and atmospheric CO_2 (Figure 13.2b). An increase in T_s increases the rate of silicate weathering which, in turn, decreases atmospheric CO_2. This decreases the greenhouse effect, thereby decreasing T_s and counteracting the initial warming. This negative feedback is thought to have been important in offsetting reduced solar luminosity early in Earth's history (Walker *et al.*, 1981) and has probably also been important in regulating Earth's climate during the Phanerozoic (Berner, 1993).

13.3 Runaway and moist greenhouses

Let us switch our attention now to examples of climate instability that may have actually occurred on Earth and its neighboring planets. We begin with our "sister" planet Venus and the runaway greenhouse. Venus is similar to Earth in several respects: it has nearly the same mass ($M_V = 0.81 M_E$); it has about the same amount of carbon stored as CO_2 in its atmosphere as Earth has tied up in carbonate rocks ($\sim$90 bar); and it has within a factor of two of the same amount of nitrogen. However, Venus is deficient in water by a factor of about 10^5 compared to Earth. What water does exist on Venus makes up only about 30 ppm of the planet's lower atmosphere. By comparison, Earth has enough liquid water to cover its surface to an average depth of 3 km.

Is Venus' lack of water primordial, or did Venus lose its water over time? This question has been debated for many years (e.g., Rasool and DeBergh, 1970; Ingersoll, 1969; Donahue *et al.*, 1982; Kasting, 1988) and is still not entirely resolved. However, the extremely high D/H ratio in Venus' atmosphere, $\sim$160 times the terrestrial value (Donahue and Hodges, 1993), suggests strongly that Venus' initial water endowment was much higher than its present inventory. Grinspoon (1987; 1993) argued that resupply of water by comets could produce this high value, but re-examination of this question by Gurwell (1995) shows that his analysis was in error.

13.3.1 Runaway greenhouse atmospheres

If one accepts that Venus did have a lot of water at one time, the question then becomes: how was it lost? The classical theory for water loss on Venus involves what is often termed a *runaway greenhouse* (Rasool and DeBergh, 1970; Ingersoll, 1969; Goody and Walker, 1972). According to this theory, the water-vapor feedback was much stronger on Venus than on Earth because of Venus' proximity to the Sun. Water outgassed from volcanos remained in the atmosphere on Venus, whereas it condensed to form oceans on Earth. Water vapor in Venus' upper atmosphere was then dissociated by solar ultraviolet radiation. The hydrogen escaped to space and the oxygen reacted with reduced volcanic gases and with reduced minerals in Venus' crust and, so, the water was lost over time.

Although this theory is still considered to be correct in general terms, the details of how Venus lost its water have undergone some revision in the past two decades. The early climate calculations neglected convection and they assumed a gray atmosphere, that is, they neglected the wavelength dependence of the H_2O absorption coefficient. Furthermore, both Venus and Earth were presumed to start out with no atmosphere, whereas we now believe (e.g., Lange and Ahrens, 1982)

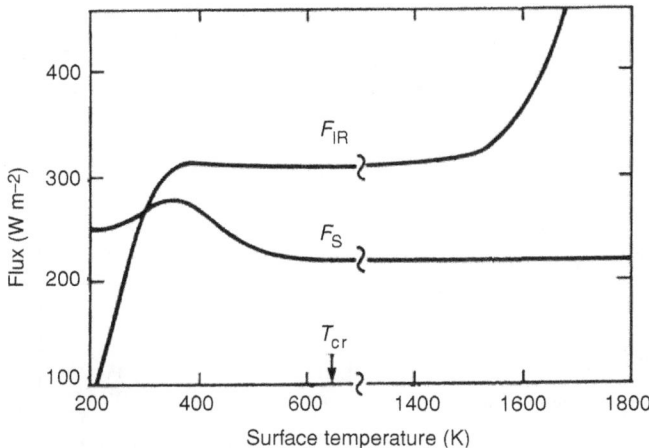

Figure 13.3. Outgoing infrared flux, F_{IR}, and absorbed solar flux, F_S, as a function of surface temperature for an Earth-like atmosphere containing 1 bar of N_2/O_2 and variable amounts of H_2O. From Kasting, 1988.

that the terrestrial planets acquired substantial atmospheres during their accretion from impact degassing of volatile-rich planetesimals.

A more recent simulation that looks at the problem in a different way was carried out by Kasting (1988). In these calculations, which were performed with a radiative–convective climate model with realistic absorption coefficients, Venus was assumed to have formed with as much water as Earth. The numerical simulation was comparable to sliding an Earth-like planet in towards Venus' orbit and calculating its surface temperature. However, for ease in computation the calculation was done in inverse mode: T_s was varied, and the climate model was used to compute radiative fluxes at the top of the atmosphere. The results are shown in Figures 13.3 and 13.4. In Figure 13.3, F_{IR} is the outgoing thermal-infrared flux, and F_S is the absorbed solar flux calculated for Earth's present solar luminosity. As the outgoing IR flux must, in reality, balance the absorbed solar flux, the ratio F_{IR}/F_S gives the "effective" solar flux, S_{eff}, which appears as the bottom scale in Figure 13.4. This is the flux required to maintain a given surface temperature. The distance of the planet from the Sun is then given by the inverse square law: $d = 1$ AU/$S_{eff}^{1/2}$ (1 AU = 149597870.691 km. An astronomical unit is approximately the mean distance between Earth and Sun).

Examination of Figure 13.3 makes it clear what is happening. As T_s increases, F_S first increases because of absorption of near-IR solar radiation by H_2O, then decreases because of increased Rayleigh scattering from the increasingly dense atmosphere. However, the runaway behavior of the system is dictated by F_{IR}. As the surface warms beyond about 360 K, the atmosphere becomes completely opaque at

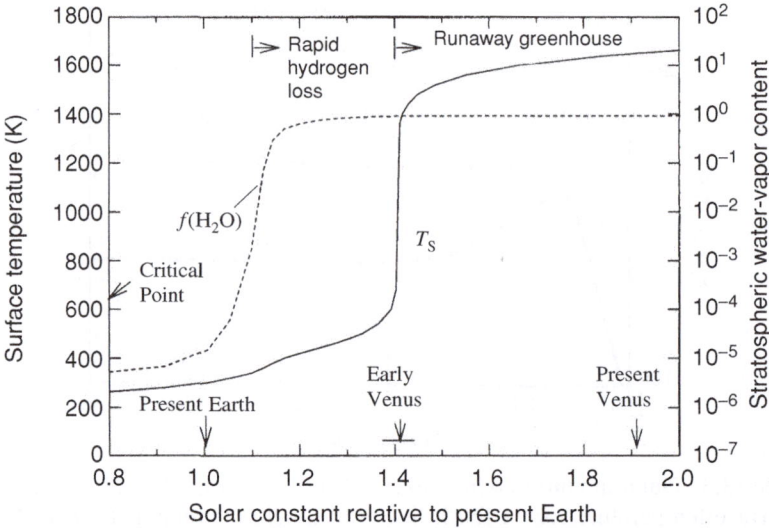

Figure 13.4. Surface temperature (solid curve) and stratospheric H_2O mixing ratio (dashed curve) as a function of effective solar constant, S_{eff}. The solar flux for present-day Venus and early Venus (at 4.6 Ga) is indicated on the figure. Water is lost rapidly for $S_{eff} > 1.1$. After Kasting, 1988.

all thermal-infrared wavelengths, so that none of the radiation from the surface or the lower atmosphere escapes to space. The outgoing IR flux originates entirely from the upper atmosphere and therefore becomes completely independent of surface temperature and pressure. When this point is reached, T_s "runs away" and does not stabilize again until it exceeds 1500 K. At this temperature the surface begins to emit radiation at visible wavelengths where H_2O does not absorb, and F_{IR} once again begins to increase. So, the climate eventually stabilizes, but it does so in a regime where liquid water, and life, cannot exist.

The remainder of Venus' story is told in Figure 13.4. The solid curve here shows T_s as a function of S_{eff}. Evidently, T_s runs away at $S_{eff} \cong 1.4$, which corresponds to an orbital distance of ~ 0.85 AU. Venus orbits at 0.72 AU, well inside this "runaway greenhouse limit"; hence, it is not surprising that Venus has lost its water. In actuality, the runaway greenhouse limit may occur at a higher solar flux than found here because clouds, which were treated very crudely here, may help to cool such a planet's surface. Clouds were parameterized in these calculations by assuming an enhanced surface albedo that allows the model to reproduce Earth's present mean surface temperature given the present solar insolation. This surface albedo was then held constant throughout the calculations. If cloudiness increases with increasing T_s, as seems likely, then runaway conditions may be harder to achieve.

13.3.2 Water loss and the 'moist greenhouse'

The water-loss story is more complicated than this, however. The critical factor that determines the rate of water loss is the water-vapor mixing ratio in the stratosphere, above the region where water vapor can condense. As T_s increases, water vapor becomes more and more abundant at low levels in the troposphere. The larger amount of latent heat released when air parcels ascend leads to a decrease in the tropospheric lapse rate and an increase in the altitude and H_2O mixing ratio at the tropopause. Thus, as mentioned earlier, the tropopause *cold trap* becomes increasingly ineffective as the surface H_2O mixing ratio begins to exceed 10–20 percent by volume. In the climate-model calculations, this happens at $S_{eff} \cong 1.1$, corresponding to an orbital radius of 0.95 AU. A planet inside this distance would not remain habitable for very long because its oceans would be lost even though liquid water was still present at its surface. This process has been termed a *moist greenhouse*. Again, the actual solar flux at which this phenomenon occurs may be higher than suggested here because of negative cloud feedback. However, the fact that Venus is so dry shows that the process of water loss can and does occur.

13.4 Runaway glaciation: the "snowball Earth"

13.4.1 Energy-balance climate modeling

The flip side of the climate stability question concerns the possibility of runaway glaciation. As recognized a long time ago (Öpik, 1965; Eriksson, 1968), runaway glaciation is possible because of the strong positive feedback loop provided by snow and ice albedo. This phenomenon was initially studied with *energy-balance climate models (EBMs)*, which are described in Chapter 3. In such models, vertical energy balance is computed at a number of different latitudes, and heat transport between latitude zones is parameterized as a diffusion process. A nice review of these models is provided by North (1975). Early EBMs (Budyko, 1969; Sellers, 1969) predicted that runaway glaciation could be triggered by as little as a two to five percent decrease in solar output. A more recent EBM calculation is shown in Figure 13.5 (from Caldeira and Kasting, 1992a). The solid curves in this figure represent stable solutions for the position of the ice line, that is, the maximum equatorward extent of polar ice. The lower axis is effective solar flux, S_{eff}, which has the same meaning as before. Three different sets of curves are shown in the figure, corresponding to three different atmospheric CO_2 levels. The right-most curve is for the present (or almost-the-present) CO_2 level of 3×10^{-4} bar. Let us spend a few moments considering what these curves mean.

For the present climate system ($S_{eff} = 1$; $pCO_2 = 3 \times 10^{-4}$ bar), four different solutions are mathematically possible: an ice-free solution, a "small-ice-cap"

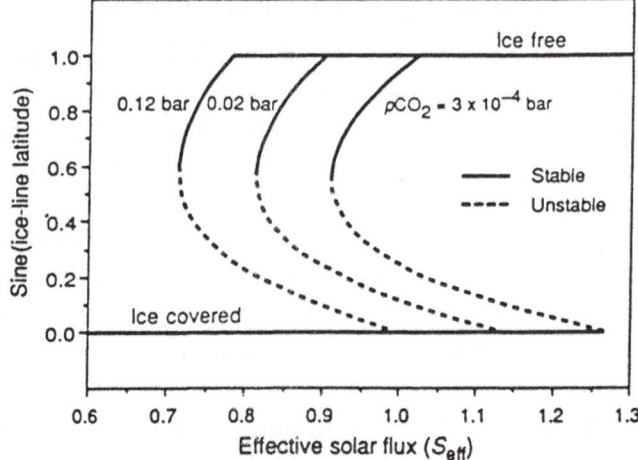

Figure 13.5. Sine of the ice-line latitude versus effective solar flux from the energy-balance climate model of Caldeira and Kasting (1992a). Solid curves represent stable solutions; dashed curves represent unstable solutions. The rightmost curve is for the present atmospheric CO_2 level; the other two curves are for higher CO_2 levels.

solution in which the ice-line is at a latitude of ~72 degrees (sine latitude = 0.95), a "large-ice-cap" solution with the ice line at 16 degrees (sine latitude = 0.28), and a solution in which Earth is completely ice-covered. The dashed portion of the curve, which includes the "large-ice-cap" solution, is unstable. If the ice sheets advanced into this region, the model predicts that the oceans would continue to freeze until Earth was completely ice covered. This would be unlikely to occur today because stable ice lines only occur poleward of about 35 degrees; hence, the "large-ice-cap" state would be difficult or impossible to reach. However, if the solar constant were about eight percent lower (or if atmospheric CO_2 were lower), the critical ice line could be reached and Earth could conceivably fall into a globally glaciated state.

In the early EBMs, which assumed constant atmospheric CO_2, the ice-covered state was semi-permanent: the only way to escape from it was for the solar constant to increase. According to Figure 13.5, a 27 percent increase in solar flux would be needed to escape from an ice-covered state today. Such an increase would eventually happen if one waited long enough. The solar constant is presently increasing at a rate of about one percent every hundred million years (Caldeira and Kasting, 1992b), so the melting point for the system would be reached in a little over 2.5 billion years.

There is, however, a much faster way to escape from the ice catastrophe, namely, build-up of volcanic CO_2. An ice-covered Earth would have a nearly non-existent hydrologic cycle, so the rate of removal of atmospheric CO_2 by silicate weathering would be close to zero. According to Figure 13.5, the critical CO_2 partial pressure at which the ice cover would melt is 0.12 bar, or about 400 times the present level. At the present rate of volcanic CO_2 release, $\sim 5 \times 10^{12}$ mol y^{-1} (Holland, 1978), this amount of CO_2 should accumulate in something under 4 million years. The actual required amount of CO_2, and the corresponding recovery time, might be somewhat shorter if the ice became dirty, as it almost certainly would. (The surface albedo was assumed to remain fixed at 0.66 in the calculations shown here.) However, the recovery could not be hastened too much because once the ice started to melt, the hydrologic cycle would be re-established and the surface would be freshened up by clean snow. When the ice did finally melt, the climate would become quite warm because the surface albedo would now be much lower and atmospheric CO_2 levels would be very high. This would cause a rapid pulse of silicate weathering, and the atmospheric CO_2 content and surface temperature would eventually decline back to normal levels.

13.4.2 Evidence for a snowball Earth

The question of runaway glaciation has recently become much more concrete because of new geologic evidence that such an event may actually have occurred three or more times during Earth's history (Hoffman *et al.*, 1998). The geologic record is rather complicated, but the two strongest pieces of evidence are the following. First, geologists have long been aware of paleomagnetic evidence for low-latitude glaciation in the Late Proterozoic Era, at ~ 0.75 Ga and again at ~ 0.6 Ga (e.g., Williams, 1975; 1993; Frakes, 1979). More recently, evidence of low-latitude glaciation has been identified in the Early Proterozoic Era as well, around 2.3 Ga (Evans *et al.*, 1997; Williams and Schmidt, 1997). The Late Proterozoic glacial evidence is particularly well documented in Australia. It consists of tillites (rock piles dropped by glaciers), striations (stripes cut into rocks by moving glaciers), and dropstones (rocks dropped into otherwise nicely laminated marine sediments by melting icebergs). The dropstones, in particular, show that Australia experienced continental-scale glaciation, not just localized mountain glaciers, at this time. And, according to paleomagnetic data, at 600 Ma Australia was situated within a few degrees of the equator (Kirshvink, 1992a). (Paleolatitudes are determined by examining rocks containing magnetite, Fe_3O_4, or some other magnetic mineral and looking at the dip angle of the remnant magnetic field relative to the bedding plane – original surface orientation – of the rocks. Earth's present magnetic field is approximately a dipole with the magnetic poles oriented about 11 degrees away from the rotation axis. Thus, the field lines are approximately perpendicular to the surface in the polar regions and parallel to the surface near the equator. The magnetic axis drifts randomly around the spin axis on time periods of hundreds of thousands to millions of years – a process called *apparent polar wander* – but it does not appear to have ever deviated by more than about 11 degrees from the spin axis during the last 200 Ma, when one can perform reasonably secure reconstructions of continental drift. Thus, unless the magnetic field was behaving very differently during the Late Proterozoic Era, the evidence for low-latitude glaciation must be taken seriously.)

The second critical piece of geologic evidence for global glaciation comes from so-called *cap carbonates* that are found directly above the 0.6-Ga and 0.75-Ga glaciations. In Namibia, where Hoffman *et al.* did their field work, the glacial deposits are overlain by several meters of fine-grained limestone ($CaCO_3$), followed by about 400 m of coarser-grained limestone. Their interpretation is that the relatively thin lower layer of carbonate may represent the initial removal of the CO_2 built up in the atmosphere during the global glaciation, whereas the thicker, coarser layer may be caused by rapid weathering of carbonate rocks on land following the deglaciation, and redeposition of these carbonate minerals in the ocean (D. Schrag, private communication).

It should be noted that an alternative explanation for low-latitude glaciation that would not require a completely frozen planet is that Earth's obliquity (the tilt of its spin axis with respect to the ecliptic plane) was originally much higher, and that it decreased rapidly between about 0.6 Ga and 0.44 Ga. Today, Earth's obliquity is 23.5 degrees, and we know that it varies from 22–24.5 degrees with a period of about 41 000 years (Berger, 1992). If the obliquity was at one time greater than 54 degrees, the equator would have been the coldest part of the planet and the poles would have been the warmest (Ward, 1974). This *high-obliquity hypothesis* (Williams, 1975; 1993) would allow the polar oceans to remain ice-free, and so would not pose as great a threat to the survival of the biota as does the snowball-Earth theory. In order for this hypothesis to be tenable, however, Earth's obliquity must have decreased by several tens of degrees in the time span of $\sim$ 150 million years. One possible mechanism for causing such a change has been identified: the so-called *climate friction* mechanism, in which the advance and retreat of the polar ice sheets causes resonant torques on Earth's equatorial bulge (Rubincam, 1993; Bills, 1994). However, the required phase relationships between the solar forcing and ice-sheet extent are peculiar (Williams *et al.*, 1998). Furthermore, this hypothesis does not explain the cap carbonates or other lines of geologic evidence discussed by Hoffman *et al.* (1998), all of which appear to be consistent with the snowball-Earth hypothesis.

13.4.3 Possible triggers for global glaciation

Let us return now to the question of climate stability, which is the underlying theme of this chapter. We have shown that global glaciation is possible theoretically and that it appears to have actually happened several different times. But we have not yet addressed the question of why it occurred. This is a bit of a puzzle because we argued earlier that Earth's climate is stabilized by the carbonate–silicate cycle. What could have gone wrong to allow this stabilizing feedback mechanism to break down?

The general answer to this question is that global glaciation requires either low solar luminosity or low greenhouse-gas forcing, or both. Solar luminosity, S, can be estimated from theoretical models of stellar evolution. A convenient analytic approximation from the work of Gough (1981) is given in Equation (13.1):

$$\frac{S}{S_0} = \frac{1}{1 + 0.4(t/t_0)}. \tag{13.1}$$

Here, t = time in Ga, $t_0 = 4.6$ Ga, and S_0 is the present solar constant, 1370 W m^{-2}. Using this formula, we find that for the most recent global glaciation ($t = 0.6$ Ga), solar luminosity was ~ 95 percent of its present value, whereas for the earlier, Huronian global glaciation ($t = 2.3$ Ga), it was only ~ 83 percent of today's value. Thus, global glaciation would have been considerably easier to trigger at earlier times. Or, to put it a different way, significantly higher concentrations of greenhouse gases would have been required to avoid global glaciation early in Earth's history when the Sun was less bright. This is simply a restatement of the well-known *faint young Sun problem* (Sagan and Mullen, 1972; Kasting *et al.*, 1988).

The more interesting question is: what could have caused greenhouse-gas concentrations to fall to dangerously low levels three or more times during Earth's history? The answers here may be quite different for the Early and Late Proterozoic glaciations. Let us begin with the Late Proterozoic Era. The climate system then was similar to today in that CO_2 and H_2O were probably the dominant greenhouse gases. Atmospheric O_2 levels were at least five percent of present at this time (Canfield and Teske, 1996) and probably higher (Canfield, 1998), so reduced greenhouse gases like CH_4 should have been limited to low concentrations, as they are today.

If these assumptions are correct, then the Late Proterozoic global glaciations were most likely triggered by low levels of atmospheric CO_2. (As discussed earlier, H_2O acts as a feedback on climate; hence, one cannot invoke changes in H_2O as a forcing mechanism.) Hoffman *et al.* (1998) showed carbon isotopic data that indicate increased burial of organic carbon directly prior to the 0.6-Ga Ghaub glaciation. Although we did not discuss the organic carbon cycle earlier, this cycle involves carbon fluxes similar in magnitude to those in the carbonate–silicate cycle and, hence, could have caused significant changes in atmospheric CO_2. Hoffman *et al.* suggested that rifting of a Late Proterozoic supercontinent created large areas of new continental shelf on which organic-carbon burial could take place, and that the resulting drawdown of atmospheric CO_2 triggered the global glaciation event.

While this hypothesis has merit, it does not explain why the carbonate–silicate cycle failed to compensate for the increase in organic-carbon burial. If the theory outlined in the first section of this chapter is correct, silicate weathering should have slowed down as the climate cooled, and so less CO_2 should have been removed by this part of the carbon cycle. Why was this negative feedback not sufficient to keep Earth from freezing?

A possible answer to this question was suggested by Marshall *et al.* (1988). These authors pointed out that if the continents were clustered at low latitudes, global climate could have become quite cold and yet silicate weathering could still have proceeded at an appreciable rate. And, indeed, some paleogeographic reconstructions suggest that this was the case during the Late Proterozoic (Kirshvink, 1992b; L. Lawver, private communication, 1999). These reconstructions are somewhat problematic and are not totally consistent with each other – Kirshvink's map shows low-latitude continents at 0.6 Ga, whereas Lawver sees this signal much more strongly at 0.75 Ga – but they do suggest that the idea is tenable. If so, then the result is paradoxical because it contradicts conventional geologic wisdom. Conventional wisdom suggests that glaciations occur when large land masses are located near the poles, where it is cold. While this may be true for ordinary, high-latitude glaciations, global glaciations appear to be favored by just the opposite continental arrangement. In any case, it is clear that the climate system can respond quite differently to continental motions when the effect of the carbonate–silicate cycle on atmospheric CO_2 is taken into account.

By contrast, the Early Proterozoic global glaciation at 2.3 Ga may have had an entirely different cause; 2.3 Ga is about the time that significant amounts of O_2 first appeared in Earth's atmosphere (Walker *et al.*, 1983; Kasting, 1993; Holland, 1994). Karhu and Holland (1996) argued on the basis of carbon isotopes that the rise in O_2 occurred somewhat later, around 2.0–2.2 Ga; however, the dates of the Huronian glacial deposits are rather uncertain, so it is not clear that any real discrepancy in timing exists. More tellingly, Roscoe (1969; 1973) (quoted in Walker *et al.*, 1983) pointed out a long time ago that in the Huronian sequence in southern Canada, glacial deposits are underlain by grayish (unoxidized) sediments and overlain by reddish (oxidized) sediments, as might be expected if the glaciations had coincided with the rise of atmospheric O_2.

Recent climate-model calculations by our group (Pavlov *et al.*, 2000) (Figure 13.6) show that this idea is quantitatively feasible. Prior to the rise in O_2, CH_4 is expected to have been much more abundant than today. Today, the atmospheric lifetime of CH_4 is about 12 years, and the CH_4 concentration is ~ 1.6 ppmv. In an anoxic early atmosphere, the photochemical lifetime is close to 20 000 years (Brown, 1999), so an equivalent surface flux of CH_4 could have supported much higher atmospheric concentrations. Methane is produced by anaerobic bacteria

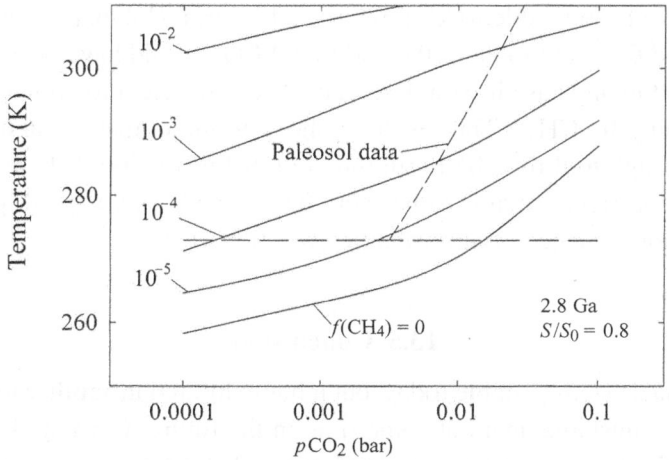

Figure 13.6. Surface temperature (solid curves) as a function of atmospheric pCO_2 and CH_4 mixing ratio, $f(CH_4)$. The horizontal dashed curve shows the freezing point of pure water. The other dashed curve shows the upper limit on pCO_2 derived from paleosols. From Pavlov *et al.*, 2000.

(methanogens) that appear to be evolutionarily ancient (Woese and Fox, 1977), so the idea that CH_4 was abundant in the early atmosphere is also consistent with our knowledge of biological evolution.

Figure 13.6 shows climate calculations performed with a one-dimensional radiative–convective model for 2.8 Ga, when solar luminosity was ∼ 80 percent of its present value. The solid curves show global mean surface temperature, T_s, as a function of pCO_2 and CH_4 mixing ratio, $f(CH_4)$. The horizontal dashed line represents the freezing point of water, and the other dashed curve is an upper limit on pCO_2 derived from paleosol data (Rye *et al.*, 1995). (Paleosols are ancient soils. The upper limit on pCO_2 is based on the absence of the mineral siderite, $FeCO_3$, from such soils.) If one accepts the paleosol constraint, then acceptable solutions to the climate problem at this time lie to the upper left of the two dashed curves. One can see, for instance, that a CH_4 mixing ratio of 10^{-3} (1000 ppmv), coupled with an atmospheric CO_2 partial pressure of 10^{-3} bar (three times the current value) could have produced a mean surface temperature in the vicinity of 295 K. This is some seven degrees higher than the current mean surface temperature and would thus be consistent with a lack of glaciation at that time. By contrast, an atmosphere with the same amount of CO_2, but no CH_4, would have had $T_s \cong 265$ K, well below the freezing point of water.

Now, suppose that this CH_4-rich atmosphere had persisted for the next 500 million years and that O_2 levels had risen suddenly near the end of that time. The atmospheric CH_4 concentration would have dropped dramatically, and CO_2 by itself may have been unable to keep the climate warm. (Methane oxidation would have produced CO_2, but most of the additional CO_2 would have been absorbed by the ocean, and in any case it would not have been sufficient to compensate for the loss of warming by CH_4.) The resulting shock to the climate system might well have been enough to throw the Earth into a global glaciation. Whether or not this explanation is correct remains to be seen, but it stands as a logical, perhaps even predictable, consequence of atmospheric evolution.

13.5 Conclusion

Earth's climate system is stable today, but it has exhibited instability in the past and it will become unstable again at some time in the future. Current climate models suggest that Earth's climate has remained generally temperate over a wide range of solar insolation as a consequence of negative feedback provided by the carbonate–silicate cycle. Solar fluxes more than about ten percent higher than the present value may, however, lead to the loss of a planet's water. Global glaciations can occur if the combination of solar flux and greenhouse-gas forcing is too low. During the few occasions when this appears to have happened, the carbonate–silicate cycle provided an escape hatch by allowing large amounts of CO_2 to accumulate in the atmosphere, thereby melting the ice. If this same cycle operates on planets around other stars, then the chances of finding a planet with a habitable, Earth-like climate appear to be relatively good.

References

Berger, A. (1992). Astromical theory of paleoclimates and the last glacial–interglacial cycle. *Quat. Sci. Rev.* **11**, 571–81.

Berner, R. A. (1993). Paleozoic atmospheric CO_2: importance of solar radiation and plant evolution. *Science* **261**, 68–70.

Berner, R. A., A. C. Lasaga, and R. M. Garrels (1983). The carbonate–silicate geochemical cycle and its effect on atmospheric carbon dioxide over the past 100 million years. *Amer. J. Sci.* **283**, 641–83.

Bills, B. G. (1994). Obliquity–oblateness feedback: are climatically sensitive values of the obliquity dynamically unstable? *Geophys. Res. Lett.* **21**, 177–80.

Brown, L. L. (1999). Photochemistry and climate on early Earth and Mars. Ph.D. Thesis, Pennsylvania State University, PN.

Budyko, M. I. (1969). The effect of solar radiation variations on the climate of the Earth. *Tellus* **21**, 611–19.

Caldeira, K. and J. F. Kasting (1992a). Susceptibility of the early Earth to irreversible glaciation caused by carbon dioxide clouds. *Nature* **359**, 226–8.

   (1992b). The life span of the biosphere revisited. *Nature* **360**, 721–3.

Canfield, D. E. (1998). A new model for Proterozoic ocean chemistry. *Nature* **396**, 450–3.

Canfield, D. E. and A. Teske (1996). Late-Proterozoic rise in atmospheric oxygen concentration inferred from phylogenetic and sulfur isotope studies. *Nature* **382**, 127–32.

Donahue, T. M. and R. R. Hodges (1993). Venus methane and water. *Geophys. Res. Lett.* **20**, 591–4.

Donahue, T. M., J. H. Hoffman, and R. R. Hodges, Jr. (1982). Venus was wet: a measurement of the ratio of deuterium to hydrogen. *Science* **216**, 630–3.

Eriksson, E. (1968). Air–ocean–icecap interactions in relation to climatic fluctuations and glaciation cycles. *Meteor. Monogr.*, **8**, 68–92.

Evans, D. A., N. J. Beukes, and J. L. Kirshvink (1997). Low-latitude glaciation in the Proterozoic era. *Nature* **386**, 262–6.

Frakes, L. A. (1979). *Climates Throughout Geologic Time*, New York, Elsevier.

Goody, R. M. and J. C. G. Walker (1972). *Atmospheres*, Englewood Cliffs, NJ, Prentice Hall.

Gough, D. O. (1981). Solar interior structure and luminosity variations. *Solar Phys.* **74**, 21–34.

Grinspoon, D. H. (1987). Was Venus wet? Deuterium reconsidered. *Science* **238**, 1702–4.
  (1993). Implications of the high D/H ratio for the sources of water in Venus' atmosphere. *Nature* **363**, 428–31.

Gurwell, M. (1995). Evolution of deuterium on Venus. *Nature* **378**, 22–3.

Hoffman, P. F., A. J. Kaufman, G. P. Halverson, and D. P. Schrag (1998). A Neoproterozoic Snowball Earth. *Science* **281**, 1342–6.

Holland, H. D. (1978). *The Chemistry of the Atmosphere and Oceans*, New York, Wiley.
  (1994). Early Proterozoic atmospheric change. In *Early Life on Earth*, ed. S. Bengtsson, New York, Columbia University Press, pp. 237–44.

Hyde, W. T., T. J. Crowley, S. K. Baum, and W. R. Peltier (2000). Neoproterozoic 'snowball Earth' simulations with a coupled climate/ice-sheet model. *Nature* **405**, 425–9.

Ingersoll, A. P. (1969). The runaway greenhouse: a history of water on Venus. *J. Atmos. Sci.* **26**, 1191–8.

Karhu, J. A. and H. D. Holland (1996). Carbon isotopes and the rise of atmospheric oxygen. *Geology* **24**, 867–70.

Kasting, J. F. (1988). Runaway and moist greenhouse atmospheres and the evolution of Earth and Venus. *Icarus* **74**, 472–94.
  (1993). Earth's early atmosphere. *Science* **259**, 920–6.

Kasting, J. F., O. B. Toon, and J. B. Pollack (1988). How climate evolved on the terrestrial planets. *Scient. Amer.*, **256**, 90–7.

Kirschvink, J. L. (1992a). Late Proterozoic low-latitude global glaciation: the snowball Earth. In *The Proterozoic Biosphere: a Multidisciplinary Study*, eds. J. W. Schopf and C. Klein, Cambridge, Cambridge University Press, pp. 51–2.
  (1992b.) A paleogeographic model for Vendian and Cambrian time, in *The Proterozoic Biosphere: a Multidisciplinary Study*, eds. J. W. Schopf and C. Klein, Cambridge, Cambridge University Press, pp. 569–81.

Lange, M. A. and T. J. Ahrens (1982). The evolution of an impact generated atmosphere. *Icarus* **51**, 96–120.

Lindzen, R. S. (1990). Some coolness concerning global warming. *Bull. Amer. Meteor. Soc.* 288–99.

Marshall, H. G., J. C. G. Walker, and W. R. Kuhn (1988). Long-term climate change and the geochemical cycle of carbon. *J. Geophys. Res.* **93**, 791–802.

Mojzsis, S. J., G. Arrhenius, K. D. McKeegan *et al.* (1996). Evidence for life on Earth before 3 800 million years ago. *Nature* **384**, 55–9.

North, G. R. (1975). Theory of energy-balance climate models. *J. Atmos. Sci.* **32**, 2033–43.

Öpik, E. J. (1965). Climatic changes in cosmic perspective. *Icarus* **4**, 289–307.

Pavlov, A. A., J. F. Kasting, L. L. Brown, K. A. Rages, and R. Freedman (2000). Greenhouse warming by CH_4 in the atmosphere of early Earth. *J. Geophys. Res.* **105** 11,981–11,990.

Rasool, S. I. and C. DeBergh (1970). The runaway greenhouse and the accumulation of CO_2 in the Venus atmosphere. *Nature* **226**, 1037–9.

Roscoe, S. M. (1969). Huronian rocks and uraniferous conglomerates in the Canadian Shield. Geological Survey of Canada, Paper 68-40, 205.

(1973). The Huronian Supergroup: a Paleophebian succession showing evidence of atmospheric evolution. *Geol. Soc. Can. Spec. Pap.* **12**, 31–48.

Rubincam, D. P. (1993). The obliquity of Mars and "climate friction." *J. Geophys. Res.* **98**, 10,827–10,832.

Rye, R., P. H. Kuo, and H. D. Holland (1995). Atmospheric carbon dioxide concentrations before 2.2 billion years ago. *Nature* **378**, 603–5.

Sagan, C. and G. Mullen (1972). Earth and Mars: evolution of atmospheres and surface temperatures. *Science* **177**, 52–6.

Schopf, J. W. (1993). Microfossils of the Early Archean Apex chert: new evidence for the antiquity of life. *Science* **260**, 640–6.

Sellers, W. D. (1969). A climate model based on the energy balance of the Earth–atmosphere system. *J. Appl. Meteor.* **8**, 392–400.

Walker, J. C. G., P. B. Hays, and J. F. Kasting (1981). A negative feedback mechanism for the long-term stabilization of Earth's surface temperature. *J. Geophys. Res.* **86**, 9776–82.

Walker, J. C. G., C. Klein, M. Schidlowski *et al.* (1983). Environmental evolution of the Archean-Early Proterozoic Earth. In *Earth's Earliest Biosphere: Its Origin and Evolution*, ed. J. W. Schopf, Princeton, Princeton University Press, pp. 260–90.

Ward, W. R. (1974). Climatic variations on Mars. 1. Astronomical theory of insolation. *J. Geophys. Res.* **79**, 3375–86.

Williams, D. M., J. F. Kasting, and L. A. Frakes (1988). Low-latitude glaciation and rapid changes in the Earth's obliquity explained by obliquity–oblateness feedback. *Nature* **396**, 453–5.

Williams, G. E. (1975). Late Precambrian glacial climate and the Earth's obliquity. *Geol. Mag.* **112**, 441–65.

Williams, G. E. (1993). History of the Earth's obliquity. *Earth Sci. Rev.* **34**, 1–45.

Williams, G. E. and P. W. Schmidt (1997). Paleomagnetism of the Palaeoproterozoic Gowganda and Lorraine formations, Ontario: low paleolatitude for Huronian glaciation. *Earth Planet. Sci. Lett.* **153**, 157–69.

Woese, C. R. and G. E. Fox (1977). Phylogenetic structure of the prokaryotic domain: the primary kingdoms. *Proc. Natl. Acad. Sci. USA.* **74**, 5088.

Glossary

Frequently used abbreviations

AMSR	Advanced Microwave Scanning Radiometer
AMSU	Advanced Microwave Sounding Unit
ARM	Atmospheric Radiation Measurement
CERES	Clouds and Earth's Radiant Energy System
DOE	Department of Energy
EBCM	Energy-balance climate model
EOS	Earth Observing System
ERBE	Earth Radiation Budget Experiment
ERBS	Earth Radiation Budget Satellite
GFDL	Geophysical Fluid Dynamics Laboratory
GISS	Goddard Institute for Space Studies
GOES	Geostationary Operational Environmental Satellites
GPS	Global Positioning System
HIRS	High-resolution Infrared Radiometer Sounder
HIRDLS	High Resolution Dynamics Limb Sounder
ICCP	International Panel on Climate Change
ISCCP	International Satellite Cloud Climatology Project
MODIS	Moderate Resolution Imaging Spectroradiometer
MSL	Microwave Limb Sounder
NASA	National Aeronautics and Space Administration
NCAR	National Center for Atmospheric Research
NCEP	National Centers for Environmental Prediction
NOAA	National Oceanic and Atmospheric Administration
NSF	National Science Foundation
SAGE	Stratospheric Aerosol and Gas Experiment
SMMR	Scanning Multichannel Microwave Radiometer
SSMT	Special Sensor Microwave Temperature
TRMM	Tropical Rainfall Measuring Mission